RESEARCH IN MARITIME HISTORY
NO. 40

THE VITAL SPARK:
THE BRITISH COASTAL TRADE,
1700-1930

John Armstrong

International Maritime Economic History Association

St. John's, Newfoundland
2009

ISSN 1188-3928
ISBN 978-0-9864973-0-8

Research in Maritime History is available free of charge to members of the International Maritime Economic History Association. The price to others is US $25 per copy, plus US $5 postage and handling.

Back issues of *Research in Maritime History* are available:

No. 1 (1991)   David M. Williams and Andrew P. White (comps.), *A Select Bibliography of British and Irish University Theses about Maritime History, 1792-1990*

No. 2 (1992)   Lewis R. Fischer (ed.), *From Wheel House to Counting House: Essays in Maritime Business History in Honour of Professor Peter Neville Davies*

No. 3 (1992)   Lewis R. Fischer and Walter Minchinton (eds.), *People of the Northern Seas*

No. 4 (1993)   Simon Ville (ed.), *Shipbuilding in the United Kingdom in the Nineteenth Century: A Regional Approach*

No. 5 (1993)   Peter N. Davies (ed.), *The Diary of John Holt*

No. 6 (1994)   Simon P. Ville and David M. Williams (eds.), *Management, Finance and Industrial Relations in Maritime Industries: Essays in International Maritime and Business History*

No. 7 (1994)   Lewis R. Fischer (ed.), *The Market for Seamen in the Age of Sail*

No. 8 (1995)   Gordon Read and Michael Stammers (comps.), *Guide to the Records of Merseyside Maritime Museum, Volume 1*

No. 9 (1995)   Frank Broeze (ed.), *Maritime History at the Crossroads: A Critical Review of Recent Historiography*

No. 10 (1996)   Nancy Redmayne Ross (ed.), *The Diary of a Maritimer, 1816-1901: The Life and Times of Joseph Salter*

No. 11 (1997)   Faye Margaret Kert, *Prize and Prejudice: Privateering and Naval Prize in Atlantic Canada in the War of 1812*

No. 12 (1997)   Malcolm Tull, *A Community Enterprise: The History of the Port of Fremantle, 1897 to 1997*

Research in Maritime History would like to thank Memorial University of Newfoundland for its generous financial assistance in support of this volume.

Table of Contents

Series Editor's Foreword

Maritime history has made some amazing strides in the past half century. From largely being the preserve of aniquarians and aficionados of ships, it has become an increasingly respected genre within the larger body of historical scholarship. Yet if there is one area that has been relatively ignored – with, as John Armstrong would likely put it, "a few honourable exceptions" – it would be the history of the trades in which merchant vessels engaged. In the early modern period we know a lot about the tobacco and timber trades, and the study of the slave trade in all its ramifications has been revolutionized by the work of a dedicated group of scholars. But virtually all the great trades of the nineteenth and twentieth centuries still await proper scholarly treatment.

There is, however, an "honourable exception" to this last generalization, for as the essays that follow demonstrate with grace and insight, we know quite a bit about the British coastal trade. That I can make this claim is due almost entirely to the work of a single remarkable scholar – John Armstrong. Although as John is always careful to remind us, his work has been entirely on the British trade, it is no less important for this focus. Moreover, although even in the United Kingdom this sector was far from homogeneous, I believe that readers of the essays in this volume will come away with an appreciation of the main characteristics of its various branches. As John has pointed out repeatedly, there remain many significant things that we still do not know. Yet what is striking to me – and I suspect that it will be to many other readers as well – is that thanks to John Armstrong's industry, imagination and hard work we know quite a bit about the British coastal trade.

For this reason, and many others besides, we are extremely happy to have this selection (and as the bibliography at the end of the book demonstrates, it is only a selection) of John Armstrong's writings in a single volume in the *Research in Maritime History* series. Having worked on this book for the past few months and reacquainted myself with his body of work, I am convinced that all maritime historians will learn a great deal from the essays included here. While each of the articles can be read and appreciated on its own, my strong recommendation is that this volume is best consumed in its totality. Taken together, the essays really are a history of nineteenth- and early twentieth-century British coastal shipping, an excellent overview of what we know and a superb guide to the questions that still lack answers. If I were reviewing the book rather than merely writing a brief foreword, I would argue that this is one of the dozen or so books to appear in the past half century that truly does belong on the bookshelves of all maritime historians.

All the essays in the pages that follow have appeared elsewhere, and we would like to thank the original editors and publishers for giving us the opportunity to bring them together. For the most part, each essay is identical to the original, but in the interests of scholarly precision I would like to note that we have introduced some changes during the editorial process. For example, we have standardized the text to fit the house style of *Research in Maritime History*. As well, the text has been edited *slightly* to remove as far as possible the typographical errors and stylistic infelicities that creep into all publications. But the most obvious difference is in the footnotes. Given the number of different places where John has published, we felt it necessary to introduce a common footnoting style. In addition, to make this book as useful as possible to the widest range of readers, we have also made some additions to the notes. In a few places, and with the author's permission, we have added works that appeared after the original essay was published. In particular, we have tried to find as many of the reprints of various works cited in the notes as possible in order to help interested readers track them down. With his encyclopaedic knowledge of the field, John has been of invaluable assistance in this process, and I thank him personally for all the time he devoted to this task.

Unlike many other re-printed collections in recent years, this book is not merely the outcome of modern xerography. Because we felt from the outset that John's work deserved better treatment, the book has been completely re-typeset in a consistent fashion. As readers will appreciate, this was a lengthy process that would not have been possible without some very important assistance (anyone who knows either John or me will know that neither of us could have done this!). In particular, we would like to thank the two most important "vital sparks" for their irreplaceable efforts. For her tireless work to see this book appear in a timely fashion, John and I would like to thank our Managing Editor, Maggie Hennessey, who has spent more time learning about coastal shipping than she ever imagined she would when she took the job. While she has never been to Connah's Quay, I am certain that Maggie now knows more about the shipping of this small port than almost any of its residents. And we would particularly like to extend our gratitude to Pamela Armstrong for solving innumerable problems that would otherwise have been intractable. Although I am sure that John would like to thank Pam for many other things as well, both of us appreciate all her hard work in the preparation of this book. Indeed, I have begun a file with e-mails and various attachments that Pam has sent (or given to John to send) at rather inconvenient hours of the day and night in answer to pleas from St. John's. I hope that both of these remarkable women are pleased with the result.

Lewis R. Fischer
St. John's
November 2009

iv

About the Author

John Armstrong <john@johnarmstrong.eu> has recently retired as Professor of Business History from Thames Valley University, London. He is a graduate of the London School of Economics and a Fellow of the Royal Historical Society. He is interested in all forms of transport history, has published on many of them, and for over a dozen years was the Editor of the *Journal of Transport History*. His main continuing academic interest has been in British maritime history, especially the British coasting trade and industrialization. As this volume suggests, there are many aspects of the coastal trade which he has addressed, but particularly from a business history perspective the routes served and cargoes carried, and the relationship between the coaster and other forms of domestic transport, such as the railways, canal system and road transport.

Over the last decade or so, he has become more interested in the early steamboats and, with David M. Williams of Leicester University, has published a number of articles on various aspects of the steamboat in the period from about 1810 to about the middle of the nineteenth century. These essays have explored the implications of a new form of technology in terms of both the new services offered and the additional risks posed. Some of this research has led into aspects of leisure travel and the beginnings of large-scale tourism, while other parts have led to examining attempts to operate long-distance steam liners. This line of research is ongoing, with a current concern being who built the hulls and engines of the very early steamboats in the United Kingdom and what the implications were for the diffusion of this new technology.

He has served on various academic bodies and committees. He was the Honorary Treasurer of the Business Archives Council for a decade or so in the 1990s after editing its journal, *Business Archives*, for half a dozen years. He has served as the Vice President of the International Maritime Economic History Association between 2004 and 2008, and was on its editorial board between 2000 and 2004. He also served on the Council of the Economic History Society in the 1980s. He has been a member of the British Commission for Maritime History (BCMH) for the last decade or so and has helped arrange and chair their flagship seminars at King's College, London, for about the same time. He is also currently Deputy Chair of the BCMH and serves on the Advisory Board for the Centre for Port and Maritime History, which links the University of Liverpool and the Merseyside Maritime Museum.

Introduction

When launching a scholarly book, it is not unreasonable to ask why the academic community should have thrust upon it yet another fat volume. Surely there already is a sufficiency of tomes on maritime history? Quite obviously, the appearance of this publication suggests little sympathy with the latter view. It is not so much the absolute number of maritime histories as the topics they address. It is my contention that little has been written on the British coastal trade, especially on the crucial period of industrialization and urbanization between about 1750 and 1914. The coastal trade was called "the Cinderella trade" a few years ago because it was a neglected area of study despite the fact that it was the largest segment of nineteenth-century British maritime commerce. It remains a relatively under-researched area of maritime history, and not just in the UK: there is relatively little written on the cabotage of any country. There are, of course, some honourable exceptions, but these are relatively rare.

The aim of this volume is two-fold. First, it endeavours to show the current state of knowledge on the British coastal trade. In so doing it indicates those areas which have not been addressed or where there is an on-going debate which could benefit from fresh research. This is the second aim: to encourage new research. In this regard the volume offers a shopping list of possible research topics which would carry the subject forward. This is based upon my belief that there may be some benefits in having these articles together in one volume. Some, it must be admitted, first appeared in rather obscure journals and thus may be difficult to access, and others saw their initial light of day in collections which are not always easy to trace. Thus, bringing them together in a single volume may ease access. It may also show that there is a critical mass of research on the British coastal trade.

A further question which might fairly be put is why has this particular set of essays been chosen? Not all of my work has been included, even on the coastal trade. So what criteria have been used for inclusion or exclusion? Obviously, the economics of producing this volume and its structural integrity dictate a maximum, especially when printing and postal charges are taken into account. The first criterion for exclusion was to omit those with a significant degree of overlap with other articles. Similarly, where an article was part of an exchange which did not make much sense without the rest of the contextualizing material, it was omitted, as were those essays that were very brief or which appeared as introductory or concluding remarks.

On the positive side, there were some criteria employed to determine inclusion. One was the period covered by the article. Although the majority of the pieces deal with the mid- to late nineteenth century, there are also essays on the early modern period and the eighteenth century; at the other end of the spectrum, one essay wanders into the interwar years. Thus, there is some coverage of a very long period, stressing the longevity of the coastal trade.

Another criterion considered when selecting pieces concerned the sources employed in constructing it. While most of the sources have been used in the past by maritime historians, virtually none had been used to analyze the coastal trade. Sources such as the Bills of Entry, Parliamentary Papers, crew agreements, newspapers, shipping registers and Admiralty records, as well as the business records of coastal shipping companies, were all used in some of the articles. This selection of essays draws upon a wide range of diverse sources.

The final, and most important, criterion was that the article should be important in its own right by making an original contribution to knowledge. Again, the range of topics covered was examined, and those which demonstrated a breadth of coverage were chosen. Some articles assessing the relative importance of the coastal trade over time, and especially relative to other forms of inland transport, such as railways and canals, have been included. There are other articles which focus on the aggregate, for example on the size of the coastal fleet at particular times and the relative importance of the coastal liner as compared to the coastal tramp trade. These can be seen as largely statistical exercises at the national level. Other essays deal with specific trades, such as coal from North East England to London, the Thames estuary and the South East. This was probably the single largest coastal trade in the eighteenth and nineteenth centuries. At the other end of the spectrum, one essay explores the coastal trade of a remarkably small port on the Dee estuary and its reason for decline. To counterpoint this essay, there is an article on the most important river in the UK, namely the Thames, and the role of the coastal trade in bringing into the capital raw materials for food and drink as well as fodder for animals. Various building materials, such as bricks, slate and stone, were brought in, and above all coal for heat, light and power.

Some of the essays in this collection explore the economics of the coastal trade. One is a contribution to a large and on-going debate over the course of maritime freight rates, especially in the second half of the nineteenth century. The whole of the debate was about deep-water freights. The essay in this collection looks at freight rates in the coastal trade as a contrast. Another paper looks at the depression in shipping in the early twentieth century and tests some of the ideas here against coastal coal freights. This broadened the debate a little. A couple of articles examine the nature of the relationship between coasters and railway locomotives. These two modes of transport have traditionally been seen as in competition, with the railway gradually taking away market share from the coaster. Several essays suggest that there might be some amendments to this story. There were many coastal liner companies which traded until the First World War and beyond. The railway did not eliminate the coaster as a carrier of high-value goods. There were also many "regular traders" plying coastal routes, such as between Liverpool and Hull, a classic railway route one might have thought. But competition between coaster and railway was not the only type of relationship. There was also cooperation between railway companies and coastal liner firms, some of it monitored by the Railway Clearing House. Traffic was shared and freight rates determined in concert. This cooperation also existed between coastal liner companies. These relationships were often known as conferences. It used to be thought that conferences originated in the 1870s in the Calcutta trade. The article reproduced here, and others, have pushed the date of the first conferences much earlier and onto coastal routes. There is also a long discussion of the role of government in the coastal trade. Given that this is sometimes characterized as an age of *laissez faire*, there was a great deal of government activity in the transport industry, and shipping was the recipient of much of it, perhaps the second most regulated industry after the railways.

This brings us to the final big question. Why does the British coastal trade deserve to be studied and hence, what is the justification for this volume? If we define coastal trade as domestic shipping, then the coaster at least until the 1870s was carrying the bulk of internal trade. That makes it worth studying. In addition, when the railway network was completed, the coastal trade did not decline but continued to grow. That too deserves explanation. When we consider the quantity of literature on the railway system, then by comparison the coastal trade looks under-researched. This is especially the case when we recall the great debate about the role of the railways in economic growth sparked off by the "new economic historians." The coaster deserves similar in-depth research.

Another significant feature of the coastal trade is its role in pioneering technical changes which spread to other maritime routes. These innovations included steam power, iron hulls, compound engines, water ballast, steel frames and hulls, triple-expansion steam engines, screw propulsion and stronger boilers. Virtually all of these were pioneered on coastal routes and subsequently spread to a wider constituency. Was the coastal shipping sector more dynamic than deep-water entrepreneurs? This might seem blasphemy to some maritime historians, but the possibility deserves more investigation.

There is another area which deserves study: the nature of the relationship between the coastal trade and the rest of the maritime economy. To what extent did the coastal sector use the same institutions as other sectors, and to what extent did they have separate institutions? For instance, were arrangements for insuring ships and cargoes the same for coastal vessels as for deep- water ships, or were they different? Did they use the same agents or was there specialization?

In a limited space I have only been able to indicate some of the areas which might reward attention. There are many other aspects of the coasting trade inviting enquiry and many sources waiting to be exploited.

John Armstrong
London
July 2009

Chapter 1
British Coastal Shipping:
A Research Agenda for the European Perspective[1]

In the UK until recently the coastal trade was largely ignored when transport developments were discussed.[2] Over the last decade or so much more research has been published on this topic, and the broad outlines of the structure of the industry, the commodities carried, the routes worked and the role of the coaster in industrialization are now known.[3] This paper will review that literature in order to establish the broad parameters of the economics of the British coastal trade in the hope that it may suggest the sort of topics and angles which need to be researched on a European-wide basis.

An acceptable definition of what is meant by coastal trade is crucial, since there is no guarantee that the idea of coastal trade in one country will be the same as in the other twenty-five or so European states. In this regard the United Kingdom has an advantage over its continental neighbours because of its geographical make up. Britain is a series of islands and therefore was completely separate from all other countries, so coastal trade was internal trade, and this is a commonly used definition. For instance, the Customs Consolidation Act of 1876, section 140, stated that "all trade by sea from one part of the United Kingdom to any part thereof shall be deemed to be coasting trade."[4] To go foreign meant venturing into deeper water and abandoning the coastline. This is the definition I have used in my work and the one most commonly used in British writing on this topic. This definition does not fit other European countries so well. They had common boundaries and thus, whereas in Britain

---

[1]This essay appeared originally in John Armstrong and Andreas Kunz (eds.), *Coastal Shipping and the European Economy, 1750-1980* (Mainz, 2002), 11-23.

[2]For example, standard texts on the development of British transport devoted a few pages to coastal shipping: H.J. Dyos and Derek H. Aldcroft, *British Transport: An Economic Survey from the Seventeenth Century to the Twentieth* (Leicester, 1969; reprint, Harmondsworth, 1974), 45-46, 52 and 208-210; and T.C. Barker and Christopher I. Savage, *An Economic History of Transport in Britain* (London, 1974), 70-72.

[3]For a recent survey of the literature, see John Armstrong (ed.), *Coastal and Short Sea Shipping: Studies in Transport History* (Aldershot, 1996).

[4]P. Ford and J.A. Bound, *Coastwise Shipping and the Small Ports* (Oxford, 1951), 48.

1

sticking to the coast ensured the ship stayed in British waters, on the mainland of Europe it was very easy to go into foreign waters by following the coastline. However, internal trade could be one definition of coasting trade.

In Britain the distinction between coasting and foreign trade was reinforced in law by the question of certificates of competency for masters and mates. Under the Mercantile Marine Act of 1850 both officers had to obtain such a certificate, but only if the ship was foreign-going.[5] A new concept, the "home trade," was introduced by the act. Masters of ships sailing in the home trade needed no qualifications until 1854 and thereafter only if their ships carried passengers.[6] Thus, from 1854 the coastal liner trade had certified officers. In addition, from 1862 coastal steamships, along with foreign-going, had to have certified engineers.[7] However, sailing ships plying in the home trade which did not carry passengers could be skippered by totally uncertified men until the First World War.[8] In 1894 the limits of the home trade were defined to include voyages to the near-continent, between Brest in France and the mouth of the river Elbe in Germany.[9] Hence, about half of both the French and German coastline and all of Belgium and the Netherlands were capable of being visited by sailing ships with uncertified masters. It would be interesting to discover how far other European countries had similar distinctions in qualification and certification for masters and mates who were foreign-going compared to the coastal trade only.

Another common phrase which might be confused with coastal shipping is "short-sea shipping." For instance, the British Shipping Federation for some years had a Short Sea Trades Committee and in the 1960s, both a Short Sea Liner Section and a Coastal and Short Sea Tramp Section.[10] This usage was to distinguish between deep-water or blue-sea voyages, such as across the Atlantic or to the Far East, and those going to the Mediterranean or the Baltic.

---

[5]Nicholas Cox, "The Records of the Registrar-General of Shipping and Seamen," *Maritime History*, II, No. 2 (1972), 179; Emrys Hughes and Aled Eames, *Porthmadog Ships* (Caernarfon, 1975), 77-78; and Clifford Jeans, "The First Statutory Qualifications for Seafarers," *Transport History*, VI, No. 3 (1973), 255.

[6]Cox, "Records," 179.

[7]*Ibid.*, 180.

[8]Basil Greenhill, *The Merchant Schooners* (2 vols., London, 1951-1957; 4th rev. ed., London, 1988), 119.

[9]J.A. MacRae and Charles V. Waine, *The Steam Collier Fleets* (Wolverhampton, 1990), 30.

[10]Leslie H. Powell, *The Shipping Federation: A History of the First Sixty Years, 1890-1950* (London, 1950), 73.

The latter were within slightly more protected waters where there was insufficient space to allow waves to build up the long swells typical of deep-water seas.[11] In this case, "short sea shipping" included the coastal trade but was more than it alone, extending to near-continental and European voyages but excluding inter-continental journeys. Thus, many ferry journeys from the United Kingdom to European destinations counted as short sea but not coastal. In mainland European states it is conceivable that journeys could commence as coastal and then continue into short sea with no real distinction.

Another defining feature of coastal trade was that in many European countries it was reserved for the ships of nationals only, or at least ships which were registered in that country. In Britain the coastal trade had been protected under the Navigation Act of 1660,[12] but was opened to foreign competition in 1853.[13] It was part of the general liberalization of Britain's trade, at a time when British shipbuilders and shippers had technological and organizational competitive advantages over other nations. In practice the liberalization had no effect before the First World War. The proportion of non-British ships plying in the coastal trade was tiny, and British shipping made up the vast majority of clearances and entrances. This was quite different from many other European countries which retained protection of their coastal trade for their own ships and seamen. This restriction may have been beneficial in that it required inspection and enforcement and may therefore have generated records for the historian. Some nations, such as Greece, still reserve their cabotage for their own nationals, though that is to be phased out in the new century.[14]

It has sometimes been suggested that it is foolish to study the coastal trade as separate from the general history of shipping and maritime activity. Among the reasons put forward for this view is that changes in technology, whether on the ship or at the dockside, affect all shipping irrespective of trades. Sailors did not differentiate between the trades, and of a more practical nature, the ships did not distinguish between coastal and overseas destinations but went where there was a cargo to be carried at a remunerative freight rate. Thus, it is a false dichotomy to separate coastal from overseas ships, as there

---

[11]Ian Dear and Peter Kemp (eds.), *An A-Z of Sailing Terms* (Oxford, 1992; 2nd ed., London, 1997), 168.

[12]Ronald Hope, *A New History of British Shipping* (London, 1990), 188.

[13]Adam W. Kirkaldy, *British Shipping: Its History, Organisation and Importance* (London, 1914; reprint, Newton Abbot, 1970), 27.

[14]Maria Lekakou and E.S. Tzannatos, "Policy Options for Increasing the Competitiveness of Intra-Greek Cargo Coastal Shipping," in Chris Peeters and Tør Wergeland (eds.), *Third European Research Roundtable Conference on Shortsea Shipping* (Delft, 1997), 471-485.

was much crossover. All of these points can be disputed, but now is not the time or place. What matters here is that we are not solely interested in the *ships,* but in the voyages they made. If we accept the definition of coastal trade as internal trade, then all voyages between two internal ports are coastal and all between one internal and one foreign are foreign voyages. We can then study those voyages which we have designated as coastal to determine their characteristics. It becomes a little more complex when we wish to investigate the ships themselves, but I will argue later that there are many ships which can be identified as almost entirely operating in the coastal trade and that as specialization of shipping type increased over time this distinction became even sharper.

The main aim of this book is to examine the relationship between the coastal trade and economic development and it is to this that I now turn. I wish to argue that the coaster was crucial in supporting and encouraging the rapid industrialization usually designated as "the industrial revolution;" that it continued to be important in fostering economic growth throughout the long nineteenth century, despite the advent of the railway; and that the coaster had special advantages over the long haul and in carrying bulky low-value goods, but also played a part in moving high-value goods. The economic advantage of the coaster was based on large capacity and low unit operating costs compared to its rival modes of transport. Let us look firstly at the coaster's role in industrialization. In Britain, as elsewhere, industrialization accompanied urbanization. The coaster played a crucial role in both movements. Urbanization meant a large increase in the physical structure of towns. This required bricks or stone, timber, slates, sand, lime and plaster for building houses, factories and offices. Granite setts were needed for roadways, and as much traffic was horse-drawn huge amounts of hay, straw and oats and other grains were required, as each horse ate about twenty pounds of grain per day. The increased population of these towns and cities required food and drink, and the main fuel for heat, light and power was coal. Many of these commodities came long distances or from specific locations. Thus, taking London alone, most of its coal came from the North East,[15] a journey of about 300 miles, granite setts and kerbs came from Aberdeen,[16] over 400 miles distant, Portland stone came from the south coast for public buildings and civic improvements such as The Embank-

---

[15]Raymond Smith, *Sea-Coal for London: History of the Coal Factors in the London Market* (London, 1961), 15-22 and 276-296; and Roger Finch, *Coals from Newcastle: The Story of the North East Coal Trade in the Days of Sail* (Lavenham, 1973).

[16]Clive H. Lee, "Some Aspects of the Coastal Shipping Trade: The Aberdeen Steam Navigation Company, 1835-80," *Journal of Transport History*, 2nd ser., III, No. 2 (1975), 95, reprinted in Armstrong (ed.), *Coastal and Short Sea Shipping*, 91.

ment, slates came from North Wales,[17] cheese from Cheshire,[18] salmon and beef from northern Scotland.[19] Some products came more locally, grain from Kent and East Anglia,[20] hay from west Middlesex, but one of the characteristics of British urban growth was the wide network of trade flows. Many of these commodities reached the cities which were on the sea or navigable rivers by coaster. It should be borne in mind that the last qualification was no problem. Eight out of the top twelve of Britain's urban centres were either on the sea or a navigable river. The east coast coal fleet is well known, as were the regular sailing packets from Aberdeen and Dundee. Cheshire cheese came by sea, as did Welsh slate and Aberdeen granite. Many of these were sent in large consignments and were of low intrinsic value. Many also demonstrated punctiform characteristics; that is, they needed to be moved from one location to one destination. In addition to the large variety of goods flowing into the towns and cities, there was also a need for a less pleasant outward flow. Night soil, horse droppings from stables and the roads,[21] the ash from boilers and domestic hearths,[22] broken glass, brick rubble, and numerous other waste or bye products had to be removed from urban areas if the death rate was not to rise to even worse proportions. Again, much of this refuse was cleared from the cities by coastal ships. Thames barges brought wheat and barley from East Anglia and took back night soil or horse manure to fertilize the ground. Cullet was carried back to Newcastle for re-using, ashes were taken out to Kentish brick makers to be used in their trade and brick rubble often acted as ballast for coasters sailing light.

Industrial growth created similar strains in terms of requiring large amounts of bulky raw materials to be moved around the country. Coal was

---

[17]Jean Lindsay, *A History of the North Wales Slate Industry* (Newton Abbot, 1974), 188-194.

[18]Robin Craig, "Some Aspects of the Trade and Shipping of the River Dee in the Eighteenth Century," *Transactions of the Historic Society of Lancashire and Cheshire*, CXIV (1963), 99-128, reprinted in Craig, *British Tramp Shipping, 1750-1914* (St. John's, 2003), 261-292 ; and Walter M. Stern, "Cheese Shipped Coastwise to London towards the Middle of the Eighteenth Century," *Guildhall Miscellany*, IV, No. 1 (1973), 207-221.

[19]See Lee, "Some Aspects," 95.

[20]Guildhall Record Office, London, Mss. 1667/1-3.

[21]Hervey Benham, *Down Tops'l: The Story of the East Coast Sailing Barges* (London, 1951; 3rd ed., London, 1986), 139-141.

[22]Patricia E. Malcolmson, "Getting a Living in the Slums of Victorian Kensington," *London Journal*, I, No. 1 (1975), 37.

pre-eminent as an industrial fuel, and large amounts of salt, china clay, iron and copper ores, timber for pit props, pig iron and alkalis also needed to be moved. These too required the transport services of the coaster, for they needed to move long distances and in large quantities. Such goods could not stand high unit costs or they were priced out of the market.

If it is accepted that urbanization and industrialization required the transport of large consignments of relatively low-value raw materials over long distances, perhaps we should now explain the dynamics of the coaster's competitive advantage. Essentially, before the railway, the only real competitor to coastal transport was land transport by packhorse, horse and cart or horse-drawn wagon. For the United Kingdom we can dismiss river and canal traffic as competitors to the coaster since sailing coasters could and did sail far up rivers; hence, river trade was often an extension of coastal trade and even some canals were accessible to coastal ships and acted in the same way. Canals were rarely built parallel to the ocean, which might have provided competition for the coaster. Rather they were built from port or navigable river to an inland town or between two inland towns. In this sense canals did not compete for the same traffic as the coaster, but instead extended the all-water routes either by small coasters entering canals directly or by unloading over the side into barges or lighters which continued the journey. Thus, canals, river transport and coasters were not competitors but complemented each other.

This leaves only horse-drawn road haulage as an alternative to the sailing coaster in the industrial revolution. Here again, there was relatively little real competition. Each mode had its economic strengths and weaknesses which determined in which routes and trades it was likely to thrive. Compare the capacities of the two modes. A horse-drawn wagon could carry up to four tons by the 1760s,[23] a cart perhaps thirty hundredweight and a packhorse a couple of hundredweight. The average coaster in the middle of the eighteenth century carried seventy tons,[24] and there were some much larger, for instance in the east coast coal trade, where Ville believes them to have been about 300-ton burthen ships.[25] Thus, whereas a consignment of thirty or fifty tons would have been easy for a single coaster, it would have required a string of wagons

---

[23]John A. Chartres, "Road Transport and Economic Growth in the Eighteenth Century," *Refresh*, VIII (1989), 7, reprinted in Anne Digby, *et al.*, *New Directions in Economic and Social History* (2 vols., London, 1989-1992). Twenty hundredweight made one ton.

[24]T.S. Willan, *The English Coasting Trade, 1600-1750* (Manchester, 1938; reprint, Manchester, 1967), 11-13.

[25]Simon P. Ville, "Total Factor Productivity in the English Shipping Industry: The North-east Coal Trade, 1700-1850," *Economic History Review*, 2nd ser., XXXIX, No. 3 (1986), 360.

and a plethora of carts. The operating costs of horse-drawn road transport were also much greater per unit carried than was the case for the coaster. Each wagon needed a human conductor and, if it was carrying four tons, about six horses.[26] Thus, to carry the previously cited fifty tons would have taken thirteen wagon drivers and about eighty horses. By comparison the crew of a coaster which could carry fifty tons would be four or five,[27] so that the labour costs of a coaster were much lower per ton of cargo than those of horse-drawn road transport. Horses were also expensive to feed, consuming about twenty pounds of corn or grain a day as well as prodigious amounts of hay.[28] By comparison the wind came free and when it took a toll on sails, spars or other gear, it was part of the implicit contract with sailors that among their duties were to repair, paint and generally maintain the fabric of the ship. A mariner who could not stitch a sail, scrape a spar or caulk a seam was not worth hiring. Horses were not merely expensive to feed, they also contracted diseases and ailments which reduced their output, required them to rest and incurred vets' bills, or in extreme cases killed them early. They were also expensive pieces of capital equipment. A mature carthorse in the late eighteenth century cost about £10, to give £60 for the full team. A broad wheeled wagon at the same time cost between £30 and £40,[29] to give a total outlay of between £90 and £100 for the whole equipage, which carried about four tons. A coastal ship's initial cost has been calculated by Craig to have been about £5 to £6 per ton in the late eighteenth century.[30] Thus, to build and outfit four tons of carrying space on a coaster might cost between £20 and £24, considerably less than the initial cost of a wagon and six horses. Although the coaster represented a larger initial outlay, for the minimum size was about thirty tons burthen, and the average about seventy to give capital costs of £150 to £420, capital costs per ton of capacity were lower than for road haulage. Thus, freight costs per ton

---

[26]Dorian Gerhold, "Packhorses and Wheeled Vehicles in England, 1550-1800," *Journal of Transport History*, 3rd ser., XIV, No. 1 (1993), 20, reprinted in Gerhold (ed.), *Road Transport in the Horse-Drawn Era* (Aldershot, 1996), 139-164.

[27]Willan, *English Coasting Trade*, 16-17.

[28]Gerhold, "Packhorses," 16; and John Tilling, *Kings of the Highway* (London, 1957), 80.

[29]Gerard L. Turnbull, *Traffic and Transport: An Economic History of Pickfords* (London, 1979), 17-18.

[30]Robin Craig, "Capital Formation in Shipping," in J.P.P. Higgins and Sidney Pollard (eds.), *Aspects of Capital Investment in Great Britain, 1750-1850* (London, 1971), 14-34, reprinted in Craig, *British Tramp Shipping*, 41-58; Willan, *English Coasting Trade*, 39; and Edward Hughes, *North Country Life in the Eighteenth Century. Volume II: Cumberland and Westmoreland, 1700-1830* (Oxford, 1965), 187.

mile by sea were much lower than by land, perhaps a quarter or even a fifth of land rates.

In addition, the coaster had particular advantages on long hauls. Such trades were quite normal, the east coast coal trade from the Newcastle region to London and the south coast being one example and Cheshire cheese from the North West to London being another. The explanation of the coaster's advantage in long hauls lies in many of its costs being terminal costs, that is incurred going into or leaving a port, and fixed irrespective of the length of the journey made. For example, harbour or quay costs were usually charged per ton of ship. Unloading, where not carried out entirely by the crew, was a function of cargo size; pilotage was a fixed sum; the costs of taking on ballast or discharging it lawfully were also incurred at the end or beginning of the voyage. These costs, fixed irrespective of the length of the journey, were obviously lowered per ton mile as the voyage length increased. By contrast, road transport's costs were much more proportional to the journey length as horse feed and wage costs were incurred relative to length of haul and not a once and for all charge, each journey.

A further consideration which pointed up the cost advantage of the coaster was longevity. A wagon horse had a working life of only five or six years, then it was too feeble to continue in harness.[31] This meant continuing costs of replacement of capital and hence high depreciation costs – though not expressed as such contemporaneously. By comparison the wagon, as the coastal ship, was capable of lasting decades, subject to relatively minor repairs and maintenance. The objection to this raised by road haulage experts is that ships sunk whereas wagons did not. However, wagons were occasionally wrecked due to overturning or running away on hills, and the rate of coaster wrecking is contentious and uncertain. The point is that whereas the coaster could be depreciated at five percent, assuming a twenty year life, for road transport, horses needed to be written off over five years, or at twenty percent per annum, and they were the preponderant capital cost in that mode.

Nor should the coaster be dismissed as much slower than horse-drawn road transport. Chartres believed large, heavily laden wagons travelled around twenty miles per day.[32] They needed to stop at night to allow drivers time to rest and feed, and more importantly for the horses to do likewise. Not allowing the horse to rest, requiring it to pull extra loads or to climb steep inclines increased fodder consumption and reduced longevity, so pushing up costs. By comparison the coaster travelled twenty-four hours a day once at sea and was capable of six or seven knots given a wind of appropriate strength and direction. Thus, in a day it could manage a journey of 140 miles or so. For in-

---

[31]Tilling, *Kings*, 63 and 81.

[32]Chartres, "Road Transport," 7.

stance, in 1687 the ketch *Edward and Jane* averaged just under 100 miles per day on its journey from London to Lancaster.[33] The coaster only needed to sail at one knot on average over the day to travel further than a road vehicle. Fast passages were achieved and even at average rates of progress a coastal voyage was quicker than a similar journey by road. The disadvantage of the coaster was its unpredictability. It could make fast passages, but it could also be locked into a harbour by lack of wind, or too much of a blow or from the wrong direction. Similarly, once at sea it was at the mercy of winds and tides and whereas the latter were predictable the former were not. By comparison, road transport was less likely to be affected by weather and more able to predict times of arrival, as Gerhold has shown for Russell's, the Exeter carriers.[34]

As a result of these different characteristics, the two modes catered for different routes and commodities. Low-value goods requiring long-distance transport had to go by coaster. Land carriage of such products was only possible for a few miles as otherwise the good was priced out of its market. Similarly, wherever the point of origin and the ultimate destination were within a few miles of water and the haul was long, say over forty miles, the coaster was the first choice because of its low cost, even when the commodity was more intrinsically valuable. Land carriage was used for long journeys when both origin and destination were well inland; for short journeys within towns, such as from quay to warehouse or factory; for long journeys when the good was very highly valued, such as bullion, specie or spices, and not very voluminous. When the good was bulky, but low weight and high value, such as Dent's stockings[35] or the serge carried by Russell and Co.,[36] it went by road. In addition, at certain times the coaster suffered because of external conditions. When Britain was at war with another maritime power its shipping was subject to predation by pirates, privateers and enemy men of war. As a result, insurance rates rose steeply and delivery times suffered as ships waited in port to make up a convoy. At this time high-value goods were likely to prefer the less vulnerable land route rather than the riskier, and now more expensive because of the increased insurance premiums, sea journey. This could be compounded where consignments were small in size and the frequency of coastal services was low. As Turnbull has shown, where the product was high value, such as

---

[33]John D. Marshall (ed.), *The Autobiography of William Stout of Lancaster, 1665-1752* (Manchester, 1967), 89.

[34]Dorian Gerhold, *Road Transport before the Railways: Russell's London Flying Wagons* (Cambridge, 1993).

[35]T.S. Willan, *An Eighteenth-Century Shopkeeper: Abraham Dent of Kirkby Stephen* (Manchester, 1970).

[36]Gerhold, *Road Transport before the Railways*.

fine linen cloth, the consignments small and the frequency of a coastal service low, it made more sense to send the parcels by road carrier, even though the price was several times higher than by sea.[37] This was compounded when war aggravated the coastal trade, and there were many years in the late eighteenth century when Britain was at war with at least one maritime power. However, in general shippers did not have a real choice between modes; it was only for expensive low-weight goods that there was any possibility of long-distance road carriage. Even when war made coastal shipping riskier and more expensive, the vast majority of low-value, high-bulk goods could not travel by road.

Thus, in the United Kingdom the coastal ship was crucial to industrialization and urbanization, and it is difficult to imagine the "industrial revolution" taking place without it. The fact that Britain was particularly well endowed with a long coastline, ample navigable rivers and very few urban centres more than ten miles from navigable water may well have aided her in being the first to industrialize. No other European country was so richly endowed. The extent to which this British model was similar to the mainland continental experience is more uncertain. Apparently Britain had an advantage in her abundance of coastline compared say, to France or Germany, but countries like Denmark, Norway or Italy had extensive coasts and should have faced similar cost conditions and demands.

The early nineteenth century saw the application of steam power to marine navigation, with Henry Bell's *Comet* plying from Glasgow in 1812.[38] In theory this should have removed one of the coastal ship's weaknesses. By allowing it to proceed at a steady pace irrespective of wind strengths or direction, it became more reliable and predictable, hence more able to compete with road transport. In practice the early steamboats required so much space for their engines and bunker fuel as to have limited range and restricted cargo capacity, so that they carried passengers, perishables, parcels and high-value, low-weight commodities. However, over time their efficiency improved, thanks to compounding; later, triple-expansion steam engines, combined with more efficient boiler designs, reduced the coal consumption drastically for any given power output. Thus, by the 1850s they were in widespread use.

The role of the coastal ship in the second half of the nineteenth century was similar to that in the industrial revolution, but with added strengths and a more effective competition. The advent of steam gave birth to the coastal

[37]Gerard L. Turnbull, "Scotch Linen, Storms, Wars and Privateers: John Wilson and Son, Leeds Linen Merchants, 1754-1800," *Journal of Transport History*, 3rd ser., III, No. 1 (1982), 47-69, reprinted in Gerhold (ed.), *Road Transport in the Horse-Drawn Era*, 75-98.

[38]Brian D. Osborne, *The Ingenious Mr. Bell: A Life of Henry Bell (1767-1830), Pioneer of Steam Navigation* (Glendaruel, 1995).

liner trade which ran to a scheduled time of departure and arrival.[39] It was fast, as the most modern ships were employed; reliable, as skippers were expected to adhere to posted times of departure and arrival; and convenient as small consignments were carried. It was also frequent, many services between major cities being twice weekly. Thus, the coastal steam liner brought a new quality of service to coastal trade, and along with improvements in dockside facilities, a new reliability and rapidity, for coastal liners represented significant aggregations of capital and hence needed to be used intensively. They also were large ships which meant great quantities of cargo to be loaded and discharged. Time was also of the essence because they had a schedule to adhere to and so would not stand the lowly place in the pecking order to which most coasters were consigned. To obviate this, many coastal liner companies bought or hired their own dedicated wharf so their vessels did not have to await a turn and installed appropriate dockside equipment for loading and unloading.

However, if steam eliminated the main weaknesses of the sailing coaster it also brought a potent new competitor in the form of the railway. Initially this was no real threat to the coaster for the early lines tended to complement the coastal routes, running from coal field to port or from inland city to port city.[40] Indeed, in two regards the early railway enhanced the coaster's trade by requiring large quantities of bulky material, such as sleepers, rails, ballast, chairs and even locomotives, to be carried around the coast, and also by unlocking a flow of goods from the interior to the coast for onward transmission by coaster. These early railway lines were also generally short distance, thus competing more with horse than with seaborne traffic. By the 1850s a national structure of long-distance through railway lines was beginning to emerge, thanks to expansion, mergers, working agreements and the work of the Railway Clearing House.[41] However, even then the railway did not prevent the coaster from continuing to expand its service. The statistics collected by government of ships entering and clearing each port in the coastal trade, despite the occasional alteration in the basis on which the figures were collected, show a steady if unspectacular growth rate of about two percent per annum on

---

[39]John Armstrong, "Freight Pricing Policy in Coastal Liner Companies before the First World War," *Journal of Transport History*, 3rd ser., X, No. 1 (1989), 180-197, reprinted in Armstrong (ed.), *Coastal and Short Sea Shipping*, 112-129.

[40]John Armstrong, "Management Response in British Coastal Shipping Companies to Railway Competition," *The Northern Mariner/Le marin du nord*, VII, No. 1 (1997), 17-28.

[41]Philip S. Bagwell, *The Railway Clearing House in the British Economy, 1842-1922* (London, 1968).

average from 1830 to the First World War.[42] Other evidence suggests a flourishing coastal trade until this date: the data on coal carried coastwise,[43] the occasional glimpses of annual cargoes carried by a coastal company, and the growth in the frequency and coverage of the coastal liner network.

How do we explain the continuing ability of the coaster to compete with the shining new technology of the railway, which some transport historians have portrayed as all-conquering in the nineteenth century? The answer might be sought in several features, and it is interesting to consider how many of these were at work in other European countries. One feature already alluded to is that the coaster segmented its market to provide differing types of service to different customer needs. This varied from the cheap but unreliable sailing ship for low-value, non-deteriorating, bulky goods, via the steam tramp for large consignments of irregular despatch and regular traders specializing in a shuttle service for one commodity, such as the east coast screw colliers,[44] to the highly reliable coastal liner, carrying mixed manufactured goods in small consignments, such as Bovril meat extract and Lever's soap,[45] as well as passengers and livestock. In this way, different levels of service were offered at different price levels to cater for a range of trades.

In addition, just like the railway, technology in the coastal ship was not static.[46] There were dramatic improvements in the technology of engines and boilers to give lower fuel consumption per unit of output and hence lower cost. The hulls were initially made of wood, then composite and later all iron; in the late nineteenth century, steel was being used. Metal hulls allowed larger ships without losing structural integrity and hence the average size of ships in the British coastal fleet increased. Overall coaster size rose from about 100 tons in mid-century to nearly 200 tons just before the First World War.[47] By then some of the steam colliers were capable of carrying cargoes of coal in

---

[42]Michael J. Freeman and Derek H. Aldcroft (eds.), *Transport in Victorian Britain* (Manchester, 1988), 172.

[43]*Ibid.*, 183-190.

[44]John Armstrong, "Late Nineteenth-Century Freight Rates Revisited: Some Evidence from the British Coastal Coal Trade," *International Journal of Maritime History*, VI, No. 2 (1994), 45-81.

[45]Armstrong, "Freight Pricing," 194.

[46]This is well dealt with by J. Graeme Bruce, "The Contribution of Cross-channel and Coastal Vessels to Developments in Marine Practice," *Journal of Transport History*, 1st ser., IV, No. 2 (1958), 65-80, reprinted in Armstrong (ed.), *Coastal and Short Sea Shipping*, 57-72.

[47]Freeman and Aldcroft (eds.), *Transport in Victorian Britain*, 172.

excess of 1500 tons.[48] In addition, innovations such as water ballast cut costs and speeded up turnaround times and the use of propellers rather than paddles made for more slender lines and greater speed. The coasting trade acted as a nursery for overseas shipping in the sense that most of the technical innovations were pioneered on coastal routes, and only when imperfections had been ironed out did they then go on to be adopted in blue-water trades.

In this way the coastal ship was continually being refined and improved to provide lower operating costs, economies of scale, greater speed and more rapid turnarounds. Nor were such improvements confined to steamers. Sailing ships too were subject to technical change. The most obvious was the move away from square to fore-and-aft rigs which cut crew requirements drastically. In addition, the use of winches, pulleys and roller reefing allowed more to be done on deck and to be carried out more quickly.[49] The design of hatches and the use of winches also allowed quicker loading and unloading and the carriage of larger objects. Thus, the coastal ship of the early twentieth century was a far more efficient and productive instrument than its predecessor of mid-century.

Given the emphasis placed upon railway development in most textbooks of nineteenth-century transport,[50] we need to explain why the coaster did not succumb to the railway. One important aspect of its success was that its rates for other than very short journeys were significantly lower than the appropriate rail rate. The coaster continued to have a competitive advantage on long hauls, say over 100 miles, because its costs were peculiarly bunched at the terminal points of its voyages irrespective of distance travelled, whereas the railway's costs were much more proportional to distance travelled as it needed to own land, construct expensive civil engineering works and guard, maintain and repair its property. In addition, the railways were not always able to offer merchants the service they needed. Goods trains gave way to passenger trains often by sitting idle in a siding. Their braking equipment was normally antiquated and incompatible, so speeds were notoriously slow. Train load traffic was a largely alien concept, so much time was lost in assembling and disassembling trains and shunting. Wagons were not always readily available and remained small in size. As a result, many movers of goods preferred to use the coaster to the railway. An extreme example of this was the Wilson Line of Hull, which ran fortnightly steamers between Hull and Liverpool, a distance of about 1000 miles when the land route was only one-fifth of this,

---

[48]Armstrong, "Late Nineteenth-Century Freight Rates," 70-71.

[49]Greenhill, *Merchant Schooners*, 80-81; and Hervey Benham, *Once Upon a Tide* (London, 1955; 2nd rev. ed., London, 1986), 47.

[50]For instance, Dyos and Aldcroft, *British Transport*; and Barker and Savage, *Economic History of Transport*.

because the railway provision between the two cities was so slow, unpredictable and untraceable.[51] It would be interesting to know how far railways in other European countries were similarly unable to offer a sufficiently superior service to offset their higher costs.

There are many other aspects of the coastal trade which require investigation, including the extent to which ships stuck to coastal trading or moved into near-continental work and the nature of the coasters' crews. Was the coasting trade a nursery for seamen in that many started there and then moved to deep-water work, or did sailors specialize in one or the other? Ownership is another important aspect. Who invested in coastal ships and why? Were the owners people who had a direct interest in the venture because they had goods to ship, hoped for employment or wished to sell goods or services to the shipping firm? Did ownership vary between sail and steam or tramp and liner? We think we know the main commodities carried in the United Kingdom, but how profitable were particular trades and indeed coastal ship ownership *per se*? A colleague recently remarked that as far as the coastal trade was concerned we were starting with a blank slate. I do not entirely agree; there is a little faint writing on the slate, but there is still much to be explained. Hopefully, this chapter has indicated some of the gaps in our knowledge of the British experience and hence where future research effort is needed. In addition, it may suggest some worthwhile possibilities for investigation on a European-wide scale.

---

[51]See John Armstrong and Julie Stevenson, "From Liverpool to Hull – by Sea?" *Mariners Mirror*, LXXXIII, No. 2 (1997), 150-168.

Chapter 2
The Significance of Coastal Shipping in British Domestic Transport, 1550-1830[1]

An examination of the literature on inland transport in Britain in the early modern period might lead one to believe that the most important method of conveying goods was by road. In the last twenty years there has been a plethora of articles on the role of road transport, including important pieces by John Chartres, Gerard L. Turnbull, Michael J. Freeman and, most recently, Dorian Gerhold.[2] The effect of this research has been to disprove the idea that road travel was at best difficult, and at worst, in winter, near impossible, when mud, inclines, potholes and other hazards prevented mobility. It is now clear that there was a network of carriers, for both short-distance local travel and long distances, which provided regular and reasonably reliable journey times for a range of commodities. Road transport grew in significance over the period. Coastal shipping was equally important, however, and the aim of this paper is to highlight its role in the transport developments which preceded the coming of the railways.

The pendulum has swung too far in one direction, and the bias in the literature in favour of road transport might be mistaken as a sign of the relative unimportance of coastal shipping. This essay sets out to correct that impression. It isolates economic variables relevant to transport and reassesses the relative importance of coastal shipping in light of those factors. Material is drawn chiefly from secondary sources. Little primary research has been done since T.S. Willan wrote his masterpiece over fifty years ago, which is a la-

[1]This essay appeared originally in the *International Journal of Maritime History*, III, No. 2 (1991), 63-94.

[2]John A. Chartres, "Road Carrying in England in the Seventeenth Century: Myth and Reality," *Economic History Review*, 2nd ser., XXX, No. 1 (1977), 73-94; Chartres, *Internal Trade in England, 1500-1700* (London, 1977); Gerald L. Turnbull, "Provincial Road Carrying in England in the Eighteenth Century," *Journal of Transport History*, New ser., IV, No. 1 (1977), 17-39; Turnbull, *Traffic and Transport: An Economic History of Pickfords* (London, 1979); Michael J. Freeman, "The Carrier System of South Hampshire, 1775-1851," *Journal of Transport History*, New ser., IV, No. 2 (1977), 61-85; Freeman, "Road Transport in the English Industrial Revolution: An Interim Reassessment," *Journal of Historical Geography*, VI, No. 1 (1980), 17-28; and Dorian Gerhold, "The Growth of the London Carrying Trade, 1681-1838," *Economic History Review*, 2nd ser., XLI, No. 3 (1988), 392-410.

mentable reflection on the neglect of the coasting trade.[3] Thus, a further aim of
this review is to indicate some areas where further research might be profitable. Chartres, Turnbull and others have shown that road transport was a normal method of moving goods as early as the sixteenth century and that its importance increased with Britain's economic growth.[4] By the eighteenth century
there existed a reasonably well-integrated network of carriers capable of moving significant quantities of freight. The price of the service was not in all
cases the only, or even the most important, consideration in choosing a transport mode. Sometimes speed, reliability of arrival time or regularity of service
was of more importance to customers. There is no simple rule as to how
prices, speed or reliability compared between different transport modes. The
effects of the seasons, the weather, the likelihood of robbery and the degree of
competition varied enormously from route to route and from year to year, or
even month to month. Thus the customer – merchant, manufacturer, farmer or
builder – would select a means of transport according to his priorities (which
were partly determined by the commodity in question). We must assume he
acted rationally within the limitations of information, habit and probabilities.
That said, I will try to comment on the part played by the coaster in goods
transport.

When the advocates of road transport have found individual merchants or manufacturers using wagons for long-distance goods haulage it has
frequently been for textile products. Thus, Wilson's linens were despatched
from Scotland to Leeds by road, as was M'Connel and Kennedy's yarn from
Manchester to Glasgow, and Horrocks' yarn from Preston.[5] George Unwin's
study of Samuel Oldknow showed him despatching fine muslins, calicoes and
shirting to his London customers from Manchester by wagon and occasionally

---

[3]T.S. Willan, *The English Coasting Trade, 1600-1750* (Manchester, 1938;
reprint, Manchester, 1967). For a research project using the nineteenth-century Customs Bills of Entry for the coasting trade, see Valerie Burton, "Liverpool's Mid-
Nineteenth Century Coasting Trade," in Burton (ed.), *Liverpool Shipping, Trade and
Commerce* (Liverpool, 1989), 26-66.

[4]Chartres, "Road Carrying;" Chartres, *Internal Trade*; Turnbull, "Provincial
Road Carrying;" and Turnbull, *Traffic and Transport.*

[5]Gerard L. Turnbull, "Scotch Linen, Storms, Wars and Privateers: John Wilson and Son, Leeds Linen Merchants, 1754-1800," *Journal of Transport History*, 3rd
ser., III, No. 1 (1982), 54-57, reprinted in Gerhold (ed.), *Road Transport in the Horse-
Drawn Era* (Aldershot, 1996), 75-97; and Clive H. Lee, *A Cotton Enterprise, 1795-
1840: A History of M'Connel and Kennedy, Fine Cotton Spinners* (Manchester, 1972),
135-137.

sending small parcels by coach.[6] This was a volatile market brooking little delay, and the commodity was of high value and relatively low weight. Much of the transport of textiles was more areal than punctiform and in these conditions road transport had the edge. There is, however, no suggestion in this literature that the roads carried bulk goods other than for the shortest distances, whereas the coaster excelled in longer hauls. Even a semi-manufactured good such as pig iron would not stand lengthy road haulage. The Horsehay Company in Shropshire was only about fifty miles by road from Chester, yet in 1775 it delivered pig iron by carting it to the Severn and thence by trow to Bristol where an agent, John James, put it on a ship that sailed around the Welsh coast to the Dee. This roundabout route minimized land carriage.[7]

In simple quantity of work performed, coastal shipping far exceeded the roads, as can be demonstrated from London's trade. Although the London carrying trade has recently been the subject of some debate, all parties are virtually agreed that the year of maximum activity was 1826 when approximately 370,000 ton-miles were worked each week.[8] This gives an annual total of about 19.25 million ton-miles. Yet at the same time about two million tons of coal was coming into London each year, the vast bulk from the mines of the North East.[9] Reckoning the distance from Newcastle to London to be approximately three hundred miles, this gives a total ton-mileage of about six hundred million. In other words, in one trade alone the coaster performed about twenty times more work than the whole of road transport to London. Admittedly the coal trade was the capital's largest in volume, yet a vast range of commodities came to London by sea – 200,000 tons of East Anglian corn alone each year in the 1820s – and these, it should be remembered, are not accounted for in our calculation.[10] To do so might add another twenty million ton-miles to the

---

[6]George Unwin, *Samuel Oldknow and the Arkwrights: The Industrial Revolution at Stockport and Marple* (Manchester, 1924), 59-68.

[7]T.S. Ashton, *An Economic History of England: The Eighteenth Century* (London, 1955; reprint, Abingdon, 2006), 71.

[8]Gerhold suggested that Chartres and Turnbull overestimated the growth of the London carrying trade and underestimated its extent in the earlier eighteenth century: Gerhold, "Growth of the London Carrying Trade," 397.

[9]T.S. Ashton and Joseph Sykes, *The Coal Industry of the Eighteenth Century* (Manchester, 1929; reprint, New York, 1967), appendix E.

[10]Great Britain, Parliament, *Parliamentary Papers (BPP)*, 1828, XVIII, 330-345.

coaster's total, assuming a minimum average haul of 100 miles.[11] When it came to the transport of bulk goods over long distances the coaster was essential and performed far more work than the road wagon. Accepting that both Chartres' and Turnbull's estimates, as well as that of Gerhold, are not comprehensive but based on a selection of the most important services between London and the provinces, it seems highly unlikely that the roads performed any more than a small fraction of the ton-mileage provided by the coaster.

Ideally, some estimate of the rate of growth of the coastal trade in this period should be attempted. However, a lack of regularly collected data, changes in the method of measurement and uncertainty about the accuracy of the sources make this a hazardous undertaking. Willan gives the number of ships leaving London coastwise in 1628 and 1683 as 352 and 1001.[12] If we assume that this increase reflected real traffic growth, it implies 3.3 percent per annum for a fifty-five year period. But this seems too large, and it probably reflects a significant degree of under-recording in the earlier year. We know that there is inaccuracy in the figures of shipping tonnage registered in the English ports given in the Musgrave Manuscript covering the period 1709 to 1782.[13] For example, it states quite categorically that no coasters belonged to the port of London, whereas the 1789 registration return shows more than 25,000 tons of coastal shipping registered there.[14] Additionally, the figures for some ports given in the Manuscript remain suspiciously constant over time, suggesting the collector of the statistics merely repeated the previous figure. If, however, these inaccuracies were cancelled out in the long run, when the recorded figures caught up with the actual changes, then the source is not invalidated as a guide to long-term growth trends.

To further confuse the problem, several calculations have been made of the total tonnage of ships in the coastal trade at various dates. Davis thought there were about 132,000 tons of shipping engaged in the coastal trade in 1686.[15] Robinson challenged this, believing the figure to be more like

---

[11]The top three ports sending grain coastwise in the period 1819-1827 were Great Yarmouth, Boston and Kings Lynn, all significantly more than one hundred miles by sea from London.

[12]Willan, *English Coasting Trade*.

[13]British Museum (BM), Additional Mss. (Ad. Mss.) 11255. While widely-used by historians of British shipping, this is accepted as a very unreliable source; see the discussion below.

[14]Great Britain, National Archives (TNA/PRO), Board of Trade (BT) 6/191.

[15]Ralph Davis, *The Rise of the English Shipping Industry in the Seventeenth and Eighteenth Centuries* (London, 1962; reprint, Newton Abbot, 1972), 397-401.

212,000.[16] The degree of difference is large and makes any growth rates calculated even more tentative. Robinson has also thrown grave doubts onto the already suspect Musgrave Manuscript. The latter claims about 230,000 tons of shipping were engaged in the coastal trade in 1776, while Robinson, using the Registers of Protections, calculates it was more like 505,000 tons.[17] This difference is also enormous and indicates how much work needs to be done to try to arrive at estimates of the number and tonnage of ships actively engaged in the coastal trade at key years throughout the period. The process is complicated. Certainly by the eighteenth century ships were flexibly deployed in whatever trade they were suited for and which offered a profit. Hence, coastal ships undertook voyages to the near-continent or even the Mediterranean.[18] Despite these complications scholars should not be deterred from further research, perhaps using port books in combination with the Musgrave Manuscript and the Registers of Protections to reassess the quantitative importance of the British coasting fleet.

Here all that can be done is to point to some approximate growth rates suggested by these sources. Davis, although primarily interested in foreign trade, gives figures for coastal tonnage of 132,000 tons in 1686, 125,000 in 1730 and 133,000 in 1737.[19] In sum, Davis believed there was no growth in the British coastal fleet between the late-seventeenth and mid-eighteenth century. Thereafter the growth rate was quite rapid – 1.8 percent per annum from 1737 to 1782. This gives a growth rate for the overall period 1686 to 1782 of about 1.2 percent per annum.

Robinson, as I have already mentioned, disagrees with Davis' figures. He suggests 212,000 tons of coastal shipping in 1686 and 505,000 in 1776.[20] Unfortunately, he gives no intermediate figures for the first third of the eighteenth century to test Davis' belief that there was stagnation or even decline in the British coastal fleet between 1686 and the 1730s. Given the importance of the coal trade in coastal shipping, Davis' slowdown is surprising, for between

---

[16]Dwight E. Robinson, "Half the Story of *The Rise of the English Shipping Industry*," *Business History Review*, XLI, No. 3 (1967), 303-308.

[17]Dwight E. Robinson, "Secret of British Power in the Age of Sail: Admiralty Records of the Coasting Fleet," *American Neptune*, XLVIII, No. 1 (1988), 5-21.

[18]Simon P. Ville, *English Shipowning during the Industrial Revolution: Michael Henley and Son, London Shipowners, 1770-1830* (Manchester, 1987), 52-62.

[19]Davis, *Rise of the English Shipping Industry*, 396-405. Since the last two figures are taken from the Musgrave Manuscript, the series could be extended to give 200,000 tons in 1765 and 284,000 tons in 1782; BM, Add. Ms. 11255.

[20]Robinson, "Half the Story," 307; and Robinson, "Secret of British Power," 6.

1686 and the 1730s coal sent coastwise from Newcastle rose by nearly forty percent, and between 1700 and the 1730s coal brought to London by sea rose by about 1.25 percent per annum.[21] This dynamism in the major commodity carried coastwise surely indicates increasing demand for shipping tonnage. Either that or the efficiency of the coastal fleet improved significantly. Obviously Robinson tends towards the former view, for his estimates give a growth rate in coastal tonnage from 1686 to 1776 of about 1.5 percent per annum.

The shipping statistics in this period are a minefield, and a major exercise of cross checking, reconciling and evaluating the primary sources awaits a brave scholar. In the meantime we might reasonably assume that growth rates were between one and 1.5 percent per annum. Significantly, these rates are much lower than those calculated by Chartres and Turnbull for the growth in the carriage of goods by road in the eighteenth century. They estimated that for the period 1715 to 1796 ton-miles worked rose ten-fold, or roughly eleven percent per annum.[22] Their calculation only applied to services to the capital, however, and it might be argued that London-bound traffic was likely to increase at a faster rate than in the provinces given London's population and pre-eminence as a centre of conspicuous consumption. Moreover, the likelihood of a constant and sustained growth exceeding ten percent per annum over this entire period is remote. The figures suggested above for the coasting trade may be rough but are probably closer to reality.

If we accept that the coaster carried much more than road transport in terms of ton-mileage, we need to explain what advantages it had over the wagon or packhorse. The most significant was price. In general the coaster was able to perform its transport service at a fraction of the rate charged by the road carriers. For example, at the height of the French wars in 1812 and 1813, coastal rates were in all probability high owing to the difficulty of obtaining seamen given the depredations of the press gang; the demand for vessels for government service;[23] and the increased cost of higher wartime insurance premiums. Yet in these years M'Connel and Kennedy were sending their yarn from Manchester to Glasgow by coaster out of Liverpool.[24] The road rate was about sixteen shillings per hundredweight whereas the sea freight ranged be-

---

[21]Michael W. Flinn, *The History of the British Coal Industry, Vol. 2: 1700-1830: The Industrial Revolution* (Oxford, 1984), 217.

[22]John A. Chartres and Gerard L. Turnbull, "Road Transport," in Derek H. Aldcroft and Michael J. Freeman (eds.), *Transport in the Industrial Revolution* (Manchester, 1983), 85.

[23]The Henleys, shipowners of London, put their coasting vessels into government service; Ville, *English Shipowning*, 52-56.

[24]Lee, *Cotton Enterprise*, 136.

tween 2s 3d and 4s 3d per cwt.[25] This suggests that transport costs by sea were between fourteen and twenty-seven percent of the road rate. A similar relationship can be established from the records of Peter Stubs, a Warrington-based file and hand toolmaker.[26] In 1802 and 1803 (also a period of war) the carriage from Liverpool to Glasgow by coaster for metal-working files was between thirteen and seventeen percent of the comparable road charge – three or four shillings as compared with twenty-three shillings. For steel rod from Greenock to Liverpool the rate was even less. Six bundles of steel rod, each weighing about a hundredweight, were sent by sea and thence upriver to Warrington. Even with the cost of cartage, warehousing and other incidental expenses the total charge was only one shilling per bundle, whereas the road carriers' charge would have been in excess of £1. Lewis Lloyd has shown that the cost of sending woollen webs by sea from Barmouth to London in the latter part of the eighteenth century was four shillings per web. By land, however, it was five times as costly.[27] Peter Barton showed that merchants could save a penny per cwt. or 1s 8d per ton, by using the Peirses' sloop for groceries from Yarmouth to Low Worsall.[28] Turnbull has suggested the rate from Leith to Hull in 1797 was fourteen shillings per bale of linen by coaster, but about two guineas by road carriage, so that the sea route was about one-third of the land charge.[29] There is little doubt that contemporaries were fully aware of these differentials in favour of sea transport.

Although the coaster was crucial in the carriage of bulky low-value goods which were essential to industrialization and urbanization – such as coal, iron ore, salt and china clay – it should not be thought that these were the only type of goods carried, or that all low-weight, high-value goods were sent by road. The coaster also conveyed high-value goods of low bulk. Reference has already been made to semi-manufactured goods of high value travelling by coastal ship – commodities as varied as Peter Stubs' consignment of bundles of steel rod, sent from Greenock to Liverpool, and thence upriver to Warrington in 1803; and the woollen webs manufactured in Merioneth's domestic textile industry between the 1760s and 1790s which were sent by sea to Liverpool,

---

[25]*Ibid.*, 95.

[26]T.S. Ashton, *An Eighteenth-Century Industrialist: Peter Stubs of Warrington, 1756-1806* (Manchester, 1939), 95.

[27]Lewis W. Lloyd, *The Town and Port of Barmouth, 1565-1973* (Harlech, 1973), 13.

[28]Peter Barton, "Low Worsall: The Shipping and Trade of an Eighteenth Century 'Port' on the River Tees," *Mariner's Mirror,* LV, No. 1 (1969), 63.

[29]Turnbull, "Scotch Linen," 65.

Chester and London.[30] Moreover, Gordon Jackson has shown that large quantities of English wool were shipped from ports on the south and east coasts to Hull in the eighteenth century.[31]

Finished goods of high value and low bulk were also shipped by coaster. For example, Archibald Turner of Glasgow insisted that his bundles of metal-working files be sent by sea from Liverpool.[32] The Carron Line was established in 1765 to ship to London the products of the iron foundry, including sugar pans, sad irons, stoves, ovens, plates and guns – all high-value manufactured goods.[33] Jackson has also demonstrated that Hull sent coastwise a whole range of manufactured textiles, including cotton and woollen cloth as well as large quantities of "ironmongery."[34] Turnbull has shown that John Wilson of Leeds found road transport a viable alternative to coastal freight for his linen cloth brought from Scotland, especially when war, its threat or its aftermath made the east coast more risky. When there was no war disruption, the preferred method of transport for textiles between major ports with regular coastal sailings (Leith, London and Hull) was by sea.[35] As early as the first decades of the seventeenth century there was a considerable coastal trade between Scotland and Boston (Lincolnshire) in linen cloth and yarn, twill and ticking.[36] Jackson has pointed out that large quantities of alcoholic beverages were carried by coaster. Hull received "British spirits," presumably gin, in large volume.[37] Peter Mathias, moreover, suggests that "beer could be traded over long distances only under conditions of water carriage direct to an urban or coastal market because of its great bulk."[38] William Stout's autobiography shows that in peacetime –admittedly a rather rare condition in the late seven-

---

[30]Lloyd, *Town and Port*, 12.

[31]Gordon Jackson, *Hull in the Eighteenth Century: A Study in Economic and Social History* (Oxford, 1972), 75-76.

[32]Ashton, *Eighteenth-Century Industrialist*, 95.

[33]Anon., *Carron Company from the Reign of George II to the Reign of George VI, 1759-1938* (Falkirk, 1938), 15.

[34]Jackson, *Hull*, 91-94.

[35]Turnbull, "Scotch Linen," 56-57.

[36]R.W.K. Hinton, *Port Books of Boston, 1601-1640* (Lincoln, 1956), 4-69.

[37]Jackson, *Hull*, 85.

[38]Peter Mathias, *The Brewing Industry in England, 1700-1830* (Cambridge, 1959).

teenth and early eighteenth centuries – he preferred to send valuable finished goods for sale in his shop in Lancaster by sea from London.[39]

The coaster also played an important part in distributing exotic foreign products from their point of importation to their final destination. Hull received by coastal ship considerable amounts of coffee, cocoa, tea, sugar and tobacco, as well as distributing "red and white port wine and Spanish wine."[40] At each of the large ports engaged in substantial foreign trade, coasters were used both to assemble export cargoes and to distribute high-value imports. In the 1750s tobacco was sent regularly from Weymouth to Portsmouth by ship.[41] Even perishables were not automatically excluded from the trade. Alan McGowan has shown that in the eighteenth century Scotch salmon was brought to London more or less regularly by sailing smack, and meat was sent from Aberdeen to the London market. Butter, which originated in Ireland, was sent coastwise from Bristol to Portsmouth in the 1750s.[42] Thus, road transport had no monopoly of high-value cargoes or even fresh produce. They could equally well be transported by coaster given the right conditions, such as a straight run down the east coast and a fast passage before the produce spoiled.

The cost advantages of the coaster as compared with road transport, originated primarily in economies of scale. It may be shown that throughout the period from 1600 the coaster carried a larger average cargo than the road haulier. For instance, Willan found that in the early seventeenth century the average size for coasting vessels on the west and south coasts was about thirty tons burthen, and on the east coast about forty.[43] Chartres, examining the ships bringing grain from East Anglia to Wiggins Key in the port of London in the 1670s and 1680s, found that their average size was forty-two tons.[44] At the same time, the coal trade was serviced by much larger ships; Willan estimated their average burthen to be about seventy-three tons in the early seventeenth

---

[39]John D. Marshall (ed.), *The Autobiography of William Stout of Lancaster, 1665-1752* (Manchester, 1967), 89-90, 169 and 172.

[40]Jackson, *Hull*, 84-85 and 89.

[41]James H. Thomas, "The Seaborne Trade of Portsmouth, 1650-1800," *Portsmouth Papers*, XL (1984), 3.

[42]*Ibid.*; and Alan McGowan, *The Ship. The Century before Steam: The Development of the Sailing Ship, 1700-1820* (London, 1980), 41.

[43]Willan, *English Coasting Trade*, 11-13.

[44]John Chartres, "Trade and Shipping in the Port of London: Wiggins Key in the Later-Seventeenth Century," *Journal of Transport History*, 3rd ser., I, No. 1 (1980), 38, reprinted in John Armstrong (ed.), *Coastal and Short Sea Shipping: Studies in Transport History* (Aldershot, 1996), 1-19.

century. This points to a much larger average load by sea compared with road transport.[45] Chartres estimated the average wagon load in the mid-seventeenth century to have been only one and a half tons.[46] Thus, between twenty and thirty wagons were required to carry the equivalent of one shipload, or nearly fifty wagons for the average load of a collier. The situation is no different if the pack train is taken into account, for in this case Chartres and Turnbull estimate the average load for both wagons and pack trains to have been only 1.2 tons in the early eighteenth century.[47]

Road haulage gained some ground on coastal shipping during the course of the next century. By the late eighteenth century the average size of ship coming into Hull was eighty-two tons.[48] This is in line with Ville's findings for 1785 which put the average coaster size for Bristol, Hull and Liverpool at between sixty-three and eighty-three tons.[49] The average size in the whole English coasting trade was perhaps seventy tons. At the same time, the average load for the collier had risen more appreciably, to around three hundred tons according to Ville, though Willan and Flinn suggest that even larger tonnages were common.[50] By the 1760s Chartres believes the average wagon load to have risen to four tons, and by 1800 he suggests that six tons was a normal load.[51] These consignment loads indicate the need for about twelve wagons to do the work of the average coaster and between forty-two and fifty for the average collier. Thus, despite a slight reduction in differentials, the coaster was still capable of carrying a much larger cargo than the wagon and as a result continued to benefit from economies of scale.

Manning levels on the coasters were such as to give them significantly lower labour costs per ton of cargo than those borne by the road hauliers. The

---

[45]Willan, *English Coasting Trade,* 11.

[46]John Chartres, "Road Transport and Economic Growth in the Eighteenth Century," *Refresh,* VIII (1989), 7, reprinted in Anne Digby, *et al.* (eds.), *New Directions in Economic and Social History* (2 vols., London, 1989-1992).

[47]Chartres and Turnbull, "Road Transport," 85.

[48]Jackson, *Hull,* 73.

[49]Simon P. Ville, "Shipping in the Port of Newcastle, 1780-1800," *Journal of Transport History,* 3rd ser., IX, No. 1 (1988), 73.

[50]Simon Ville, "Total Factor Productivity in the English Shipping Industry: The North-East Coal Trade, 1700-1850," *Economic History Review,* 2nd ser., XXXIX, No. 3 (1986), 360; Willan, *English Coasting Trade,* ii; and Flinn, *History of the British Coal Industry,* 172.

[51]Chartres, "Road Transport," 7.

small, typical coaster of say thirty tons usually had a crew of about three to give a ratio of ten tons per man. Williams, looking at the late sixteenth century, believed a crew of five was normal for a thirty-five chaldron ship, giving a ratio of one man to nineteen tons.[52] In 1632, *Endeavour* of Shields, a large ship of 140 tons burthen, had a crew of nine men, a ratio of about 15.5 tons per man.[53] Willan suggests that in the case of colliers, tonnage per man rose from about twenty-four to over thirty between 1665 and 1729.[54] Compare this with the average wagon which, according to Chartres, carried one and a half tons and at minimum required one driver, though the evidence suggests that two men not uncommonly formed its complement.[55] Thus, the labour costs per ton were likely to be between six and twelve times as high on the roads as for the average coaster. Even greater economies applied to the collier: in this case it is reasonable to suggest that land-based labour costs were ten to twenty times higher. Since the carriage of large quantities by road required a multiplication of wagon, horses and drivers, all costs were similarly multiplied, thus denying any reduction in unit costs for large loads.

Additionally, the sailing ship bore no costs for power, being dependent upon the free, if fickle, wind. The road haulier, on the other hand, had to find feed for his horses, no small expense given that for the large wagons, teams of four or even six were common. Gerhold has few doubts that this was the single largest cost of a carrier.[56] While it is true that the coaster bore a cost for propulsion, via its sails, spars and rigging, unlike the road haulier the shipowner treated these as fixed rather than variable costs. They are, in consequence, much harder to apportion accurately.

Another way in which costs were kept down on the coaster was through discharging crew at the end of the trip, thus incurring only limited labour costs when ships were in port or otherwise inactive. Ville has demonstrated that the crews of the Henley ship *Freedom* were remunerated by the

---

[52]Neville J. Williams, *The Maritime Trade of East Anglia, 1550-90* (Oxford, 1989), 147.

[53]Davis, *Rise of the English Shipping Industry*, 60.

[54]Willan, *English Coasting Trade*, 16-17.

[55]Chartres, "Road Transport," 7.

[56]Dorian Gerhold, "Packhorses and Wheeled Vehicles in England, 1550-1800," *Journal of Transport History*, 3rd ser., XIV, No. 1 (1993), 20, reprinted in Gerhold (ed.), *Road Transport in the Horse-Drawn Era: Studies in Transport History* (Aldershot, 1996), 139-164.

trip during the period 1791 to 1816.[57] John Press confirms the prevalence of the practice in the coasting, as distinct from the foreign-going trade, where it was becoming rarer.[58] It continued to be used in the coasting trade into the early twentieth century. In some trades the wages of the seamen were tied to the price of the product carried so that labour costs moved in line with freight earnings. Thus, in 1770 shipowners in Cumberland agreed that seamen's wages should be *pro rata* with the price of coal; that is "to Dublin 2 guineas or the average price of three tons of coal."[59]

An additional method of keeping costs down was to provide the captain with an incentive to ensure parsimony in disbursements. If the master sailed "on shares" he had every reason to keep expenditure down since expenses were paid from his share of freight earnings; if in addition he were a part-owner of the vessel, he had a reason for keen interest in the profitability of its voyages. His best efforts would be put to securing fast turnaround in port. Davis demonstrated that the share system was in use as early as the late seventeenth century. He quotes the example of the *Ann* plying along the Channel coast in 1685: the division of her earnings was one-third to the owners and two-thirds to the master, out of which he paid the crew's wages, the victualling of the vessel and all incidental running expenses.[60] A similar arrangement applied to *Two Brothers* of Wisbech trading on the east coast in the 1730s.[61] This method of motivating the skipper was still in use on smaller sailing schooners and smacks in the early twentieth century, while master-owners were commonplace in the coasting trade into the nineteenth century.[62]

Given that quite a high proportion of the coaster's costs were "terminal" – a point to which I shall return – it was beneficial to the shipowner to minimize these. One way was for the vessel not to enter an official port, but instead to be run onto a sloping beach and discharge its cargo over the side at

---

[57]Simon P. Ville, "Wages, Prices and Profitability in the Shipping Industry during the Napoleonic Wars: A Case Study," *Journal of Transport History*, 3rd ser., II, No. 1 (1981), 48-51, reprinted in Armstrong (ed.), *Coastal and Short Sea Shipping*, 21-34.

[58]John Press, "Wages in the Merchant Navy, 1815-54," *Journal of Transport History*, 3rd ser., No. 2 (1981), 37. Willan, *English Coasting Trade*, 17, also believes that coastal crews were normally paid by the voyage.

[59]Edward Hughes, *North Country Life in the Eighteenth Century. Volume II: Cumberland and Westmoreland 1700-1830* (Oxford, 1965), 196.

[60]Davis, *Rise of the English Shipping Industry*, 133.

[61]*Ibid.*

[62]Willan, *English Coasting Trade*, 37.

low water into carts or wagons. Many of the small, wooden, flat-bottomed coasters were quite capable of this, as the paintings of Constable and Ibbotson and the prints of Cooke testify.[63] Savings came from avoiding port charges and perhaps pilotage, as well as the expense of dock labour, since the crew and the recipient's employees could be called upon to work the cargo. Additionally, of course, this allowed the coaster to serve small communities distant from large ports while keeping expensive land carriage to a minimum.

There were other methods of cost minimization widespread among coasting owners. One was to employ boys or apprentices who did not qualify for a full wage. This was often perceived by the boys as a training position, and they tolerated it in prospect of future employment as a fully-paid able seaman, perhaps in the more lucrative foreign trades.[64] It was also relatively common for master-owners to recruit their own family members to form part or all of the crew. Family could be expected to sacrifice current earnings in prospect of future returns, such as the inheritance of a vessel.[65]

There were also additional ways of earning income beyond the carriage of goods. Throughout the early modern period the sailing coaster owner was willing to carry passengers as well as freight, and there is evidence to suggest that the service was widely-used. William Stout sailed from London to Lancaster by ketch in 1698, accompanying masts, sails, rigging, anchors and cables which he hadbought for friends building a ship at Wharton.[66] John Wesley was a frequent user of coastal vessels when he toured the country on his preaching missions in the mid-eighteenth century.[67] From the point of view of the owner and captain of the coaster, passengers provided an alternative source of income while incurring little additional cost.

In order fully to compare costs between the coaster and the road wagon, some element of fixed costs needs to be calculated. This should take into account the initial capital cost of the ship, on the one hand, and the cart or wagon and horses, on the other, and then allow

---

[63]McGowan, *Century before Steam*, 43 and 53; Basil Greenhill, *The Ship. The Life and Death of the Merchant Sailing Ships, 1815-1965* (London, 1980), 15; and John H. Farrant, "The Harbours of Sussex as Part of an Inland Transport System in the Eighteenth and Nineteenth Centuries," *Sussex Industrial History*, XV (1985-1986), 2.

[64]Willan, *English Coasting Trade*, 17-19; and Williams, *Maritime Trade*, 150.

[65]Willan, *English Coasting Trade*, 36: "the great number of Passengers who go to and from Newcastle, etc."

[66]Marshall (ed.), *Autobiography of William Stout*, 122-123.

[67]John Wesley, *Journal* (London, 1906), 263.

for the depreciation of these assets over their average life. This element should then be added to the variable cost to arrive at total cost. This raises two questions: what was the cost per ton of building and fitting out a coaster, and how long could the owners reasonably expect to keep the ship in profitable employment? The former obviously varied over time, with the place of construction, size and trade in which the ship was to ply. Coasters were among the smallest vessels, the least sumptuously appointed compared with the haughty East Indiaman, and were usually built in the outports where costs were low.[68] Hence, they were likely to be at the cheaper end of any spectrum. Davis published some estimates for the cost of constructing the hull, which varied from £2 8s to £6 17s per ton in the late seventeenth century, £2 11s to £5 11s in the early eighteenth, and £3 18s to £9 9s for the late eighteenth century.[69] These figures are given some support by the 350-ton collier *Blessing*, built in 1714 at Newcastle, which cost £4 11s 5d per ton.[70] But the expense of fitting-out with the necessary masts, spars and ropes has also to be taken into account. This would vary with the sophistication of the fittings. Robin Craig reckoned the cost at between fifty and one hundred percent of initial building costs.[71] If we assume that most coasters were fairly basically equipped, a figure of about £3 15s per ton may not be far wrong for the late seventeenth century, rising to £4 in the early eighteenth century and to £6 at its end. This is borne out by Hughes' observation that in 1763 a 150-ton burthen collier cost just over £7 per ton to build in Cumberland.[72]

Turning to ship longevity, Davis estimated twenty to twenty-five years to be a normal lifespan for wooden sailing ships in the early modern period, and worked on a depreciation rate of four percent per annum, which Craig endorsed.[73] If we accept these estimates, the depreciation per ton-year for the coaster works out at three shillings in the late seventeenth and five shil-

---

[68]Davis, *Rise of the English Shipping Industry*, 367-368.

[69]*Ibid.*, 373-375.

[70]Willan, *English Coasting Trade*, 39.

[71]Robin Craig, "Capital Formation in Shipping," in J.P.P. Higgins and Sidney Pollard (eds.), *Aspects of Capital Investment in Great Britain, 1750-1850: A Preliminary Survey* (London, 1971), 143-144, reprinted in Craig, *British Tramp Shipping, 1750-1914* (St. John's, 2003), 41-58.

[72]Hughes, *North Country Life, Vol. II*, 187.

[73]Davis, *Rise of the English Shipping Industry*, 376; and Craig, "Capital Formation," 140-141 and 155.

lings in the late eighteenth century. But the exact method of apportioning these costs raises considerable difficulties. Should they be divided among the number of trips, by the mileage worked or in some other way? For current purposes I shall merely compare the coaster's cost with the equivalent for the road haulier.

Turnbull gives a figure of between £30 and £40 each for the broad-wheeled wagon in the mid to late eighteenth century.[74] The value of the draught horses must be added, however, and Turnbull cites £10 as the going price for a mature carthorse at this period. A minimum of six were needed for a four-wheel wagon.[75] Thus, an outlay of at least £90 to £100 was required for a vehicle which would carry between four and six tons. If we assume an average life for both the horse and wagon of about twenty-five years,[76] we arrive at a depreciation rate of four percent, or £3 12s to £4 per annum. Assuming the wagon's capacity was between four and six tons, this gives an annual depreciation per ton per annum of between twelve shillings and £1. These figures, as those for the coaster, can only be taken as very approximate. While it is to be hoped that further research will refine them considerably, they do indicate the importance of economies of scale enjoyed by the coaster over the wagon, in this case in relation to construction costs and depreciation. In sum, the vital difference between the two modes of transport rests upon the greater carrying capacity of the coasting vessel. Of course, these calculations assume the coaster was not wrecked at an early age, not an unusual occurrence, though premature losses are balanced out by the longevity of some vessels well beyond the average of twenty-five years. Again, more research is required to determine precisely the average life expectancy of a coaster, and hence to refine these calculations.

The coaster had a special cost advantage over land-based transport when it came to the long-haul. "Water transport was the only key to unlock distant markets," as Mathias has stated.[77] As early as the first years of the seventeenth century Boston was receiving frequent calls from coasters from Kirkcaldy, Dysart and Leith carrying white salt, rye, linen cloth and yarn and twill.[78] Even the shortest of these voyages is three hundred miles. Examples of long-hauls abound, such as the coal trade from the North East to London

---

[74]Turnbull, *Traffic and Transport,* 17.

[75]*Ibid.,* 18.

[76]This almost certainly overstates the working life of a cart horse, which Gerhold estimates to have been less than ten years.

[77]Mathias, *Brewing Industry,* 140.

[78]Hinton, *Port Books,* 4-69.

(about three hundred miles) and the south coast, the trade in Cheshire cheese from the North West to London (about seven hundred miles), or the carriage of salt from the Mersey to the east coast Scottish ports (between five and six hundred). In the 1730s iron ore was being shipped from Barrow and Argyll for the Sussex iron industry and pig iron to Hampshire and Sussex from Bonawe on Loch Etive, while Barmouth imported lambs' wool from Dover and London.[79] Many of the coaster's costs were incurred as a fixed cost per voyage, irrespective of the length of haul. If expenses were spread over a relatively short journey, the cost per ton-mile was likely to be high. The longer the haul, however, the lower per mile these charges became, and this was the basis of the coaster's cost advantage.

The costs which the coaster bore irrespective of haul were those generally associated with entry into and exit from ports. These included pilotage charges, port and light dues, the cost of lighterage to and from shore, warehousing and the cost of unloading when not borne by the crew. Chartres has shown for Wiggins Key that these could be substantial, especially where a number of these services were controlled by the same individuals or partnerships and where there was a form of cartelization preventing the ship's master from shopping around to play one wharfinger off against another.[80] In the late eighteenth century ships entering Portsmouth harbour were charged by the Corporation "harbour dues of groundage, anchorage and bushelage," as well as "wharfage dues for goods shipped or landed."[81] The road haulier bore no comparable expenses, and terminal costs were limited to loading and discharging, which were likely to be quite small absolutely given the small average load of the wagon.

It has been argued that the sailing coaster was too slow, especially for expensive goods where inventory costs were high if large cargoes were tied up on long voyages. But it is questionable whether the coaster was, in fact, very much slower than road transport. Chartres has suggested that the speed of the great stage wagon rarely exceeded twenty miles per day before 1800,[82] and he implies that in the seventeenth century, when wagons were less technically advanced, the speeds were likely to have been less. Gerhold, moreover, be-

---

[79]Tom Evans, "The Coastal Trade in Iron Ore for Sussex and Hampshire in the Eighteenth Century" *Sussex Industrial History*, XV (1985-1986), 3; and Lloyd, *Town and Port*, 12.

[80]Chartres, "Trade and Shipping," 34 and 46.

[81]Thomas, "Seaborne Trade," 13.

[82]Chartres, "Road Transport," 7.

lieves two miles per hour was the optimum speed.[83] Even the more specialist carriers, such as Pickfords in the early nineteenth century, considered forty miles a day as fast.[84] Speedier services were introduced on some routes in the late eighteenth century, but these were intended essentially for the light parcel trade, and the additional speed carried with it extra costs.[85] Measured against these statistics the average sailing coaster performed well. Speeds of five knots were not difficult to maintain, and over a twenty-four hour period this would amount to about one hundred miles. These speeds were, indeed, achieved in practice. In 1687, when William Stout sent goods valued at £200 in the ketch *Edward and Jane* from London to Lancaster, they took seven days on the passage, thus averaging just under one hundred miles per day.[86] Although the distance by sea was three times the road route, the goods took no longer to reach Stout than had they been sent by road carrier. Ville has demonstrated for a number of ships plying between Shields and London in the 1790s that the duration of a voyage was between three and ten days, giving an average haul per day of thirty to one hundred miles.[87]

The real problem with the coaster was not that it was necessarily slower than the road wagon, but rather that its speed was much less predictable. If wind and tide were favourable it could make a fast passage, but the wind was not to be relied upon. A contrary wind could keep the coaster locked in harbour for days, or unable to enter, and at sea, if the wind was not in the right quarter, the master would make slow progress as he ordered a tacking course. The nature of the rigging, mostly square-rigged rather than the fore-and-aft rig of the schooners, smacks and ketches of the nineteenth century, made the vessel less able to take advantage of winds. A contrary tide also made for low speeds. In this sense the ship might be seen as the fabled hare against the road wagon's tortoise. The wagon was slow, lumbering and cumbersome, but it could be relied upon to arrive within a few hours of its rough timetable. The ship might make a much faster journey, but on the other hand, it might be very much delayed. Hence, where goods were required by a set time, for instance to be loaded into a foreign-going ship leaving at a specific date, or where fashions were volatile, the merchant might choose to use the

---

[83]Dorian Gerhold, unpublished paper presented at Coventry Polytechnic, 16 March 1991.

[84]Turnbull, *Traffic and Transport*, 64.

[85]*Ibid.*, 66-67.

[86]Marshall (ed.), *Autobiography of William Stout*, 89.

[87]Ville, "Wages, Prices," 40.

road wagon for its reliability and predictability rather than the potentially faster and undoubtedly cheaper, but less reliable, coaster.

If reliability was one of the coaster's weaknesses, its regularity has also been criticized compared with road transport. Turnbull, in his study of John Wilson of Leeds, suggested that one reason for sending linen from Scotland to Leeds by road was the irregularity of the service from some smaller Scottish ports. Merchandise might be delayed for weeks awaiting a boat to a suitable destination, and this was a particular problem where expensive goods representing a high capital investment were concerned.[88] Wilson chose to pay the much dearer road transport charges for transporting high-value linen since his profit margins could bear the expense. Before the advent of the steam vessel and liner services the coaster was low in the regularity stakes. But it should not be forgotten that even in the pre-steam age some coastal shipowners did manage to keep to regular sailing days. Regular services, justified by the volume of trade, were more common between the major ports. Otherwise, the wagon with its smaller consignment size had the advantage over the ship and may have been more appropriate to the transport needs of lesser communities.

The onset of war is usually seen as inhibiting coastal trade, as the fear of ships being seized by enemy privateers or naval ships rose and craft were delayed in harbour awaiting the formation of a convoy. As a result, freight charges rose to take account of the greater risk, diminishing the advantage of the coaster. However, this disadvantage should not be overstated. In the eighteenth century the assertion of British naval supremacy resulted in reduced risk of capture.[89] Not all coastal ships were unprotected. The Carron Company boasted of its "heavily armed ships" in the 1780s, and claimed there was no need to await a convoy since its ships could sail regularly every week.[90] Thus, although there were interruptions to service as a result of war, the shipping community responded by building fast or well-armed vessels.

Turnbull has suggested that insurance premiums on expensive cargoes, especially in times of high risk, such as war, could push the coasting rate above the road rate.[91] But many merchants did not insure their cargoes, thus

---

[88]Turnbull, "Scotch Linen," 62.

[89]Ville, "Wages, Prices," 42; and David J. Starkey, *British Privateering Enterprise in the Eighteenth Century* (Exeter, 1990).

[90]Their armaments were listed as twenty eighteen-pounders on *Paisley*, twelve twelve-pounders on *Forth* and fourteen carronades on *Galatea*; Anon, *Carron Company*, frontispiece, 15-16. Lewis W. Lloyd has identified about twenty Barmouth square riggers which were armed with a few six- or nine-pound guns; Lloyd, *The Unity of Barmouth* (Caernarfon, 1977), 21.

[91]Turnbull, "Scotch Linen," 62-65.

saving the premium, and they took their chances with the coaster. The low level of risk incurred by individual shippers is exemplified by William Stout. He calculated that in the nine years 1688-1697 his total losses by privateers and other war-related causes were negligible. Bad debts in the same period cost him more than £200. Throughout this period he continued to send his heaviest goods from London to Lancaster by sea.[92] Hence, although fear of war losses may have diverted the most valuable goods from the coast to the road, the risk was not so great as to cause significant loss when compared with the ordinary vicissitudes of business.

The fear of total loss of the cargo as a result of shipwreck or foundering has sometimes been cited as an additional hazard of coastal transport which had no equivalent on land and which therefore made the coaster less attractive to merchants.[93] Severe storms did cause high losses, but the risk is exaggerated, in part owing to the publicity given to these occasional events. Flinn recently suggested the loss rates of colliers were less than one percent in 1703 and 1710, and much lower in the period 1720-1750.[94] Thus, as Stout found for war-related loss, the reality was much less severe than the mythology. Though Flinn's conclusions are based on research done many years ago, unless further evidence establishes otherwise there are good grounds for believing that loss rates were quite low.

Another factor which enhanced the coaster's value as a method of transport was its ability to penetrate many miles inland using the extensive network of navigable rivers, thus avoiding the need for lengthy, expensive cart or wagon haulage.[95] Because the average coaster was still relatively small, there were few difficulties of access along most of Britain's rivers. In the seventeenth century tanners from South Yorkshire regularly shipped hides from London to Hull and up the Humber to Turnbridge and Bawtrey. Jackson has shown that in the eighteenth century coasters were going via the Humber and its tributaries as far as Selby, Stockwith, Thorne and York, between thirty-five and sixty miles beyond Hull.[96] Welsh rivers such as the Mawddach and Dyfi were navigable by small shallow coasting ships, and where that was difficult, as on the Conway, small boats plied further upriver bringing down mineral and

---

[92]Marshall (ed.), *Autobiography of William Stout*, 119.

[93]Turnbull, "Scotch Linen," 63-64.

[94]Flinn, *History of the British Coal Industry*, 174.

[95]Leslie A. Clarkson, *The Pre-Industrial Economy in England, 1500-1750* (London, 1971), 122.

[96]Jackson, *Hull*, 94.

timber to tranship into coasting ships sailing for Liverpool.[97] Where the coaster was too large or too unwieldy to get to its final destination, it was quite normal for the goods to be off-loaded straight into lighters or barges which would then complete the voyage. Chartres has shown for Wiggins Key in the late seventeenth century that charges for unloading barrels of oil could be halved in this way.[98] Wool coming into Hull by coaster was put into smaller river craft for Leeds, Halifax and Wakefield, each of which was more than fifty miles upstream from Hull.[99] The main commodity carried on the Thames and Severn Canal in the 1790s was coal which had come coastwise and then upriver to the canal.[100] In 1732 the Peirse brothers turned Low Worsall on the river Tees into a thriving "port," although it was ten miles upstream from Stockton and twenty-six miles from the sea: they ran small sloops from there with corn and lead and transhipped at Stockton. In the 1750s they operated sloops of forty tons burthen direct from Low Worsall to ports on the east coast such as Newcastle, Sunderland and Blythnook.[101] River improvements added to the usefulness of the coaster by extending its effective range. When the canal network was constructed in the late eighteenth century, it too facilitated the use of coastal shipping. At the juncture of coast and canal the ship's cargo was discharged direct into a canal barge, thus continuing cheap transit by water. Some of these riverine ports became very important as transhipment points, for example Runcorn, Selby and Goole.[102]

The coaster was used in a similar way to redistribute foreign imports and to assemble overseas exports from their origins. At major ports foreign-going ships discharged direct over the side into coasters – avoiding the need to dock and thus minimizing port charges and land transport costs. In a like manner many goods for export were marshalled by coaster from a multitude of small harbours and transhipped direct into ocean-going vessels at larger ports. Both Hull and London were perceived as lighterage ports, and there a high proportion of foreign goods never touched land.

---

[97]Lewis W. Lloyd, *Sails on the Mawddach: The Account Book of David Evans, Boatman* (Harlech, 1981), 8-9; and Gareth H. Williams, "The Building of a Conway River Boat, 1685," *Maritime Wales*, No. 5 (1980), 5-6.

[98]Chartres, "Trade and Shipping," 37.

[99]Jackson, *Hull*, 120.

[100]H.R.H. Prince Naruhito, *The Thames as Highway* (Oxford, 1989), 64-65.

[101]Barton, "Low Worsall," 61-62 and 68.

[102]J. Douglas Porteous, *Canal Ports: The Urban Achievement of the Canal Age* (London, 1977), chaps. 4 and 7.

As far as ship technology is concerned, it is too easy to be seduced by the large paradigm shifts and to ignore smaller incremental changes. Thus, while a revolution took place in the nineteenth century with steam propulsion, the use of iron and steel hulls, and the screw propeller, there were important changes in technology affecting coastal shipping in the earlier period; albeit not so dramatic, they improved the coaster's efficiency. Ville and Hausman, although they disagree about the details, concur in the general outlines of this improvement in performance. During the eighteenth century average cargo size increased, possibly by fifty percent, thus maintaining the coaster's ability to carry much larger cargoes than the wagon and benefit from greater economies of scale.[103]

In addition, the average number of trips that the coaster made in any one year rose – Ville suggests that collier voyages doubled during the century.[104] The key effect of this was that fixed costs could be spread over a larger number of voyages with the possibility that lower freight rates might follow. It may have been achieved through the coaster making more voyages in the winter months. Hughes suggests winter voyages were rare in the early eighteenth century,[105] but Chartres concluded otherwise from his research on the East Anglian grain trade into London in the late seventeenth century. He found that the peak months for this trade were February to May, and this he explained by consumer demand. Grain was ready for market from February onward, so the voyages were undertaken despite bad weather.[106] The fact that colliers made few winter trips before the eighteenth century may indeed have had more to do with the difficulties of getting coal to the staithe than to any inherent weakness in the coastal vessel. Thus, to explain the increase in coaster voyages we may well have to look to land-based transport improvements, possibly colliery wagonways. Certainly this would not be incompatible with Ville's identification of sources of productivity growth, many of which were related to improvements in shore-based loading and discharging facilities during the eight-

---

[103]See William J. Hausman, "Size and Profitability of English Colliers in the Eighteenth Century," *Business History Review,* LI, No. 4 (1977), 460-473; Simon P. Ville, "Note: Size and Profitability of English Colliers in the Eighteenth Century – A Reappraisal," *Business History Review,* LVIII, No. 1 (1984), 103-117; Hausman, "Profitability of English Colliers in the Eighteenth Century: Reply to a Reappraisal," *Business History Review,* LVIII, No. 1 (1984), 121-125; and Ville, "Total Factor Productivity," 358-370.

[104]Ville, "Total Factor Productivity," 360-361.

[105]Edward Hughes, *North Country Life in the Eighteenth Century: Volume I: The North East, 1700-1750* (Oxford, 1952), 203.

[106]Chartres, "Trade and Shipping," 44-45.

eenth century.[107] Ville highlights improvements effected by the construction of staithes and spouts allowing coal to be loaded direct into the coaster rather than the slower, more cumbersome and more labour intensive keelman's boat. Unloading was expedited by basket and pulley systems, and in the late eighteenth century cranes, all of which replaced the shovel. The net result of such improvements was to speed up turnaround times and so provide the customer with a faster service and allow the coaster to make additional journeys per year. These, together with reduced cost, were material advantages in the competition with road transport.

Changes in the design of the vessels had the effect of reducing their cost of operation and increasing their reliability. These were mainly changes to the method of rigging, notably a move away from the square-rigged ship and greater use of fore-and-aft rig to facilitate faster passages whatever the strength and direction of prevailing winds. In addition, fore-and-aft rigging was usually more economical in labour, as many sail-handling tasks could be done from the deck.[108] From the early seventeenth century the most famous type of fore-and-aft rigged coaster, the schooner, appeared,[109] and during the course of the eighteenth century the ketch increasingly shifted from square to fore-and-aft rigging. The effect of this trend was to reduce costs, and so possibly prices, and to improve the coaster's reliability.

This is but one instance of a trend which had been occurring for centuries and was to continue: the constant evolution of ship types to cater to specific trades or local conditions. In this sense the coaster should not be considered as homogenous. Around the country particular types of craft had evolved to take advantage of prevailing wind strengths and directions, or they were of shallow draught to accommodate estuarial and small port trade. Examples are the Humber keel, the Severn trow, the Kent hoy and the Mersey flat.[110] Thus, the coaster was in constant evolution and adaptation to optimize the service it could provide.

Another important area of enquiry in the coastal trade is organization and ownership. There are three key players here: the master mariner, the owner and the merchant. Ideally, we wish to know the role of each and to what extent one individual filled more than one role. The ownership system

---

[107]Ville, "Total Factor Productivity," 362-365.

[108]McGowan, *Century before Steam*, 51. Willan confirms that manning levels per ton fell in the period 1600-1750; *English Coasting Trade*, 16-17.

[109]Basil Greenhill, *The Merchant Schooners* (2 vols., London-1951-1957; 4th rev. ed., London, 1988), 6-14.

[110]Frank G.G. Carr, "Surviving Types of Coastal Craft of the British Isles," *Mariner's Mirror*, XX, No. 2 (1934), 137-151.

followed the sixty-fourth share pattern, and relatively small investments were possible. Initially, it might be posited that shipowning was seen as an extension of the normal activities of a merchant –a form of vertical integration. In order to eliminate the uncertainty of finding a suitable vessel in which to send goods, a merchant entered shipowning and managing, thus giving him more control over the shipment of products. Williams' work suggests that the local merchants of Lynn in the late sixteenth century owned many of the vessels which brought coal into the port; they also advanced the finance to buy the coal in the North East to the ship's captain.[111] Thus, there was an integration of shipowning, managing and merchanting. The same is true for the grain trade from the East Anglian ports in which the merchants of Lynn were also dominant. Williams tells us that they either "entered into bonds with merchants in other English ports for the due delivery of a definite quantity of grain by a certain date" or shipped "whenever convenient" knowing their shipmasters or resident factors could always bargain for a good price there.[112] Thus, shipowning and managing were seen as integral parts of the merchant's functions, whether to deliver specific contracts or to send grain to markets. Where port books suggest that different arrangements prevailed, for example in the case of Boston's trade with Scotland during the early seventeenth century when cargoes were commonly recorded as the master's property, this is a warning to careful interpretation of commercial arrangements.[113] The possibility must be considered that captains used their own money to purchase salt and linen or, alternatively, acted as agents for the shipowners.[114]

This pattern of organization was still quite normal in the eighteenth century. Robin Craig and Rupert Jarvis have shown that in Liverpool in the late eighteenth century local merchants dominated shipowning,[115] while Ville indicates that the Henleys, as well as carrying cargoes for other merchants, regularly bought coal at Newcastle or Shields and then resold it at the port of destination, the difference in price being their income for freight and reward

---

[111]Williams, *Maritime Trade*, 146-147.

[112]*Ibid.*, 156.

[113]Hinton, *Port Books*, 4-37.

[114]Williams, *Maritime Trade*, 146-147, concluded that the latter was true for Lynn in the sixteenth century, and it seems likely that the same was true for Boston in the next.

[115]Robert Craig and Rupert Jarvis, *Liverpool Registry of Merchant Ships* (Manchester, 1967), 201-202.

for risk-taking.[116] Jackson has shown that shipowning in Hull was an extension of merchanting activities, though this was it seems less common than in Liverpool.[117] The pattern continued into the nineteenth century for G. and J. Burns (one of the most important names in coastal shipping in that and the next century), which started out as a Glasgow merchant firm and for some time carried its own goods.[118] Another classic example was the Carron Line, which was started in 1765 to carry the products of the Carron Ironworks from Scotland to London. From its inception the line carried for other merchants, and this side of the business assumed ever greater importance.[119]

A further arrangement involved the master as owner and merchant. This was more feasible in the coasting as distinct from the foreign-going trades owing to the relatively small size of coasting vessels which made it possible for masters both to own their vessels and trade on their own behalf. Williams highlighted instances of this in East Anglia in the late sixteenth century. Captains bought coal in the North East, shipped it to Lynn and sold it on the quay there, then loaded grain for Newcastle upon Tyne and sold that when they arrived.[120] However, he stresses that this sort of operation was very much a minority of total trade. Jackson has also established that Hull shipmasters adopted the practice, and Willan, too, suggests it was relatively common outside the coal trade where capital requirements were lower owing to the smaller average size of ships. He cites stone shipments from Weymouth in 1738 as being dominated by owner-masters. These owner-masters also carried cargoes for third parties when freight rates were attractive.[121]

Even when an employee, the master was a pivotal figure. Quite apart from the technical side of navigation, knowing the winds, tides, shoals and lights, he also needed labour management skills to command his crew, and he acted as trusted agent of the owners, collecting freights or selling cargoes on their behalf, making numerous disbursements and often having the privilege of shipping some goods on his own account.

---

[116]Ville, "Note," 116-117.

[117]Jackson, *Hull*, 140-141.

[118]Anthony Slaven and Sydney G. Checkland (eds.), *Dictionary of Scottish Business Biography, 1860-1960. Volume 2: Processing Distribution and Services* (Aberdeen, 1990), 266-267.

[119]Anon., *Carron Company*, 15; and Ian Bowman, "The Carron Line," *Transport History*, X (1979).

[120]Williams, *Maritime Trade*, 147.

[121]Jackson, *Hull*, 140; and Willan, *English Coasting Trade*, 42-44.

The development of specialization and professionalism is important. When did the professional master having no financial interest in the cargo appear? And when, for that matter, was shipowning essentially separated from merchanting? Jackson suggests the first use of the term shipowner in Hull was in 1766, but then it was tacked on to the occupation of merchant. It was first used independently in 1773. He traces the growth of this group in terms of a life-course progression. They were shipowning master mariners who, when retired from the sea, lived on their shares in ships, and hence became professional shipowners.[122] Ville argues persuasively that by the late eighteenth century the Henleys had become specialist shipowners.[123] Thus, by the late eighteenth century, on the east coast at least, shipowning was becoming the prime activity of some individuals. This was part of a larger development whereby acting as a carrier for other people's goods in return for a freight rate, or chartering out vessels for specific voyages, ceased to be an offshoot of the major business of merchanting and became instead a commercial activity in its own right.

Changes in the structure and organization of coastal shipping are intimately connected with commercial possibilities which are, in turn, tied to needs of industry and to domestic consumer demand. In conclusion, then, it should be noted that the developments outlined above occurred against a backdrop of population expansion and the development of domestic manufacturing. The latter has recently been characterized as a distinct stage of "industry before industrialization" and designated "proto-industrialization."[124] It is noticeable, however, that for all the emphasis economic historians have placed on the enlargement of domestic markets and savings in transaction costs, none has appreciated that the coasting trade had a vital part to play in the widening and integration of markets and the reduction of transport costs during the seventeenth and eighteenth centuries.[125] Indeed, the earlier emphasis on road transport is echoed even in the most recent work. Thus, many of the questions raised in this paper gain a new importance and urgency if we are to understand

---

[122]Jackson, *Hull*, 141-142.

[123]Ville, *English Shipowning*, 2-13.

[124]Peter Kriedte, Hans Medick and Jürgen Schlumbohm, *Industrialization before Industrialization: Rural Industry in the Genesis of Capitalism* (Cambridge, 1981).

[125]On proto-industrialization, see Leslie A. Clarkson, *Proto-Industrialization: The First Phase of Industrialization?* (London, 1985); Donald C. Coleman, "Proto-Industrialization: A Concept Too Many?" *Economic History Review*, 2nd ser., XXXVI, No. 3 (1983), 435-448; and Franklin F. Mendels, "Proto-Industrialization: The First Phase of the Industrialization Process," *Journal of Economic History*, XXXII, No. 1 (1972), 241-261.

the role of transport in the pre- and proto-industrial economy, and to give coastal shipping its proper place as the pre-eminent form of commodity transport before the coming of the railway.

Chapter 3
The British Coastal Fleet in the Eighteenth Century:
How Useful Are the Admiralty's Registers of
Protection from Impressment?[1]

In 1988, the *American Neptune* published an article by Dwight E. Robinson on the British coastal fleet in the eighteenth century.[2] It was a pioneering study in that before then very few pieces had been published on any aspect of the British coastal trade in this or any other maritime history journal. It reminded readers of the great importance of the British coastal trade, both quantitatively and qualitatively, from the earliest days of Britain and especially of its role as a nursery for seamen and in boosting Britain's naval power through training seamen in ship handling and navigation. It was also innovative in its methodology, as Robinson had to deal with the problem of double recording of ships and find a method to eliminate such double counting. He chose to use a computer database for this task, a relatively novel solution at the time. Furthermore, the article provided a quantitative assessment of the size of the British coastal trade in 1776, and was able to break this down by the nature of the cargo carried, between domestic coasting and that to near-continental ports, and also provided a ranking of which ports were the largest owners of coasters and therefore probably the most heavily engaged in operating coastal ships. This was important work and came up with revealing findings.

Robinson took maritime historians to task for ignoring "a comprehensive source of data" that would allow us to estimate "the extent and nature of the British domestic coasting fleet" in the eighteenth century.[3] Given that the article was published over a decade ago, and that no subsequent book or article has appeared tackling the question of the size of the British eighteenth-century coasting fleet or drawing on the documentation to which Robinson referred, maritime historians appear to be either incorrigibly slow and lazy or totally disinterested in the British coasting fleet. Our object is to redeem the community of maritime historians. There are good reasons why the records in ques-

---

[1]This essay appeared originally in *American Neptune*, LX, No. 3 (2000), 235-251 (with John Cutler).

[2]Dwight E. Robinson, "Secret of British Power in the Age of Sail: Admiralty Records of the Coasting Fleet," *American Neptune*, XLVIII, No. 1 (1988), 5-21.

[3]*Ibid.*, 5.

tion have not been extensively used, and why Robinson overstated their comprehensiveness and value.

The records that Robinson used were the British Admiralty's thirty-eight "Registers of Protection from Being Pressed," held in the National Archives in Kew on the western outskirts of London.[4] These covered the period 1701 to 1815, according to Robinson, and were drawn up during war, or when war threatened, to record which ships and men had been given exemption from having their crews press-ganged into service on one of His Majesty's ships. These were in addition to the letters of protection that individual ships had to carry. Impressment was similar to shanghaiing, in which force or trickery was used by the recruiting party to seize men for the navy. In practice, press gangs were mostly only interested in men with experience of the sea. They were not keen to press landsmen, except in the hottest of presses, and many groups were immune from seizure, among them foreigners, fishermen, apprentices, those aged under eighteen and captains of merchant vessels. Inevitably, some men were pressed who should not have been, and there was vigorous activity on their behalf by friends and relations, personal lobbying, correspondence and through intermediaries to gain their release.[5] Often this brought success. In cases, too, naval personnel displayed humanity, releasing people on compassionate grounds such as elderly parents to support or a large number of dependent children.[6] Impressment is now seen as a reasonably fair and efficient way of recruiting suitable people for the navy in times of stress, and although there was much local opposition, if individuals were incorrectly pressed, they stood a good chance of being released. These Registers were the centralized record of those people given immunity to impressment. Robinson stresses that these Registers constitute a valuable source for estimating the size of the British coasting fleet, referring to "the rich stores of documentary evidence" they contain.[7] Yet, when looked at objectively, these Registers turn out to be much less than "comprehensive" and to have only a limited value in assessing the British coastal fleet.

Robinson explains that the Registers exist from 1701 to 1815, although he admits that this period is not covered comprehensively. There are

---

[4]Great Britain, National Archives (TNA/PRO), Admiralty (ADM) 7/363-400.

[5]George Selement, "Impressment and the American Merchant Marine, 1782-1812," *Mariner's Mirror*, LIX, No. 4 (1973), 409-418; G. Hinchcliffe, "Impressment of Seamen during the War of the Spanish Succession," *Mariner's Mirror*, LIII, No. 2 (1967), 139 and 141; and Norman McCord, "The Impress Service in North-East England during the Napoleonic War," *Mariner's Mirror*, LIV, No. 2 (1968), 171.

[6]Hinchcliffe, "Impressment of Seamen," 140-141.

[7]Robinson, "Secret of British Power," 6.

Registers for only seventy-one years, with considerable gaps being evident, such as the period between 1713 and 1739 and that between 1759 and 1769. In practice, then, just over one-half of the 115-year period is covered in one or more of the Registers. This casts some reasonable doubt on Robinson's "comprehensive" tag. The gaps can be explained largely by the periods of peace. When no war was taking place, and none threatened, no need existed for protection from impressment, and hence no register was drawn up. Where a register survives for a particular year, there is no guarantee that it covers the whole year. Many start or stop in the course of a calendar year, as did the one register that Robinson analyzed for 1776 and 1777. The period covered by this register was from 21 September 1776 to 30 June 1777; neither calendar year was comprehensively covered, and the register did not span a whole twelve-month period. This presents the obvious problem of whether, if a register covers only a nine-month period, there is a need to inflate the figures calculated by one-third to allow for the missing three months.

There is a further complication. Even for those years where a register exists, it does not necessarily contain details of crew who were protected. Other groups of workers could also be pressed, and needed to obtain protection from impressment. Some volumes, for example, give the names of land-based maritime workers in shipyards and dockyards. For instance, "four sail-makers employed by Eliz & Thos Gillesphy at Wapping" and "eighteen riggers employed by Thos. Smith of Liverpool."[8] These were self-evidently employed in boatbuilding or repairing and of no direct relevance to the coasting fleet. Other entries are of landsmen engaged in dock activity, such as "six men employed under the direction of the Port of London Committee at the Canal in the Isle of Dogs."[9] Some contain lists of East India Company employees who were not necessarily employed at sea and with no mention of a ship. An example is "four men employed by the E[ast] India] Company."[10] The point is that there are four volumes marked "Yards" that contain no mention of ship names or tonnages and consequently are of no help in measuring the British coastal trade.

Other volumes are of no utility in assessing the importance of the British coastal fleet. One volume is entirely devoted to "Southern whale fishing containing the names of employees with jobs such as "harpooner," "line coiler" and "boat steerer," who were obviously engaged in hunting cetaceans and of no relevance to coasting.[11] Another ledger is labelled "firemen" and

---

[8]TNA/PRO, ADM 7/377, 27 August 1790, 21; and 10 November 1794, 28.

[9]*Ibid.*, ADM 7/380, 28 January 1814, 3.

[10]*Ibid.*, 12 April 1814, 7.

[11]*Ibid.*, ADM 7/389, 10 November 1779, 12.

contains just that, lists of firefighters who had been given freedom from impressment. An example is "I do hereby certify that the thirty firemen employed by the Westminster Fire Office in extinguishing fires, are entered in the books of this office, pursuant to Act of Parliament."[12] This tome is irrelevant to the coastal fleet. One volume deals exclusively with protections for fishermen and contains no information on the coastal trade or fleet.[13]

Figure 1: Typical Eighteenth-Century Coastal Traders: A Schooner and a Smack Lying at Fresh Wharf, London Bridge.

Source: From an engraving by E.W. Cooke.

A number of the Registers record only the protections granted to apprentices and the occasional foreigner. These give a few details about them – their names, place of residence, date of indenture and of being recorded, and

---

[12]*Ibid.*, ADM 7/390, 30.

[13]*Ibid.*, ADM 7/388.

names of their masters – but contain no reference to ships or trade.[14] There is no certainty that these were ship's apprentices; internal evidence suggests that some could have been apprenticed to land-bound trades rather than maritime ones. Either way, these seven volumes have no relevance to the coastal fleet. A further pair of Registers has no value when measuring the extent of the coastal trade because they contain only lists of names of men protected.[15] They do not make any mention of the ships or trades, and one is not informed of the occupation of the protected individual. The only other piece of information that is included is the reason for their protection. The individuals in question are exempted as being either too young for impressment (under eighteen) or too old (at least fifty-five years old). These Registers do not provide any significant information on the extent or nature of the British coastal trade.

In researching these records, we identified sixteen volumes as being of no relevance at all to the British coastal trade, thus eliminating about forty percent of the volumes, leaving twenty-two of the Registers still of potential interest to an examination of the British coastal fleet. Sadly, not all of these remaining volumes are of value to the historian of British coastwise trade. Some volumes catalogue predominantly or almost entirely ships in overseas trades. For instance, one gives the name of the ship, the master, the tonnage of the ship, the number of the crew and the routes on which they operated.[16] The drawback of this volume is that the majority of the trades mentioned are foreign, with the West Indies, North America and Europe being the most common destinations mentioned. There are several volumes where this occurs, and while one or two have a few ships assigned to a coastal trade (an example is "*Fortune,* Robert Hunter, 203 tons, Cork and London") the total of such ships in the register is only about fifty and cannot be considered a comprehensive listing.[17] Although of some interest for inquiries about ship size and trades worked, there is no data for calculating the size and nature of the British coasting fleet for these years. There are seven volumes that fall into this category and that must be removed from the list of Registers useful to maritime historians of the British coasting fleet.[18]

This is a disappointment for any scholar keen to measure the British coastal fleet. Some of the volumes do not distinguish between ships in the coastal trade and those in foreign trades. For example, one register contains

---

[14] *Ibid.*, ADM 7/393-399.

[15] *Ibid.*, ADM 7/391-392.

[16] *Ibid.*, ADM 7/373.

[17] *Ibid.*, ADM 7/374 and 376.

[18] *Ibid.*, ADM 7/372-376, 381 and 387.

the names of many ships, their masters, and their crews, but gives no indication of whether an individual ship is in the coastal or foreign trade or indeed what routes she plied.[19] "Protection for Rd Blackstone the Master or the Mate or Carpenter belonging to the ship *Nicholas* of N Shields" is a case in point.[20] Another register has all the requisite details except in which trade the ship was employed: "John Bay, *Triumph,* 150 tons, no crew, Chester" or "William Barker, *Unity,* 157 tons, 14 crew, Whitehaven."[21] In short, these particular Registers will not help answer any queries about the British coasting trade. This eliminates a further three volumes as being valueless in calculating the size of the British coastal fleet.

In all, twenty-six volumes were discarded as shedding no light on the coastal fleet, leaving only a round dozen as possible sources. Given that this is less than one-third of the original number, it does suggest that this source is not quite as valuable as Robinson stated. Indeed, the majority of the volumes are of no relevance to an inquiry on the size and extent of the British coastal fleet.

Robinson suggested that these records would allow us to measure not merely the number of ships in the British coastal trade but also the aggregate tonnage of ships involved. Since the home port of each ship was given, it would be possible to work out the relative importance of various towns in terms of their ownership of coaster numbers and tonnage. It is not quite as simple as that. Some Registers do not indicate the tonnage of the ships except in very rare cases. For example, protection is extended to the crew of a "Barge called the *Hopewell* of Blackney employed in carrying Corne between yt place and London," but no tonnage is assigned to this barge.[22] Thus, we are provided with fascinating glimpses of the coasting, fishing and shellfish trades, but the data are neither sufficiently systematic nor full enough to allow an estimate of the aggregate numbers or tonnage in the coastal trade. Even where the register allows the coastal ships to be distinguished, some rarely give the home port of the ship and thus do not allow a geographic analysis to be made of the ownership of the British coastal fleet.[23]

From this information, we concluded that a large number of the Registers cannot be used to provide any information on the size and extent of the British coastal fleet, as Robinson claimed. To be of maximum use, we need

---

[19]*Ibid.,* ADM 7/363.

[20]*Ibid.,* ADM 7/152.

[21]*Ibid.,* ADM 7/371.

[22]*Ibid.,* ADM 7/363-364.

[23]*Ibid.,* ADM 7/363-370, 381-382 and 384-386.

seven pieces of information to be present in the register: whether a ship was employed in the coastal or overseas trade; the name of the ship; her tonnage; the master's name; crew size; home port; and the specific cargo or commodity she was carrying. It might be objected that the ship's name and that of her master are not required to work out the aggregate number and tonnage of the British coasting fleet. This is true, but these two pieces of data are crucial when it comes to resolving double-counting problems. Similarly, we do not need the home port, crew size or type of commodity carried to arrive at the overall total of ships and their tonnage in the coasting trade. However, the lack of any one of these pieces of data seriously reduces the depth of analysis that can be undertaken. We can somewhat reclassify our list of requirements in the Registers to say we must have four pieces of information: whether the ship was in the coastal or foreign trade; her name; her master's name; and tonnage. The other three types of data are bonuses that allow us to extend the analysis.

Figure 2:   A Collier Brig Discharging into a Lighter, Probably in the Thames.

Note: These were the workhorses of the coal trade from North East England to London and the south coast.

Source: See figure 1.

On this basis, how many Registers contain the essential minimum data and how many have some of the enhanced data? Further, how many years are covered by these basic or enhanced Registers? Appendix table 1 lists those volumes that have information on the British coastal trade, together with the years covered, the average number of entries on each page relating to coastal ships, the number of pages in each volume allocated to each year, an estimate of the total number of coastal entries (calculated by multiplying columns 3 and 4 together) and the number of months in each year that are covered in the register. From this, it will be seen that only fifteen Registers contain the relevant information, and that they cover, intermittently, the 111-year period between 1701 and 1811. During this time, however, there is no coverage of sixty-four years, with these lacunae concentrated in the first half of the eighteenth century. This leaves forty-seven years dealt with in the Registers. This is far from being comprehensive, as Robinson claimed.

Looking in more depth at Robinson's analysis of the British coastal trade, we see it is based on one year, or to be more precise nine months in 1776 and 1777. For that period, he found 10,275 entries relating to the coastal trade, whereas our rough estimate in table 1 of the number of relevant entries is 8855. This suggests our estimates are about sixteen percent too low and need to be inflated by such an amount to be accurate. A glance at column 5 shows that the number of entries per year varies enormously, from 130 in 1711 to 12,768 in 1744. Given that Robinson calculated there were over four thousand vessels in the "pure" coastal trade (that is, entirely in domestic trade) of Great Britain in 1776 and 1777, either there were drastic fluctuations in the size of the British coastal fleet from year to year or, more likely, those years for which there are less than four thousand entries are nowhere near a comprehensive listing of the British coastal fleet and thus can be discarded for this purpose. This opinion is reinforced when we recall that Robinson started from a base of 10,275 entries that he had to deflate by a ratio of 1.77, caused by double-counting of ships, as a large number of them appeared twice in the register. Given that the number of entries in column 5 of table 1 is the gross total before eliminating double-counting, it might be objected that we need a minimum of ten thousand entries there. This is to assume the double-counting is always at the same level, which may not be correct. It seems safer, until we have investigated the double-counting problem for one or two more years than Robinson did, to eliminate only those years with less than four thousand entries. This removes thirty-four years from the forty-nine, to leave only thirteen years that are worth investigating to see what light they shed on the size and nature of the British coastal fleet. In addition, our elimination of these low-entry years has restricted the time period covered considerably. No years before 1740 and none after 1793 are covered. The Registers can shed light only on the later part of the eighteenth century, which is some way from Robinson's original claim, and then, for at most thirteen separate years.

In order to test some of the above ideas, it was decided to create a database for 1810 where there were only about 650 entries by our calculation. Accordingly, all of the data for 1810 was put in an "electronic spreadsheet." There were seven pieces of information on each entry, and 726 entries. Given that we estimated 648 entries, this suggests our estimate is about twelve percent below the actual number, which is roughly in line with the result we had for 1776 and 1777. This gives us confidence in our estimates and a known margin of error in the low teens. A number of problems arose at the data-entry stage that is common in historical research. Because the Registers are handwritten, some entries had faded, the ink has blurred in others, some of the information could not be deciphered and an approximation had to be made for the name of the ship, port or captain. In six cases, no tonnage was given for the ship, and in a few cases, the port name remained indecipherable. These were so few as not to affect seriously the subsequent analysis. The basis on which ships were categorized into the various trades –"coasting," "coal and coasting," "timber and coasting" – was also rather arbitrary as the same ships appeared with different designations in the same year. This probably reflected their voyage patterns and indicates the flexibility of deployment, in which many ships were employed in a number of different trades within a year rather than being devoted to one route or commodity.[24] In a few cases, this flexibility extended to ships being employed in both the coasting trade and fishing.

In the course of our research, once the data had been entered, it was sorted alphabetically by the name of the ship, so that all ships with the same name appeared together. This was done to allow removal of ships that were registered for protection more than once in the year.

In order to arrive at the number and tonnage of ships in the coastal fleet, we needed to ensure that there was no double-counting, and hence the requirement to strike from the database any duplicate entries. We accomplished this by comparing the ships' names, port of registry, captain's name and tonnage. Where three or more of these variables coincided, the second and subsequent entries were pruned from the database as being probable repetitions. Some ships were registered a large number of times. A handful appeared five times, over forty appeared four times and an even larger number were listed three times. The original database contained 726 ships of an aggregate tonnage of 85,938 and a total manpower of 7439. When the repeat entries were removed, this brought the dimensions down to 422 ships, 49,689 tons and a total manpower of 4256. This represented a ratio of 1:73 between the unpruned and pruned databases, comparing favourably with that used by Robinson for 1776 and 1777 in the original article in which he found the crude

---

[24]Simon Ville, "The Deployment of English Merchant Shipping: Michael and Joseph Henley of Wapping, Ship Owners, 1775-1830," *Journal of Transport History*, 3rd ser., V, No. 2 (1984), 16-33.

database to be 1.77 times the pruned version. These two figures are genuinely close enough to each other to provide mutual support and vindicate the method.

A database of some 420 ships cannot represent the whole British coasting fleet in 1810 – if it did we could ignore the coastal trade as being of no great significance at this time. However, if we assume our database represents the characteristics of the overall fleet and is a random sample from it – and we have no reasons for thinking the contrary – then we can analyze some of the characteristics of the British coasting fleet in 1810. The aggregate statistics suggest an average ship size of about 118 tons with a crew of ten and the average manning level nearly twelve tons per man. The range of ship sizes was large, from the twenty-five-ton *Elizabeth* to *Chapman* at 555 tons.

The database also allows a hierarchy to be drawn up of those ports that owned coastal ships (appendix table 2). It is a strange mixture of what might be expected and some surprises. The top two ports are not surprising. London was by far the largest city in Britain in the eighteenth century as well as being the capital, and the major export and import centre. It is understandable that it should have the lion's share of coasters, with over sixteen percent of the fleet. In a similar vein, Leith, the port for Edinburgh, the capital of Scotland, served a role akin to London's north of the border. The coal-exporting ports, Newcastle and North Shields, with nearly nine percent of the fleet between them, are placed fifth and eleventh in order. This points up the importance of the coal trade to the British economy. Whitby, tenth in the table, may also owe its position in part to the coal trade, since it was renowned as a collier-owning port.[25]

Liverpool's eighth place signifies the early days in this port's rise to the second position in the trade of the United Kingdom and reveals the burgeoning traffic of the Irish Sea. Given the trade in fish, cattle and granite, and its role as an embarkation port for some of the chains of the Scottish isles, Aberdeen's sixth position is understandable, if rather higher than might have been expected. Dundee, which placed seventh, also suggests that the long-distance trade in primary products was an important component of the coastal trade.

The two surprises in the top ten, Berwick and Chepstow, are placed third and fourth, each with about eight percent of the total fleet. Berwick had a regular packet service carrying the surplus of a rich agricultural hinterland to London. Grain, eggs and salmon were shipped south. Turnpike road improvements linked the port to local market towns to gather this harvest. The high position in this table suggests either that this trade has previously been under-estimated or that the shipmasters of Berwick were particularly assiduous in

[25]Ralph Davis, *The Rise of the English Shipping Industry in the Seventeenth and Eighteenth Centuries* (London, 1962; reprint, Newton Abbot, 1972), 62-63.

seeking protection. Chepstow is even more difficult to explain. It had a trade in coal from the nearby coalfield, and in bark and timber, but its small population and lack of national role make it look like a statistical fluke rather than a reflection of Chepstow's importance as a port.[26]

The table of ports also allows us to make some generalizations about the concentration of coaster ownership. Sixty-eight ports feature in the database. The top three owned one-third of the total tonnage, the top five owned forty-eight percent and the top ten owned nearly two-thirds. This leaves one-third of all coasters owned by fifty-eight ports. This may not quite conform to the 80:20 rule, but is very close: the top fourteen ports (20.6 percent of the total) owned seventy-four percent of all coasters, while eighty percent of the ports owned twenty-six percent of all coasters. This suggests a high degree of concentration.

What is clear from the number of entries, even in the unpruned database, is that the 1810 register is not a comprehensive listing of all the ships in the British coastal fleet at that time. The exercise vindicates Robinson's deflator of 1.77, but, in reality, it tells us nothing about the absolute size of the coasting fleet in either numbers or tonnage.

We therefore decided to create another database of one of the years that had a large number of entries and that was more likely to be comprehensive. We picked 1741 for this purpose for two reasons. First, we estimated it to have one of the largest total number of entries, 11,828, and second, it was a third of a century earlier than the year that Robinson analyzed. Some comparisons might be drawn as to what had happened to the size of the British coastal fleet over the middle of the eighteenth century.

The methodology was identical to that adopted for 1800, except, of course, the data inputting was much more extensive and time consuming. There was also one drawback to the Registers for 1741. They did not give the home port of the ships, so no geographical analysis was possible. This was a significant drawback, but we stuck to 1741, as all of the other years in the 1740s for which there were large numbers of entries also suffered from this weakness. They, too, did not record the home port of the ships entered in the register. We did not wish to take another year in the 1770s because that would be too close to the year Robinson analyzed, and we should not get the long-term perspective, so 1741 remained the chosen year.

Once the data were entered in the database, we had another check of our estimate of the number of entries in each register. We had estimated the number of entries for 1741 at 11,828. In fact, we ended up with 9834 ships in our database. This suggests that we overestimated the number of coastal entries

---

[26]Tony Barrow, "Corn, Carriers and Coastal Shipping: The Shipping and Trade of Berwick and the Borders, 1730-1830," *Journal of Transport History*, 3rd ser., XXI, No. 1 (2000), 6-27; and Grahame E. Farr, *Chepstow Ships* (Chepstow, 1954).

by about twenty percent. This was probably caused by there being fewer entries of coastal ships per page than we estimated, as entries for fishing vessels and those engaged in overseas as well as in the coasting trade were all jumbled up on the same page. This total was the number of unpruned entries and included any multiple entries. Just as for 1810, we needed to order the entries by ship name and eliminate all multiple entries. This was quite a considerable task because multiple entries were rife. For instance, *Adventure* of seventy tons, Ben Robinson, appeared five times in the register in 1741, *Loyalty* of ninety tons, John Taylor, appeared three times, and *Good Intent* of sixty tons, John Sandwell, also appeared three times. One ship appeared eight times, but most only appeared once in the year.

All such multiple entries were removed from the database to produce a pruned version. This brought the number of ships down to 5865, compared to the nearly ten thousand in the original unpruned version, and gave a conversion factor of 1.68. This was not too far away from the 1.72 we discovered in 1810 and the 1.77 calculated by Robinson for 1776. It suggests that somewhere between 1.6 and 1.8 would be an accurate conversion factor for these Registers in order to eliminate multiple counting.

The results of the 1741 database, after pruning out these multiple entries, worked out at 5865 ships of a total of 604,330 tons and with an aggregate crew size of 24,030. When these figures are used to calculate average ship size, the average number of crew in each ship, and the average number of tons handled by one crew member, the results are 103 tons, a crew of four and twenty-five tons per sailor. These figures can be compared to those we derived for 1810 and those Robinson calculated for 1776.

Year	Average Tonnage	Average Crew	Tons per Man
1741	103	4	25
1776	113	7	17
1810	118	10	12

This causes some problems. The figures for average tonnage are the least problematical. Given that we do not know which tonnage figure was employed, registered or deadweight, and taking into account the degree of doubt that the figures were recorded consistently, leaving a margin for error, and that the range of ships was from a small sloop or ketch to a collier brig of several hundred tons, these seem plausible. Ville, for instance, calculated the average tonnage in the coasting trades in 1785 as varying considerably between ports. For Newcastle, he estimated 231 tons; for Sunderland, 137 tons; for Bristol, eighty-three tons; for Liverpool, sixty-eight tons; and for Hull, sixty-three tons.[27] A straight arithmetic average gives a result of 116 tons, very close

---

[27]Simon P. Ville, "Shipping in the Port of Newcastle, 1780-1800," *Journal of Transport History*, 3rd ser., IX, No. 1 (1988), 73.

to the figures for 1776 and 1810. Ideally, we should weight the average size by the tonnage of ships at each port, but Ville does not provide the needed data. The figures suggest a small increase in average ship size over time, which is also consistent with growth in trade over the period.

The figures for average crew size are much more of a problem. There is such a great difference in the three figures as to call them into doubt. Also, the trend of steep increases in crew size is not what one might intuitively expect, given the slight rise in ship size. Like crew size, the tonnage handled by one mariner also looks suspicious because its trend is downward over time and, if these figures are to be believed, one sailor was coping in 1810 with half the ship tonnage handled in 1741. This also seems to run counter to some other previous findings. Willan found that, in the first third of the eighteenth century, large coastal colliers of two to three hundred tons carried crews of about nine or ten men; that would represent about twenty to thirty tons per man.[28] This is very similar to the figure for 1741 calculated here. One would have expected that the average ship size in the coasting fleet was much smaller than that of colliers, the vessels that were usually the largest and hardiest of those engaged in the coasting trade. Davis, for example, cited the example of an eighty-ton ship *Diligence* in the coal trade from Whitehaven to Dublin with a crew of seven men and two boys. This results in a tonnage of about eight tons per man.[29] Willan pointed out that 119 ships leaving Chichester coastwise in 1732 averaged 9.5 tons per man.[30] A figure of twenty-six tons per man for the whole coasting fleet seems far too high, and good even for large coastal collier brigs.

The figure may be explained by the nature of the records used. The number of crew recorded in the Registers is not the actual crew or even the preferred number of sailors, but those given protection from impressment. It might be that the number was at some variance to the crew normally employed, and the naval officers extended protection only to what they considered to be minimum numbers of sailors required to sail the ship in ideal conditions, rather than the normal number employed. The figures may understate the actual crew size and overstate the tonnage handled per mariner.

The figure for 1810 looks too large for the crew size and hence too low for tonnage per man. It could be that the "sample" of the British coastal fleet represented by the 422 ships registered as protected was not representative of the whole fleet. The sample size looks reasonable. If the fleet was about the same size as in 1741 or 1776, our sample would be between about seven

---

[28]T.S. Willan, *The English Coasting Trade, 1600-1750* (Manchester, 1938; reprint, Manchester, 1967), 16.

[29]Davis, *Rise of the English Shipping Industry*, 358.

[30]Willan, *English Coasting Trade*, 16-17.

and ten percent; thus, the sample size is not likely to be too small to be unrepresentative. It might be that the sample is unrepresentative in some way, or that fishing boats, which carried a larger crew, have been recorded wrongly as coasters. Alternatively, it could be that the captains of coastal vessels had learned to overbid, that is, to ask for protection for a larger crew than they needed, and the number recorded is larger than the actual crew. Because the Registers record only the size of the crew protected, not the actual crew employed, there must be some doubt as to their validity when measuring crew size or tonnage handled per man.

What is clear from these various attempts to use the Registers is that they are by no means a simple source to use and are riddled with complications and uncertainties. Many years do not record all of the coastal fleet, and even those years with large numbers of entries may not be comprehensive. Without more information on who received protection and who did not, and detailed knowledge on the basis for extending protection, we are working in the dark. We can absolve the community of maritime historians from most of the implicit opprobrium heaped on their heads by Robinson. The usefulness of the Registers is very limited, and any results obtained by analyzing them must be qualified by caveats and uncertainties.

There are a number of different measures that could be used to determine the aggregate size of the British coastal trade in the eighteenth century, and it is important to distinguish between them and compare only like with like. The main distinction, apart from that between numbers of ships and their tonnage, is the difference between the fleet and the trade. The former refers to the total number of ships in the coasting trade, counting each only once, the latter to the total number of entries or clearances in the trade where the same ship entering a given port several times in a year is counted on each occasion. The fleet size multiplied by the average number of voyages made by a ship in a year (if it could be ascertained) would give the size of the trade. The Admiralty Registers of Protection used here say nothing about the trade but have been used to measure the fleet size. There now exist a number of sources that can make a contribution to the fleet size. Ralph Davis estimated the size of the British mercantile fleet, including ships engaged in both the overseas and coasting trades. His figures ranged from 340,000 tons in 1686 to 323,000 tons in 1702 to 608,000 in 1775. Of these figures, about 80,000 were coastal colliers and 52,000 other coasters in 1686, and his estimate of coasters in 1702 was about 125,000 to 130,000, although he never made this totally explicit.[31] These figures for coastal trade came under attack from Robinson as far too small. Robinson thought Davis' figures for the coastal trade in 1686 should

---

[31]Davis, *Rise of the English Shipping Industry*, 401, 403 and 427.

have been at least doubled to 270,000 tons plus.[32] Davis conceded they under-stated reality but thought they under-recorded by two-fifths, to give a total in 1686 of about 220,000 tons of coasters.[33]

An important source used by historians of shipping is the Musgrave Manuscript in the British Library.[34] This was cited by Willan and drawn on by Davis, but with grave reservations on both their parts as to its reliability and accuracy.[35] What is clear is that there is an enormous difference between the figures given in the Musgrave Manuscript and those derived from the Registers of Protection by Robinson and the current exercise. For 1741, we have a fig-ure of 604,000 tons from the Registers and for 1776, 505,000 tons. For 1744, the closest year in the Musgrave Manuscript to 1741, Musgrave gives 123,000 tons and for 1776, 228,000 tons, although this excludes London, which could have accounted for another fifty percent, to give a total around 330,000 tons. The figures from the Registers are respectively five times and two and a quar-ter times the Musgrave figures. These are large differences and difficult to reconcile. It could be that some of the ships described as coasters in the Regis-ters were not, in fact, in those trades, but were refugees from the overseas trade; it is difficult to see how that could be estimated. It might be suggested that, despite our best efforts, some double-counting persisted in the pruned database. This might seem to be supported by the fact that for 1741 the ratio between pruned and unpruned ship numbers was lower (1.68) than in the 1776 and 1810 databases (1.77 and 1.73 respectively). However, this variance does not extend to the tonnage figures, where the ratio for 1741 was 1.88, a little higher than the 1776 and 1810 figures. The double-counting of tonnage has been eliminated to the same extent as in the other years. In any case, the small errors in removing double-counting are not going to account for the significant differences.

Another complication with the two figures from the register is that they indicate that the British coastal fleet declined over the middle third of the eighteenth century by about twenty percent, roughly equivalent to about one-half of one percent per annum. This contradicts the trend in the Musgrave Manuscript that shows rapid growth, nearly doubling in thirty-two years, a rate of growth over two and a half percent per annum.

---

[32]Dwight E. Robinson, "Half the Story of *The Rise of the English Shipping Industry*," *Business History Review*, XLI, No. 3 (1967), 307.

[33]Ralph Davis, "Well, Maybe Three Fifths," *Business History Review*, XLI, No. 3 (1967), 308-311.

[34]British Library, Additional Manuscript 11,255.

[35]Willan, *English Costing Trade*, appendix 7, 220-225; and Davis, *Rise of the English Shipping Industry*, appendix A.

Despite Robinson's comments, the effect of analysis of the Registers of Protection seems to be to complicate the picture we have of the British coastal trade. The calculations carried out in this article for the year 1741 support Robinson's thinking on the size of the British coastal fleet in the eighteenth century more than those of Davis or Willan. Davis' revised estimate of approximately 220,000 tons was much smaller than Robinson's 1967 figure of about 270,000, but both are tiny compared to the total suggested by the Registers of about half a million tons. Given that the large difference stems from using two different sources, new sources or research methods are needed to try to resolve this. One way forward might be to compare the figures in the Musgrave Manuscript for individual ports with local sources to see if the former holds up. Another might be to look at the trend in coastal ship ownership in a number of ports to see if they support the steady growth of the Musgrave Manuscript or the decline between 1741 and 1777 shown in the Registers. Two other areas might help to shed some light on the size of the British coasting fleet, the extent of flexible deployment, and the number of voyages made each year in particular trades. The former might help explain an apparently large coasting fleet, if it contained many vessels in the overseas trade at some seasons. The latter would allow comparisons with the total output of the coastal sector since, presumably, fleet size multiplied by the average number of voyages should equal the aggregate output.

There obviously remains much more work to be done on the eighteenth-century British coastal trade. The Registers are much more complicated and time consuming to work with than Robinson suggested, and may be more deeply flawed than he allows.

Appendix Table 1
ADM 7 "Registers of Protection" with Information on the British Coastal Fleet

Volume Number	Years Covered	Average Entries/ Page	No. Pages	No. Coastal Entries	Months Covered	Deflated by 1.72	Deflated by 1.77
363	1701	1	386	386	11.00	224	218
	1702	1	156	156	3.00	91	88
364	1711	2.5	52	130	3.00	76	73
	1712	2.5	187	468	9.00	272	264
365	1740	31	252	7812	9.66	4542	4414
	1741	34	302	10,268	10.50	5970	5801
366	1741	30	52	1560	1.50	907	881
	1742	29	330	9570	12.00	5564	5407
	1743	23	145	3335	5.00	1939	1884
367	1743	29	192	5568	7.00	3237	3146
	1744	36	303	10,908	10.00	6342	6163
368	1744	31	60	1860	2.00	1081	1051

Volume Number	Years Covered	Average Entries/ Page	No. Pages	No. Coastal Entries	Months Covered	Deflated by 1.72	Deflated by 1.77
	1745	27	386	10,422	12.00	6059	5888
	1746	25	88	2200	3.00	1279	1243
369	1746	25	278	6950	9.30	4041	3927
	1747	22	264	5808	8.00	3377	3281
370	1747	25	122	3050	4.00	1773	1723
	1748	24	58	1392	2.00	809	786
	1754	21	51	1071	1.00	623	605
	1755	13	303	3939	4.33	2290	2225
381	1755	8	114	912	7.33	530	515
	1756	4	160	640	12.00	372	362
382	1770	33	144	4752	3.00	2763	2685
	1771	23	42	966	2.00	562	546
	1776	23	72	1656	2.00	963	936
	1777	23	313	7199	6.00	4185	4067
383	1778	26	150	3900	5.00	2267	2203
	1779	26	52	1352	6.00	786	764
	1780	26	39	1014	12.00	590	573
	1781	26	16	416	2.00	242	235
384	1787	28	88	2464	1.00	1433	1392
	1790	28	240	6720	6.00	3907	3797
	1791	29	124	3596	4.00	2091	2032
	1793	25	81	2025	2.33	1177	1144
385	1793	19	193	3667	8.00	2132	2072
	1794	13	153	1989	12.00	1156	1124
	1795	13	46	598	12.00	348	338
	1796	13	31	403	12.00	234	228
	1797	13	25	325	12.00	189	184
	1798	13	15	195	11.50	113	110
	1799	13	10	130	11.50	76	73
	1800	13	8	104	11.66	60	59
	1801	13	5	65	8.50	38	37
386	1803	8	110	880	9.00	512	497
	1804	8	54	432	12.00	251	244
	1805	8	62	496	11.50	288	280
	1806	8	34	272	11.50	158	154
	1807	8	48	384	12.00	223	217
	1808	8	69	552	12.00	321	312
	1809	8	54	432	12.00	251	244
	1810	8	81	648	12.00	377	366
	1811	11	25	275	3.66	160	155
400	1781	20	57	1140	8.50	663	644
	1782	15	75	1125	12.00	654	636
	1783	15	10	150	0.66	87	85

Appendix Table 2
Protected British Coastal Fleet by Port of Ownership, 1810

Port	Tons	Percent of All ports	Ships	Percent of All ports	Crew	Percent of All ports
London	8088	16.28	49	11.61	619	14.54
Leith	4357	8.77	35	8.29	432	10.15
Berwick	4223	8.5	35	8.29	463	10.88
Chepstow	4034	8.12	28	6.64	245	5.76
Newcastle	3239	6.52	19	4.5	203	4.77
Aberdeen	2506	5.04	21	4.98	322	7.57
Dundee	2164	4.36	24	5.69	257	6.04
Liverpool	1716	3.45	24	5.69	143	3.36
Bridport	1328	2.67	12	2.84	108	2.54
Whitby	1267	2.55	10	2.37	94	2.21
North Shields	1197	2.41	6	1.42	68	1.6
Belfast	979	1.97	6	1.42	83	1.95
Margate	921	1.85	11	2.61	79	1.86
Lerwick	890	1.79	9	2.13	66	1.55
Arbroath	862	1.73	8	1.9	60	1.41
Banff	696	1.4	9	2.13	78	1.83
Poole	535	1.07	3	0.71	29	0.68
Rye	531	1.07	2	0.47	13	0.31
Plymouth	483	0.97	4	0.95	60	1.41
Inverness	453	0.91	4	0.95	49	1.15
Dartmouth	448	0.9	4	0.95	30	0.7
Ilfracombe	416	0.84	4	0.95	15	0.35
Lynn	396	0.8	4	0.95	29	0.68
Stockton	389	0.78	3	0.71	24	0.56
Littlehampton	375	0.75	3	0.71	28	0.66
Colchester	346	0.7	5	1.18	31	0.73
Greenock	332	0.67	2	0.47	35	0.82
McDuff	332	0.67	4	0.95	30	0.7
Dover	321	0.65	3	0.71	15	0.35
Jersey	299	0.6	4	0.95	33	0.78
Ulverstone	289	0.58	3	0.71	18	0.42
Whitstable	283	0.57	4	0.95	29	0.68
Barmouth	273	0.55	2	0.47	18	0.42
Faversham	262	0.53	4	0.95	19	0.45
Kirkcaldy	226	0.45	2	0.47	15	0.35
Runcorn	226	0.45	2	0.47	24	0.56
Weymouth	211	0.42	4	0.95	24	0.56
Bridgewater	188	0.38	2	0.47	11	0.26
Workington	181	0.36	1	0.24	13	0.31
Perth	176	0.35	2	0.47	25	0.59
Falmouth	163	0.33	5	1.18	10	0.23
Dublin	161	0.32	2	0.47	10	0.23
Teignmouth	161	0.32	2	0.47	10	0.23

Port	Tons	Percent of All ports	Ships	Percent of All ports	Crew	Percent of All ports
Glasgow	159	0.32	2	0.47	13	0.31
Penarth	130	0.26	1	0.24	16	0.38
Pwllheli	125	0.25	1	0.24	7	0.16
Chester	124	0.25	1	0.24	10	0.23
Guernsey	120	0.24	2	0.47	26	0.61
Rochester	120	0.24	1	0.24	7	0.16
Newport	118	0.24	1	0.24	6	0.14
Findhorn	116	0.23	1	0.24	12	0.28
Portsmouth	110	0.22	1	0.24	8	0.19
Larne	107	0.22	1	0.24	2	0.05
Ardrossan	105	0.21	1	0.24	12	0.28
Hastings	104	0.21	1	0.24	14	0.33
Hull	103	0.21	1	0.24	8	0.19
Montrose	91	0.18	1	0.24	11	0.26
Shetland	84	0.17	1	0.24	6	0.14
Porthsay	80	0.16	1	0.24	8	0.19
Ramsgate	78	0.16	1	0.24	7	0.16
Ipswich	76	0.15	1	0.24	4	0.09
Exeter	70	0.14	1	0.24	8	0.19
Greenwich	61	0.12	1	0.24	30	0.7
Bristol	46	0.09	1	0.24	6	0.14
Ipswich	45	0.09	1	0.24	6	0.14
Ripley	45	0.09	1	0.24	2	0.05
Gloucester	38	0.08	1	0.24	2	0.05
Scarborough	25	0.05	1	0.24	5	0.12
Port not given	224	0.45	2	0.47	28	0.66
Unreadable	265	0.53	3	0.71	25	0.59
	49,689	100	422	100	4256	100

Port	Tons	Percent of All ports	Ships	Percent of All ports	Crew	Percent of All ports
Glasgow	199	0.32	2	0.47	15	0.31
Penrith	130	0.20	1	0.24	16	0.?
Bedbeth	195	0.25	1	0.32	7	0.16
Chester	104	0.25	1	0.24	10	0.23
Guernsey	110	0.24	2	0.47	26	0.6?
Rochester	129	0.26	1	0.24	7	0.16
Newport	28	0.24	1	0.24	6	0.14
Lisburn	123	0.56	1	0.24	12	0.25
Portsmouth	0	0.00	1	0.24	8	0.1?
Luton	97	0.??	1	0.??	2	0.05
Atkinson	105	0.??	1	0.??	12	0.??
Fleetway	104	0.21	1	0.24	14	0.?
Hull	103	0.21	1	0.24	8	0.18
Montrose	91	0.18	1	0.024	11	0.23
Sheffield	84	0.17	1	0.24	7	0.14
Banbury	80	0.16	1	0.24	8	0.19
Ramsgate	78	0.15	1	0.24	6	0.16
Ipswich	76	0.15	1	0.28	4	0.09
Exeter	71	0.14	1	0.24	8	0.15
Greenwich	61	0.12	1	1.36	60	0.2
Bristol	46	0.09	1	0.24	6	0.14
Norwich	45	0.04	1	1.18	4	0.14
Ripley	45	0.09	1	0.28	4	0.09
Gloucester	48	0.56	1	0.24	7	0.05
Scarborough	45	0.27	1	1.21	7	0.12
Portland Bill	204	0.26	1	0.87	24	0.40
Dunstable	207	1.??	1	0.??	24	0.56
	49,2??	100	?1?	100	11??	100

Chapter 4
Management Response in British Coastal Shipping Companies to Railway Competition[1]

This essay will examine the ways in which British coastal shipping businesses reacted to competition from railways. It is divided broadly into five sections. The first sketches the role of coastal shipping before the advent of the railways and explores the impact of steam on short-sea shipping. The second analyzes the part played by the short-distance early railways, which were perceived initially as at best minor threats to coastal shipping. Indeed, many were seen as beneficial because they enhanced the flow of goods to and from ports. The third section examines the threat from the long-distance national rail lines that began to appear in the 1840s. The fourth considers the range of responses, including attempts at intra- and inter-modal collusion; a search for technological improvement; a more positive market segmentation; and a re-appraisal of pricing methods. Finally, I will evaluate the success of these responses in securing market share for the coaster.

As a method of moving goods and people, coastal shipping has a long history. In the early modern period it was an important industry in Britain, even if estimates of its significance vary enormously.[2] With an extensive coastline and many navigable rivers, Britain was particularly reliant on coasters to move coal, grain, ore and a wide range of agricultural and extractive goods.[3] Despite having been virtually ignored by some recent historians, coastal ship-

---

[1]This essay appeared originally in *The Northern Mariner/Le Marin du nord*, VII, No. 1 (1997), 17-28.

[2]Ralph Davis, *The Rise of the English Shipping Industry in the Seventeenth and Eighteenth Centuries* (London, 1962; reprint, Newton Abbot, 1972), 395-406; Dwight E. Robinson, "Half the Story of *The Rise of the English Shipping Industry*," *Business History Review*, XLI, No. 3 (1967), 303-308; Davis, "Well Maybe Three Fifths," *Business History Review*, XLI, No. 3 (1967), 308-311; and Robinson, "Secret of British Power in the Age of Sail: Admiralty Records of the Coasting Fleet," *American Neptune*, XLVIII, No. 1 (1988), 5-21.

[3]Simon P. Ville, *Transport and the Development of the European Economy, 1750-1918* (London, 1990), 101-103; and John Armstrong, "The Significance of Coastal Shipping in British Domestic Transport, 1550-1830," *International Journal of Maritime History*, III, No. 2 (1991), 63-94.

ping was crucial to British industrialization and its growing trade.[4] Coasters linked the various regions into something approaching a national economy, carrying not only bulky, low-value products but also manufactures, such as linen, cheese, iron goods, and beer and spirits. They were ubiquitous, as a perusal of directories for port cities or early local newspapers reveals.

The value of the coaster was much enhanced with the advent of steam. Although there were earlier experiments, the first commercial steam-boat service in the UK was inaugurated in 1812 by Henry Bell, who ran *Comet* on the Clyde and its estuaries between Glasgow and Gourock.[5] This service, which spawned many others, began thirteen years before the pioneering Stockton and Darlington Railway (SDR) and eighteen before the Liverpool and Manchester (LMR).[6] In other words, steam was exploited much earlier on water than on rails. The advantages of steam to water transport were enormous. Although largely confined to rivers, estuaries and coastal routes, steam provided a predictability that sailing ships lacked.[7] No longer subject to tides and fickle winds, steamers could proceed at steady speeds in almost all conditions, in marked contrast to the sailing coaster, which could and did make some fast passages but could also be becalmed for days or trapped in harbours by contrary winds. On tidal rivers, sailing vessels frequently could make no headway and hence could only use half the working day. Steamboats brought speed and punctuality to transport well before the railways, allowing both more regular exchanges of information and reductions in inventories.

Yet there were also drawbacks to the early steamboat. One was safety, for the new technology posed novel problems.[8] Unreliable boilers and a

---

[4]One historian with little to say about coasting during the industrial revolution is Rick Szostak, *The Role of Transportation in the Industrial Revolution: A Comparison of England and France* (Montréal, 1991). For a corrective, see John Armstrong and Philip S. Bagwell, "Coastal Shipping," in Derek H. Aldcroft and Michael J. Freeman (eds.), *Transport in the Industrial Revolution* (Manchester, 1983), 142-161.

[5]Brian D. Osborne, *The Ingenious Mr. Bell: A Life of Henry Bell (1767-1830), Pioneer of Steam Navigation* (Glendaruel, 1995), 19-25; and H. Philip Spratt, *The Birth of the Steamboat* (London, 1958), 87-88.

[6]Maurice W. Kirby, *The Origins of Railway Enterprise: The Stockton and Darlington Railway, 1821-1863* (Cambridge, 1993); and Thomas J. Donaghy, *Liverpool and Manchester Railway Operations, 1831-1845* (Newton Abbot, 1972).

[7]Sarah Palmer, "Experience, Experiment and Economics: Factors in the Construction of Early Merchant Steamships," in Keith Matthews and Gerald Panting (eds.), *Ships and Shipbuilding in the North Atlantic Region* (St. John's, 1978), 233-235.

[8]See, for example, Great Britain, Parliament, House of Lords, *Parliamentary Papers (BPP)*, 1839, XXV, "Report on Steam Vessel Accidents."

poor understanding of safety procedures led to fairly frequent explosions that attracted press attention and parliamentary enquiries. Moreover, the steamboat's ability to make good progress when other ships were stationary could itself be a problem. For instance, while it could maintain its usual speed in fog, this could lead to collisions if lookouts were not alert. Still, such dangers did little to constrain the demand for steam. Much more important in this regard were the inefficient boilers and the voracious demands of the engines for fuel, both of which prevented early steamers from making long journeys economically. Even on short coastal, estuarine and river passages there was little room for cargo by the time the engines and boilers had been accommodated and the bunkers filled.

As a result, early steamboats were restricted in their routes and cargoes. Compared to sailing vessels, they had to charge a premium to cover the additional capital and fuel costs. Only those goods which benefited most from rapid and predictable transport could bear the higher freight rates. The early steamers thus concentrated on post, parcels, passengers, perishables and livestock, all of which were either light and lacked bulk or benefited from speedy carriage by virtue of rapid deterioration or high value. By the late 1820s there was a range of short steam services to complement the more ubiquitous, cheaper, but less predictable sailing coaster, which continued to carry bulky goods and general cargoes. It was into this environment that the early railways intruded.

Early railways were not perceived as threats to coastal shipping. On the contrary, they were seen as beneficial since they promised to increase coastal traffic. Many of the earliest, such as the colliery lines in the North East, were no more than economical methods of moving coal from pitheads to the nearest navigable waterway, be it river or coast, where it could be loaded into coastal colliers for shipment south.[9] In this respect they were feeders to the coastal shipping network. Some of the early short-distance public steam railways were essentially extensions of this principle. For example, the SDR was intended to link the inland collieries of South Durham and Northeast Yorkshire to the Tees, where coasters could take the cargo to London to compete with coal from further north on the Tyne and Wear. The importance of coal to this railway is shown by its seal, which depicts horse-drawn coal carts. To facilitate this coastal coal trade, the leading financiers of the SDR in 1831 established Middlesborough as a reliable deep-water port. As a result of the construction of the line, the number of colliers clearing the Tees with cargo rose from ninety-seven in 1826 to 2415 in 1832. It is hardly surprising that the line was considered a boon to coastal shipping. On the opposite side of the country, promoters of the LMR were motivated by the extensive raw cotton trade between the two centres. The greater speed and ease of communication

---

[9]Much of this paragraph draws on evidence in Kirby, *Origins*, esp. chapter 2.

by rail also opened the Manchester market to Irish farm produce, especially
with the rapid, reliable steamboat able to carry perishables like butter, cheese
and livestock.[10] Again, coastal traffic was enhanced by this railway, and the
tale was repeated with many other early rail lines. Few paralleled the coast;
most ran from the interior to a port or connected two inland towns, as did the
Cromford and High Peak.[11]

Early railways also created traffic for coasters by increasing demand
for materials and components. Railway construction involved gargantuan quantities of ballast, bricks, sleepers and chairs, rails, locomotives and rolling
stock. Many of these could be obtained locally, but some had to come from a
distance and were transported for a substantial part of their journey by coaster.
For example, Edwin Pratt points out that the sleepers and rails for many of the
early northeastern lines came from Sussex and Hampshire as back cargoes for
coastal colliers.[12] Maurice Kirby explains that the SDR used "wood blocks to
act as foundations for the chairs [which were] imported via the Tees from
Portsea in Hampshire." Similarly, rails came from both the Bedlington Iron
Co. in Northumberland and the Neath Abbey Ironworks in South Wales; chairs
also came from the latter.[13] These were delivered to the Tees by ship, for rails
were heavy and road carriage was a last resort. In the construction of the
LMR, similar components were needed, and many were brought by sea. The
first locomotives built by the Stephensons in Newcastle were dismantled,
transported overland to Carlisle and brought by boat to Liverpool, where they
were rebuilt.[14] An initial order of 160 tons of wrought iron rails was filled by
John Bradley of Stourbridge, and further supplies were shared between this
firm and Michael Longridge's Bedlington Iron Works.[15] An examination of the
Bills of Entry reveals the extent of this construction-related coastal traffic. To
take one instance, the Clyde Bills for 1850 show that 225 tons of iron rails
entered Glasgow from Newport, and that Glasgow despatched by sea 750 tons
of railway iron to Dumfries and 200 tons to Runcorn. Twenty-five tons of

---

[10]Robert E. Carlson, *The Liverpool and Manchester Railway Project, 1821-1831* (Newton Abbot, 1969), 22-24.

[11]D.J. Hodgkins, "The Origins and Independent Years of the Cromford and High Peak Railway," *Journal of Transport History*, 1st ser., VI, No. 1 (1963), 39-53.

[12]Edwin A. Pratt, A *History of Inland Transport and Communication in England* (London, 1912), 199-200.

[13]Kirby, *Origins*, 44.

[14]Stephen Greaves, *The Liverpool and Manchester Railway: Newton's Story* (Newton, 1980), 1.

[15]Carlson, *Liverpool*, 191-192.

wooden sleepers were sent by sea from Arbroath to Grangemouth and 1995 pieces of sleeper wood were sent from Aberdeen. Grangemouth also received one cargo of "scotch fir sleepers" from Arbroath, one consignment of railway sleepers from Lossiemouth and two from Speymouth. In return, Grangemouth sent 448 tons of iron railway chairs to Rye, 346 tons to London, 159 tons to Hastings and 115 tons to Lynn. It also sent iron railway sleepers: 380 tons to Hastings, 293 tons to Rye, 180 tons to London and 120 tons to Whitstable.[16] One port thus sent over 2000 tons of railway iron by sea inside Britain in a single year. These examples could be multiplied, but the point is that railway construction created considerable coastal traffic. Once operational, railways channelled goods onto the coasters. The Manchester and Leeds Railway, for instance, sent large quantities of textiles to London for marketing in the early 1840s. They went by rail to Hull and by steamer to London, a journey that took thirty-six hours. Previously, the goods had been sent by canal, a trip of about five days.[17] It is clear that in many ways early railways complemented coasters, as many ran from the interior to a port and hence generated traffic for coastal shipping.

The mania of the mid-1840s inaugurated an era of truly national railways. In 1840 the basic backbone of about 1500 miles of mainline was in place. By 1850 all large urban centres had been linked, and the country boasted about 6500 miles of track.[18] There was not only a quantum leap in track mileage but also a change in the way railway managers thought, as strategic and national concerns became paramount. Long-distance trunk routes were their objective rather than local or regional lines serving limited markets. Among the new mainlines were the Great Western, Midland, London and North Western and London and South Western. Lines now linked London to Aberdeen, as well as to Plymouth, Holyhead and Glasgow. The establishment of the Railway Clearing House (RCH) in 1842 further facilitated long-distance traffic by providing an impartial mechanism to allocate ticket receipts when passengers or goods required transport by more than one company.[19] The practice of allowing one company's trains onto another firm's tracks, along with more powerful locomotives and continuing mergers, created a truly national network. As a result, in the 1850s railways began to pose a real threat to long-

---

[16]Clyde Bills of Entry for 1850.

[17]*BPP*, 1840, XIII, q. 4427.

[18]T.R. Gourvish, "Railways, 1830-70: The Formative Years," in Michael J. Freeman and Derek H. Aldcroft (eds.), *Transport in Victorian Britain* (Manchester, 1988).

[19]Philip S. Bagwell, *The Railway Clearing House in the British Economy, 1842-1922* (London, 1968).

distance coastal traffic. The railways captured long-distance passenger traffic completely at an early date and retained it. Until the 1850s the railways earned more revenue from passengers than from freight. Despite emphasizing the health and recreation aspects of a sea trip, the coaster could not compete on speed or comfort with railway travel for passengers.

Initially, though, the threat was minimized, for the strategies employed by the railways did not include creating a demand for mass travel by radical reductions in transport charges. Rather, they were keen to capture existing traffic and to cater for high-value products at premium prices: "The country's trunk lines established themselves by specialising in high tariff business, quality rather than quantity...they sought to capture high-value merchandise from the road carriers by cutting rates just sufficiently to ensure it."[20] For example, despite the construction of the South Devon Railway in the late 1840s, the ball clay trade remained essentially the province of coasters through the port of Teignmouth. Railway interests did not want bulky, low-value goods like clay and coal.[21]

The nature of the railway challenge lay mostly in non-price features. One of the most important was speed. While the steam coaster could make a steady eight or nine knots (about nine or ten miles per hour), steam locomotives could sustain speeds several times as high. When goods needed rapid transit, railways could provide it much more efficiently than coasters. Perishable cargoes, such as meat or fish, could be catered for by special express trains given precedence over ordinary goods traffic and nearly on a par with passenger trains. As Geoffrey Channon has shown for the Aberdeen-London route, by the mid-1850s railways were carrying the vast majority of fresh meat, with steamers relegated to a minor role. This was achieved by "special express goods trains," which took twelve hours less than the steamers and which were by the 1860s operating nightly.[22] Similarly, fresh fish, which had previously been brought from the Yorkshire coast by steam packet to towns along the North Sea, like Newcastle, London, Sunderland and Hull, and even up the Humber and its tributaries, began to be conveyed by rail from the

---

[20]T.R. Gourvish, *Railways and the British Economy, 1830-1914* (London, 1980), 26; and Robb Robinson, "The Evolution of Railway Fish Traffic Policies, 1840-66," *Journal of Transport History,* 3rd ser., VII, No. 1 (1986), 35-36.

[21]L.T.C. Rolt, *The Potters Field: A History of the South Devon Ball Clay Industry* (Newton Abbot, 1974), 127-129.

[22]Geoffrey Channon, "The Aberdeenshire Beef Trade with London: A Study in Steamship and Railway Competition, 1850-69," *Transport History,* II, No. 1 (1969), 12 and 23.

1840s.[23] Initially, rates were high and the emphasis was on premium traffic. But from the mid-1840s, lower rates were introduced to accommodate less expensive varieties, like haddock and plaice. Traffic subsequently grew in volume and revenue. Aberdeen salmon, fresh fruit, vegetables and seasonal produce were commodities the railways desired and captured. They were not at first concerned with increasing the quantity of goods to maximize revenue as much as maximizing the carriage of high-value freight. This was a strategy similar to that adopted by the early steamboats several decades earlier.

By the 1850s, however, railway strategy had changed, as companies began to compete for long-distance, lower-value bulk freight and were willing to cut freight rates to attract it. This was a result of a number of factors. The RCH made it easier to sort out financial implications when wagons traversed the lines of a number of companies. Amalgamations gave the firms a more national outlook and reduced the number of distinct enterprises.[24] Improvements in locomotive technology made the running of a large number of heavily loaded wagons over long distances more feasible, and improved methods of goods handling, monitoring and billing obviated some of the problems of logistics and paperwork. In addition, by the late 1840s or early 1850s, the railway companies, like the overall economy, were depressed and hence sought new revenues to bolster their lagging profits and poor dividends.[25] They became more aggressive in searching for freights they had previously shunned. An example is the coal trade from the Northe East to London. In 1850 the railways had carried less than two percent of this trade, which was still dominated by the coastal ship, initially the collier brig but increasingly the screw-propelled steam collier. By 1857, however, the railway's share had risen to twenty-eight percent, and by 1867 railways carried more coal to London than did the coasters.

Having outlined the threat from the railways, we need to consider how the coastal shipping industry responded. The initial reaction to competition was often to cut costs. Even before the railway was fully operational the mere threat was often enough to cause the coastal sector to try to reduce rates to retain customers. In this way shipowners lowered the economic rent earned and brought prices more closely in line with costs. That they were able to earn any economic rent was an indication of the superiority of steamboat travel over

---

[23]Robb N.W. Robinson, "The Fish Trade in the Pre Railway Era: The Yorkshire Coast, 1780-1840," *Northern History*, XXV (1989), 202-234; and Robinson, "Evolution," 34-35.

[24]Gourvish, *Railways*, reports that 187 amalgamation acts were put before parliament in the 1860s alone. While in 1850 the top fifteen railway companies accounted for three-quarters of the turnover, in 1874 the top ten had the same share.

[25]*Ibid.*, 27-28.

previous methods in terms of speed, reliability and punctuality. It was only the railways, also utilizing steam, which could match them in punctuality and could exceed their speed, at least for passenger and special fast-goods trains. Even so, by the 1840s the cost structure for coastal steamers was such that they could offer a cheaper service than railways over long distances, although they could not compete on speed.[26]

It would be wrong to suggest that only competition from the railways caused the coasters to collaborate. Indeed, as early as 1832 there was a pool comprising a number of steamboat owners plying between Glasgow and Liverpool with mutually-agreed rates, coordinated schedules and a form of revenue-sharing.[27] In 1836 there was a similar pool on the east coast among operators running between Scottish ports and London.[28] Thus, even before the railway began to compete in earnest coasters were demonstrating the validity of Adam Smith's aphorism about businessmen and the suppression of competition. But two things happened after railways became national. The first was an attempt to increase the degree of collaboration between coastal steamboat companies in liner trades. The second was to extend cooperation between the two modes, as railway companies developed conferences into which they admitted some "rival" liner firms.

By their nature pools and conferences were usually kept hidden from the public since the parties wished to give the impression of competition. As far as objective economic information is concerned, the outside observer cannot easily discern the difference between a perfectly competitive market and one where collusion occurs. For example, in both cases the prices for similar services will be identical. If the market were competitive this would be because both firms had cut their charges to the point where marginal revenue equalled marginal cost; where the market is collusive, the result will be the same, albeit for different reasons. As a result, except where formal institutions were established to administer such agreements, few relevant records are likely to have survived. Nonetheless, we know that there were a large number of these agreements in the nineteenth century, covering a range of routes and involving a large proportion of the coastal liner companies.[29] The increased

---

[26]B.R. Mitchell with Phyllis Deane (comps.), *An Abstract of British Historical Statistics* (Cambridge, 1962; reprint, Cambridge, 1976), 113.

[27]National Maritime Museum, Greenwich (NMM), CST/38/3.

[28]Aberdeen University Library, Ms. 2479/1, Aberdeen Steam Navigation Co., Directors' Minute Book.

[29]John Armstrong, "Conferences in British Nineteenth-Century Coastal Shipping," *Mariner's Mirror*, LXXVII, No. 1 (1991), 55-65.

competitive capability and strategy of the railways played a part in this up-surge, although it is difficult to determine how big a part.

Perhaps more important was the innovation of intermodal collaboration. Once railways had long-distance routes, they established the RCH, and because there were often multiple routings between two cities, they began to establish conferences to reduce price competition and to share revenues.[30] Examples, such as the Octuple Agreement (1851), the Humber Conference (1855) and the English and Scotch Traffic Agreement (1856), spring to mind. But because coastal liners enjoyed a cost advantage over railways and hence could charge lower rates, the latter thought it prudent to allow coastal liners into some of the conferences.[31] The railways retained a speed advantage for passenger and express freight trains, but not for ordinary goods trains, which had low priority. The two modes thus complemented each other in this respect: the coaster could offer cheaper travel, the railway faster. That the coaster was cheaper was reflected in these agreements, for whenever rates were agreed and printed the coaster's rates were significantly less than the railway. The functions of these rail-coaster conferences varied in detail, but ran the gamut from mutually-agreed freight rates which could not be altered without all-party consent to agreements about the frequency of departures, revenue-sharing and sometimes even the charges for ancillary services, such as cartage.

That such collusion was not contrary to the interests of the coastal companies is shown by the number of complaints from railway representatives at RCH meetings about the low share of traffic – as varied as meat from Aberdeen to London and baled textiles from Dundee and Perth to the capital – carried by rail.[32] The advantages to the coaster appear to have been three-fold. First, by fixing rates which reflected differential costs, coasters were given a price advantage which allowed them to retain the lion's share of bulk traffic, such as baled jute, cloth and yarn, which did not require rapid transit. For instance, on the Dundee-London route from 1870 to 1879 – the only period for which continuous figures have been found – coasters carried on average over

---

[30]Bagwell, *Railway Clearing House;* and Geoffrey Channon, "Railway Pooling in Britain before 1900: The Anglo-Scottish Traffic," *Business History Review,* LXII, No. 1 (1988), 74-92.

[31]John Armstrong, "Railways and Coastal Shipping in Britain in the Later Nineteenth Century: Cooperation and Competition," in Chris Wrigley and John Sheppard (eds.), *On the Move: Essays in Labour and Transport History Presented to Philip Bagwell* (London, 1991), 76-103.

[32]Great Britain, National Archives (TNA/PRO), RAIL 1080/509, minute 857; and RAIL 1080/511, minute 2378.

ninety percent of baled jute.[33] This suggests that the conference did not disadvantage the coaster. Second, fixing rates allowed both sides to implement
higher charges than if collaboration had not occurred. For instance, on the
Dundee and Perth to London route, the pre-1870 sea rate was fifteen shillings
per ton. In 1870 this was raised to twenty-five shillings, a massive increase of
sixty-seven percent. This rate remained in force until at least 1879.[34] Given
that the 1870s was a decade of falling prices, coasters would seem to have
made good returns. Railways advanced their rates at the same time, but by a
smaller percentage. It thus appears that the two modes charged higher prices
than if there had been no collaboration and so raised profits. This gives a clue
to the third advantage for coasters: their ability within a conference to beat off
any opposition from "outside ships." When such competition threatened, the
coastal companies had the financial strength to enable them to withstand a period of intense rate cutting and eventually to see off the opposition. Most conferences were sufficiently flexible to allow the shipping lines a temporary departure from agreed rates to fight such external competition.

The owners of coastal ships responded to railway competition. Another was the deliberate
copying of the methods used by railways to charge for freight. Railways used
eight categories, depending on the value of the commodity: A, B and C were
used for cheaper products and 1-5 for more valuable goods, with five being the
highest rate. The first three groups were often "station to station," while the
latter five were normally "collected and delivered," meaning the railway company was responsible for cartage at both ends. In addition, the railways created
a profusion of "exceptional" rates, which by the late nineteenth century had
become the norm.[35] These special rates gave the railways flexibility and hence
bargaining room while being concealed behind published tariffs. At a time of
widespread criticism, railways needed published rates as a smoke screen while
actually charging much lower rates, without setting a precedent enforceable by
the Railway Commissioners.

The owners of coastal liners based their charges on the same eight
groupings as the railways and assigned exactly the same commodities to each
class. In addition, they fixed an exceptional rate to undercut most of those established by the railways.[36] It might seem strange that shipping firms knew

---

[33]*Ibid.*, RAIL 1080/509-512.

[34]*Ibid.*, RAIL 1080/510, minute 2154.

[35]Peter J. Cain, "Traders versus Railways. The Genesis of the Railway and
Canal Traffic Act of 1894," *Journal of Transport History*, 2nd ser., II, No. 2 (1973),
66-67.

[36]NMM, CST 39/15-16.

about these supposedly confidential exceptional rates, but that they did know seems unarguable. At the more conspiratorial end of the explanatory spectrum, it is possible that coasting firms had spies either in the RCH or the railway companies. But a more plausible explanation is that the information came from on-going collaboration in one of the many pools or conferences. Regardless, by aping the railways' method of charging, shipping firms made it easy for potential customers to compare and hence to see the price advantage of the coaster.

A quite different approach toward railway competition was to upgrade technology. Shipping technology changed rapidly in the nineteenth century.[37] Improvements in boilers allowed higher pressures and more efficient coal use, while higher pressures paved the way for compound and then triple-expansion engines that consumed less coal. Screw superseded paddle; iron replaced wood, to be in turn supplanted by steel, which permitted stronger yet lighter hulls and larger ships. Less expensive water ballast replaced stones, bricks and pig iron. Cargo-handling was improved by the use of winches, derricks and other gear, and the design and placing of hatchways provided easier access. There were also improvements in dockside infrastructure. While these advances are well known and require no elaboration, the important point is that all had economic consequences. The ability to reduce coal consumption diminished costs; increased the steamer's economic range; facilitated manoeuvrability; and increased speed. Larger ships provided economies of scale in capital and crewing costs, and so allowed lower freight rates. Water ballast reduced operating costs and speeded up turnaround times, while the various aids to cargo-handling gave the customer quicker service and the owner more revenue-generating voyages per annum. All these improvements lowered costs while allowing better service.

Yet what is interesting is whether technological improvements favoured the railway or the steamboat. On balance the costs of moving large quantities of bulky goods fell much more on the coaster than the railway. The large coastal collier in the late nineteenth and early twentieth centuries was capable of carrying 1000-1500 tons of coal.[38] Compare this with the average train load, estimated by Peter Cain as sixty-three tons between 1880 and 1900

---

[37]Much of this paragraph is based on evidence in J. Graeme Bruce, "The Contribution of Cross-Channel and Coastal Vessels to Developments in Marine Practice," *Journal of Transport History,* 1st ser., IV, No. 2 (1959), 65-80, reprinted in John Armstrong (ed.), *Coastal and Short Sea Shipping: Studies in Transport History* (Aldershot, 1996), 57-72.

[38]John Armstrong, "Late Nineteenth-Century Freight Rates Revisited: Some Evidence from the British Coastal Coal Trade," *International Journal of Maritime History,* VI, No. 2 (1994), 70-71.

and about ninety-six tons by 1911.[39] Even allowing for this increase and for the fact that mineral trains may have been longer than average, the coaster had an enormous advantage. Where the railway had an edge was in its greater speed. The power output of locomotives was significantly increased, which allowed faster speeds and higher pulling power. Where railway managers chose to maximize the former – for passenger trains carrying parcel traffic or with a dedicated meat or fish van attached – it extended rail's advantage. At the minimum, changes in technology kept the coaster in the same position relative to the railway, and at best improved its cost structure.

A further response of coastal shipping firms grew from an evolving perception that distinct customers required different services. Strategies to build on this insight led to market segmentation.[40] If a customer wanted the lowest possible transport price, was in no hurry for the goods and did not care precisely when they would be delivered, sailing ships were available. They offered low freight rates, a function of no fuel costs, low manning requirements and no loading or unloading costs (at least as long as the crew were required to carry out these tasks). In some cases harbour or port charges could be evaded by sailing up a sloping beach at high tide, awaiting the ebb and discharging over the side. The schooners and ketches in the coastal trades have been described as "floating storehouses." If low costs were vital, or goods needed slow and careful loading (such as bricks, clay pipes, chimney pots or other frangible products), a factor that was likely to lead to high demurrage charges, sailing vessels were most suitable. Since managers preferred full holds to maximize revenue, this made sail especially suitable for bulky goods.

If shippers wanted predictable arrivals, they needed steam. The coastal tramp as operated by firms such as Coppack Bros. of Connah's Quay provided a more reliable and usually faster service, since it could make a steady eight or nine knots in virtually any weather, twenty-four hours a day.[41] Like the sailing coaster, it required full holds but was a little dearer because of higher operating costs. Since by definition tramps did not provide regular service, a shipper might have to wait for a suitable ship. But the telegraph and a dense network of agents gave the tramps good employment while providing shippers reasonable assurance of the availability of an appropriate craft.

---

[39]Peter J. Cain, "Private Enterprise or Public Utility? Output, Pricing and Investment on English and Welsh Railways, 1870-1914," *Journal of Transport History*, 3rd ser., I, No. 1 (1980), 16.

[40]The next three paragraphs are based largely on John Armstrong, "Coastal Shipping: The Neglected Sector of Nineteenth-Century British Transport History," *International Journal of Maritime History*, VI, No. 1 (1994), 182-185.

[41]Tom Coppack, *A Lifetime with Ships: The Autobiography of a Coasting Shipowner* (Prescot, 1973).

Where there was a need for frequent consignments of a bulk good, like coal, china clay or iron ore, coastal shipping firms provided "regular traders." They were often built for a particular trade or commodity and might be owned by specialty merchants, such as Cory or Charrington, coal dealers who had their own colliers for the east coast coal trade. These firms found an advantage in backward vertical integration because they had a large regular demand for the coal and could therefore keep their vessels fully employed. The transaction costs of hiring outside vessels were internalized, saving money and providing a greater certainty of supply. These firms negotiated large contracts with electricity-generating stations, gas works and other large industrial users, such as the Gas Light and Coke Co., Metropolitan Water Board or City of London Electricity Supply Co., which were located on or near the sea or a navigable river.[42] There was thus a large regular demand that was relatively unaffected by seasonality. The degree of specialization is demonstrated by the up-river colliers, which had hinged masts and funnels to allow easy passage under London's bridges. These regular traders provided fast turnaround and reliability for large-scale consumers who could not afford an interruption in supply.

At the top end of the market were customers who wanted fast, reliable, scheduled collection and delivery. They moved high-value goods, such as manufactures, and were willing to pay a premium (in coastal shipping terms) to have them delivered quickly and on schedule. Lever's soap, Bovril meat extract, cigars, pianos and books all travelled by this method. They used large, well-appointed steamers incorporating the latest technology, owned by companies such as Tyne-Tees, Aberdeen Steam and Powell, Bacon and Hough. In addition, they usually owned or leased a dedicated berth at the ports they served to minimize delays in docking and moving cargo. These ships usually carried passengers and livestock as well as cargo, and so eschewed dirty, dangerous and malodorous goods, such as coal, dynamite and kips. They carried mixed cargoes and usually required no minimum quantity, thus making them ideal for small but frequent deliveries of low-bulk goods.

In short, one response by the owners of coasters to competition from railways was to diversify the services offered, ranging from inexpensive but unreliable for bulky, low-value raw materials, to fast, scheduled reliable services for high-value, low-bulk manufactures. As the railways attacked the coastal trade and adopted a more national strategy, the coastal shipping industry endeavoured to retain, and even attract, customers by tailoring its services more closely to the economic characteristics it perceived important to them. In this way it retained a large share of the internal trade of the UK.

---

[42]Elspet Fraser-Stephen, *Two Centuries in the London Coal Trade* (London, 1952), 99-100.

Perhaps the ultimate expression of intra-modal collaboration occurs when two or more firms in the same industry decide to merge. This may reflect competitive pressures within the industry or from competing modes, or perceptions that economies of scale may be available at a higher level of aggregate output. Sharing berths, agents or office administration might offer some savings, and better coordination of previously independent schedules might provide a faster service and hence attract additional traffic. There would also be a degree of risk-spreading due to a greater number of ships, and there might be savings in insurance for a larger fleet. Simply reducing competition might be a sufficient motive, since it would allow freight rates to be raised and might well yield higher profits. Whatever the theoretical gains, many railway companies thought them sufficiently real to indulge in corporate marriage. The ten largest companies (of a total of about 120) in both 1870 and 1913 accounted for nearly three-quarters of aggregate revenue, which was £50 million in 1870 and £140 million in 1913.[43] The coastal shipping companies followed a similar strategy, albeit more slowly and with a lag of several years.

Before 1870 most amalgamations in coastal shipping were on a small scale, usually when an existing concern absorbed a potential or actual competitor on its route. In some cases these were almost certainly "greenmails," where a firm placed a ship onto a route in order for it to be bought out. The motive in some cases was purely pecuniary, while in others it was because the owners of the competing ship wished to force a way into what were usually private companies with a closed, limited shareholder base. Since when shareholders wished to dispose of shares they often had to be offered first to existing investors, if a coastal shipping company appeared to be paying steady dividends outsiders often tried to force their way in using this strategy. There are numerous examples of one-ship businesses being absorbed by a successful coastal liner company.

From the 1890s a greater urgency and formality emerged; in the next two decades there were a number of large mergers and takeovers. In 1893, for example, C.R. Fenwick, W.M. Stobart and William France combined to become the largest managers of colliers.[44] This seems to have concerned Cory, whose ships were managed by Fenwick. In 1896 Cory, long a major player in the London coal trade and the owner or part-owner of a number of steam colliers, organized the merger of eight separate firms into a public limited-liability company. The businesses absorbed included Lambert Bros., J. and C. Harrison and Green Holland and Sons, all long-established coal merchants and col-

---

[43]Peter J. Cain, "Railways, 1870-1914: The Maturity of the Private System," in Freeman and Aldcroft (eds.), *Transport,* 103-104.

[44]J.A. Macrae and Charles V. Waine, *The Steam Collier Fleets* (Wolverhampton, 1990), 53.

lier owners.[45] As a result, the new firm owned about thirty ships and handled about seventy percent of seaborne coal carried to London. These two mergers caused the coastal coal trade from the North East to resemble a duopoly.

Mergers also occurred among coastal liner companies. In 1903 the Tyne Steam Shipping Company merged with the Tees Union Shipping Company to create a near-monopoly over liner traffic from the North East to Hull and London. In 1910 the Powell Line, operating between Liverpool and London and calling at Bristol, Plymouth and Southampton, merged with the Bacon Line, also offering similar services out of Liverpool. In 1912 the merged firm also absorbed the Hough Line. These manoeuvres were minuscule compared to what was to occur after 1917 when the Powell Bacon and Hough Line was absorbed by the Royal Mail Steam Packet Group and changed its name to Coast Lines. Between 1917 and 1920 Coast Lines absorbed thirteen separate coastal and Irish Channel firms, and engulfed another seven during the 1920s.[46] By the onset of the Great Depression the British coastal liner trade was close to being a monopoly.

How effective were these strategies? It is much easier to answer in the aggregate. On this level it may be argued that since coastal shipping still played an important part in British internal transport in 1914, the policies must have worked.[47] The volume of transport provided by coasters expanded, and they were the most important carriers of long-distance and high-bulk cargocs; in total, they probably provided about the same ton-mileage as railways. The registered tonnage of entries and clearances of ships with cargo in the coastal trade grew steadily before 1913, and the network of coastal liners was dense and comprehensive. In aggregate terms, coastal shipping firms retained their positions.

Yet there is always the counterfactual. If coastal shipping firms had pursued alternative strategies, would they have done even better, perhaps increasing their shares of traffic or driving some railway companies out of business? The latter seems highly unlikely. By the late nineteenth century the railways were financial giants, beside which even the largest coastal liner firms were financial pygmies. If the coasters had waged war against the railways, the latter had the financial muscle to survive a period of intense price and service

---

[45]Raymond Smith, *Sea Coal for London: History of the Coal Factors in the London Market* (London, 1961), 343.

[46]Peter Mathias and Alan W.H. Pearsall (eds.), *Shipping: A Survey of Historical Records* (Newton Abbot, 1971), 37-39 and 94.

[47]John Armstrong, "The Role of Coastal Shipping in UK Transport: An Estimate of Comparative Traffic Movements in 1910," *Journal of Transport History*, 3rd ser., VIII, No. 2 (1987), 164-178, reprinted in Armstrong (ed.), *Coastal and Short Sea Shipping*, 148-162.

competition which could only have led to the withdrawal of the coastal company. In any case, such a scenario would not have pitted the coastal firms against the railways, but one coastal firm against an individual railway, let us say Cory versus the North Eastern Railway for the carriage of coal from the North East to London. Again, the disparity is huge. When it went public in 1897, Cory was capitalized at £2 million; the North Eastern Railway at the same date had a capitalization of over £60 million.[48] The conclusion is that the coastal firms were irritants as far as goods were concerned but had no impact on the large and lucrative passenger traffic. The railways tolerated them while they were not too much of a problem and were mainly carrying low-value, bulky commodities. But if a coastal firm had tried seriously to impinge on the traffic of a particular railway, the latter would have retaliated, likely with great success. The railways had the formal institutions to facilitate cooperation and might have closed ranks in the face of a severe coastal threat. The coaster and railway co-existed in the late nineteenth century because it was in the interests of both not to upset this delicate balance.

---

[48]Thomas Skinner (comp.), *Stock Exchange Year Book* (London, 1897), 1464; and Robert J. Irving, *The North Eastern Railway Company, 1870-1914: An Economic History* (Leicester, 1976), 288.

Chapter 5
Conferences in British Nineteenth-Century Coastal Shipping[1]

Much is known of the extent, methods and significance of conferences as a means of regulating competition in overseas shipping.[2] They were introduced among liner companies running to a fixed schedule in a particular trade. Tramp conferences were unlikely to succeed because such ships did not ply one regular route but rather worked whatever cargo and route was available, making mutual pricing a nightmare. Tramps also were often operated by merchants to carry their own goods and were not interested in collaborative action. Conferences in overseas trade appeared with the advent of the long-distance steamer which ensured reasonably reliable arrivals and departures. Sailing ships depended too much on fickle winds and tides to run to a strict timetable. Conferences were most likely to succeed where valuable or perishable commodities needed regular and rapid transit. For instance, the tea trade from the Far East, carried so famously by clipper ships until the 1870s, was confined to steamers belonging to the Far Eastern Freight Conference after its inception in 1879. Other cargoes could await the unscheduled but cheaper freighter.

The earliest conference in foreign trade was established in August 1875 for shipping from the United Kingdom to Calcutta. One had been projected in 1869 for the North Atlantic but failed to materialize, and some sailing ship brokers attempted to form rings to fix freight rates on the more regular routes. From that beginning conferences spread rapidly into most trades[3] so

---

[1]This essay appeared originally in *Mariner's Mirror*, LXXVII, No. 1 (1991), 55-65.

[2]For a general discussion of conferences, see Francis E. Hyde, *Blue Funnel: A History of Alfred Holt and Co. of Liverpool, 1865-1914* (Liverpool, 1956), 56-75; Sheila Marriner and Francis E. Hyde, *The Senior: John Samuel Swire, 1825-98: Management in Far Eastern Shipping Trades* (Liverpool, 1967), 135-171; J.H. Dyos and Derek H. Aldcroft, *British Transport: An Economic Survey from the Seventeenth Century to the Twentieth* (Leicester, 1969; reprint, Harmondsworth, 1974), 296-372; and David R. MacGregor, *The China Bird: The History of Captain Killick and the Firm He Founded* (London, 1961), 175-178.

[3]Hyde, *Blue Funnel*, 57; Dyos and Aldcroft, *British Transport*, 269; and John Orbell, *et al.*, *From Cape to Cape: The History of Lyle Shipping* (Paul Harris, 1978), 49.

that by 1895 they were "in all the major trades from Great Britain, with the exception of the North Atlantic,"[4] and by 1913 they "regulated most of the cargo exported from the UK" except coal.[5]

Although there were variations, the common features of conferences have been identified. A schedule of freight rates was agreed to which all members adhered. Deferred rebates were pioneered in 1877 by the Calcutta conference to encourage shipper's loyalty, since the rebate was conditional upon the merchant sending all goods only via conference ships in the qualifying period. This worked sufficiently well for it to be adopted by many later conferences. In most cases conference ships ran to a regular schedule of sailings to ensure that goods were not kept waiting for transportation. Within the conferences there were also often confidential pooling agreements which usually took the form of "joint purses" where all receipts were divided between the participants on some pre-arranged basis. This meant close cooperation in terms of scrutiny of the companies' accounts and possibly the more successful firms paying some of their income to the less successful.

The logic of the conferences was to reduce excessive competition among shipping companies, allowing them to maintain or raise freight rates and ensure reasonably full cargo holds and a fair profit margin. One advantage to the shipper was stable long-term freight rates. Removing short-run fluctuations allowed the merchant to plan ahead knowing future transport costs and increasing his confidence in the transport medium. Additionally, the shipper knew there would be a steady supply of vessels in the trade rather than irregular gluts and scarcities, ensuring his goods would not wait long for a ship and possibly deteriorate. Conferences were thus believed to be beneficial to both shippers and shipowners in removing fluctuations in capacity, freight rates and passenger fares. There is some evidence to suggest that conferences allowed shipowners to raise their rates and increase their profitability. However, if shipowners endeavoured to earn excessive profits, competition soon appeared in the form of tramps or ships belonging to firms not involved in the conference, eager to cream off some of the trade. The verdict of both a contemporary court case and a Royal Commission, as well as latter-day historians, was that such conferences were legal and did not exploit shippers.[6]

Conferences were not unknown in foreign coastal trades. In the early 1870s there were a number of pools and conferences on the China coast including traffic up the Yangtse and to Shanghai and Tientsin. These included

---

[4]R.H. Thornton, *British Shipping* (Cambridge, 1945), 278.

[5]Dyos and Aldcroft, *British Transport*, 269.

[6]*Ibid.*, 271; Hyde, *Blue Funnel*, 61-63, 68 and 72-73; Marriner and Hyde, *The Senior*, 148-150; and Great Britain, Parliament, *Parliamentary Papers* (*BPP*), Royal Commission on Shipping Rings, 1909, XLVII.

joint-purse agreements. Before 1902 there was also a South African coastal conference. Certainly the Chinese coastal conference is important because it predated the Calcutta and thus might claim the title of earliest known conference. Also, it inspired others in overseas trade.[7] Given that the conditions for a conference were regular steam communication for commodities which did not brook delay and initial intense competition leading to some form of agreement, it is surprising that little attention has been given to the British coastal trade, where these prerequisites occurred about half a century before the Calcutta or China coastal conferences. The main reference to coastal conferences is by Dyos and Aldcroft who state that "early Glasgow steamship owners fixed rates by Conference" but that by 1913 conferences were in widespread use "except for the British coastal trade."[8] The aim of this article is to demonstrate that conferences and pooling agreements were widespread in the British coastal trade before and after the Calcutta and China coast concords and to suggest that they may have been one source of inspiration for later agreements, for some of the pioneer investors and managers of the early coastal steamboats were later involved in the foundation of trans-oceanic steamship lines such as Cunard and the Peninsular and Oriental Line.[9] Their previous experiences in the coastal trade may well have influenced their strategies in overseas trade.

Because of their inefficient engines, early steamboats were used in the river, estuarial and coastal trades. They consumed large quantities of coal so they could not undertake voyages far from suitable refuelling, which ruled out deep-sea routes. Because of the capital expenditure on the engines and boilers and the expense of fuel, their costs were much higher than the sailing ships' but they could offer the shipper a faster and more reliable service. Hence, they charged premium rates which helped cover their high expenditure and tended to carry low-bulk, high-value goods which needed rapid transit: parcel trade, food and drink, such as butter, cheese, ale, fish, livestock and meat, and small quantities of manufactured goods. They also carried passengers.[10] Here were exactly the prerequisites for conferences: steamships engaged in a regular liner trade carrying high-value goods not willing to await cheaper, slower transport.

The earliest example of a coastal conference found by this author bears out Dyos and Aldcroft's assertion, for it was made in 1832 between two

---

[7]Marriner and Hyde, *The Senior*, 136-137; and Francis E. Hyde, *Shipping Enterprise and Management, 1830-1939* (Liverpool, 1967), 82.

[8]Dyos and Aldcroft, *British Transport,* 269.

[9]Michael J. Freeman and Derek H. Aldcroft (eds.), *Transport in Victorian Britain* (Manchester, 1988), 268.

[10]Derek H. Aldcroft and Michael J. Freeman (eds.), *Transport in the Industrial Revolution* (Manchester, 1983), 163-165.

groups of Glasgow merchants and their Liverpool agents who owned two separate steamboat lines trading between the Clyde and the Mersey: the Glasgow
and Liverpool Steam Shipping Co. and the City of Glasgow Steam Packet
Co.[11] This group included many who became famous in later coastal and overseas shipping firms, including James and George Burns, Thomas and James
Martin, David McIver and Robert Napier. The object of the agreement was to
arrange orderly sailings, usually two per company per week on different days;
to fix common freight rates and passage fares which would only be altered in
concert; and to stipulate how any additional shipping requirements should be
met if trade increased. Additionally the companies agreed to operate a "joint
purse" whereby if less than £60,000 was earned in any one year the proceeds
were shared sixty percent to the Glasgow and Liverpool and forty percent to
the City of Glasgow. If aggregate annual earnings exceeded £60,000, the extra
was divided equally between the two firms after the 60:40 split. The agreement
was to be effective initially for eight years. This contains all the elements expected in a conference, except perhaps deferred rebates. This was not used
because there was no need to maintain shippers' loyalty since many of the
partners to the agreement were large shippers. The two shipping firms were
not concerned to reduce capacity on the route but rather to ensure there were
sufficient ships available for the existing traffic and any increase. There were
seven clauses laying down the responsibilities for putting additional ships on
the run if trade increased, and penalty payments were to be incurred by a company unable to perform the stipulated number of sailings. This departure from
the normal concern of such conferences needs to be explained. The individuals
involved, as merchants, were more concerned to obtain regular transport for
their goods than to maximize their profits in the shipping line. Also, because
of the newness of coastal transport by steamer, and as there was no direct rail
link, they anticipated, quite correctly, that there could be a boom in demand
for the steamer's services. This concern on the part of local merchants and
manufacturers to ensure adequate local transport was one of the main motives
for innovation and investment in canals, railways and steamboats[12] and acted as
a brake on exploitative transport prices. It was only when user and owner became totally separated that any idea of profit maximizing could be contemplated.

---

[11]This paragraph is based on Great Britain, National Maritime Museum
(NMM), CST/38/3.

[12]Aldcroft and Freeman (eds.), *Transport in Victorian Britain,* 165-168; J.R.
Ward, *The Finance of Canal Building in Eighteenth-Century England* (Oxford, 1974);
S.A. Broadbridge, "The Sources of Railway Share Capital," in M.C. Reed (ed.), *Rail-
ways in the Victorian Economy: Studies in Finance and Economic Growth* (Newton
Abbot, 1969), 209-210; and Gordon Jackson, "The Shipping Industry," in Freeman
and Aldcroft, *Transport in Victorian Britain,* 265.

It is not clear how long this agreement lasted or if it was renewed, but by 1848 the two shipping lines had carried their working agreement to its logical conclusion for they were then subsidiaries of the Clyde Steam Navigation Co., which in its turn formed a conference for the Clyde to Liverpool trade with two new entries: the Glasgow and Liverpool Royal Steam Packet Co. and the Glasgow and Liverpool Shipping Co. (GLS).[13] The substance of the agreement was broadly similar to that of 1832. The ships were to be despatched "in regular rotation," common freight rates and passenger fares were fixed and only to be modified collectively and there was a joint earnings scheme, this time based on profits rather than income which entailed a detailed specification of what were considered legitimate expenses and which were disallowed. Each of the lines was to receive a quarter of net profits, i.e., each had an equal share. Given that the two lines of the Clyde Steam Navigation Co. owned three steamers between them and that the GLS owned two screw steamships whereas the Glasgow and London Royal Steam Packet Co. owned only one, this seems surprisingly generous to the latter. Like the 1832 agreement, that of 1848 emphasized that this was no merger as each of the companies was to remain independent and manage its own vessels. The agreement was to be effective for a minimum of ten years and then terminable only by written notice. The main difference between it and the earlier contract was that there was less concern in 1848 to ensure that sufficient vessels were put on the route. This is explicable partly by the greater number of vessels owned by the parties to the conference (six) and partly that the shipowners now perceived themselves as professional shipowners rather than primarily merchants and only secondly as transport providers. Also, by 1848 the steamboat had been plying on the Clyde for over thirty-five years and the trade was no longer in the rapid growth stage of the product life-cycle. Additionally, a more potent competitor, the railway, was now a commercial and technical possibility.

Conferences on the west coast were not confined, despite the opening remarks of this article, to steamer companies. In 1849 the GLS, which was one of the signatories to the 1848 Clyde-Mersey steamer conference, entered into a separate agreement with James McArthur and John Scott to operate their eleven sailing vessels in concert in the Liverpool-to-Glasgow trade.[14] Both concerns were carrying non-perishable bulky products, such as pig iron and chemicals from the St. Rollox works of Charles Tennant. They agreed to despatch their ships in regular rotation in the proportion three for the GLS and two for McArthur and Scott; common freight charges were drawn up and adhered to; and all freight earnings were to be pooled and then divided in the proportion of five-fourteenths to McArthur and Scott and nine-fourteenths to

---

[13]NMM, CST/38/6.

[14]*Ibid.*

the GLS, reflecting fairly closely the relative economic strength of the two parties. The agreement was to have a currency of five years from March 1849. Such co-operation in the early days of steam was not confined to the coastal trade. Sarah Palmer has shown for passenger travel on the Thames from the late 1820s to the 1840s that "unrestricted competition rarely seems to have persisted for long."[15] The speed and frequency of service were more important than fare levels, and steamboat companies engaged in non-price competition via the quality of the service as much as, if not more than, passenger fares on which "some kind of agreement" was usually made.

Thus, the comments of Dyos and Aldcroft about the tendency for the early Glaswegian steamboat owners to collaborate are entirely vindicated, as is their belief that rings were not unknown among the sailing ship fraternity. However, the unstated suggestion, implicit in their statement, that conferences were confined to Glasgow is quite incorrect. In the late 1830s and early 1840s there was a very similar ring on the east coast. As early as 1836 the Aberdeen Steam Navigation Co. was in correspondence with Ogilvie and Crichton in Leith, who ran steamers to London, and Matthew and Nicol in Dundee, who ran to the same destination, demonstrating that there was an informal agreement between them to alter passenger fares and freight rates in concert. Indeed, they endeavoured to ensure that the rates to London were the same from all three ports, so discriminating marginally in favour of the long haul and ensuring there was no advantage to merchants in moving their goods overland to seek the cheapest port to the metropolis. As well as collaboration in pricing, this informal east coast conference also acted like an employers' organization, setting the wages of their crews in concert, especially the engineers.[16] The aim was to prevent mutual poaching of engineers, a common problem in early factory industry.[17] There was particular difficulty in attracting staff capable of operating steam machinery because it was a relatively new and rapidly expanding sector, and hence supply lagged behind demand. As a result, engineers commanded high wages and easily obtained new positions; hence the steamer companies' attempt through collective action to prevent mobility and wages being bid up.

---

[15]Sarah R. Palmer, "The Shipping Industry of the Port of London, 1815-1849" (Unpublished PhD thesis, University of London, 1979), 149-151.

[16]Aberdeen University Library (AUL), MS 2479/1, Aberdeen Steam Navigation Co. (ASN), Directors' minute books, 18 and 25 May 1836, and 29 November and 7 December 1837.

[17]T.C. Barker, *The Glassmakers. Pilkington: The Rise of an International Company, 1826-1976* (London, 1977), 83-85; and Sidney Pollard, *The Genesis of Modern Management: A Study of the Industrial Revolution in Great Britain* (London, 1968), 199-200.

The informal collaboration ended in 1839 when a formal agreement was entered into by four steam shipping companies: the Aberdeen and London Steam Navigation Co., the Dundee, Perth and London Shipping Co., the General Steam Navigation Co., which had acquired the London and Edinburgh Steamship Co. in 1836, and the London, Leith, Edinburgh and Glasgow Shipping Co., all of which ran liner services to London carrying passengers, parcel traffic, livestock, fish and a range of other goods, many perishable and of high value. Where they operated from different ports there was no need for rotated sailing, but the two lines from Edinburgh "were openly running in amity" in 1837.[18] All four agreed "a uniform rate of fare between London and these several stations" and kept each other informed of freight rates in order to prevent any one port becoming a cheap route to the capital. Like the west coast example, the main aim of this pool was to reduce competition between the shipping companies to increase the probability of profits. The new element was the attempt to control some costs, such as the wages of the crew. This had the effect of reducing labour turnover and hence recruitment and training costs and also of ensuring that individual conference members were not uncompetitive through overly high wage costs. The records do not indicate the longevity or fate of this agreement. It was certainly still in existence in mid-1845, and seventeen years later, in December 1862, the Aberdeen Steam Navigation Co. was still corresponding quite regularly with the Leith and Dundee companies to compare the conditions of service of the crew and to ensure that their emoluments were in line with those of the other Scottish lines.[19] Even if the full-blooded conference was not continued, and there is no evidence either way on this, elements of it certainly endured for over twenty years.

This was not the only trade on the east coast to be worked collaboratively rather than competitively. In the mid-1860s the Tyne Steam Shipping Co., one of the precursors of the Tyne-Tees Steam Shipping Co., held discussions with the General Steam Navigation Co., and "an agreement was reached...to share the trade on the London route."[20] Although the details of this co-operation have not survived, it seems reasonable to suppose common

---

[18]Sarah Palmer, "'The Most Indefatigable Activity:' The General Steam Navigation Co., 1824-50," *Journal of Transport History,* 3rd ser., III, No. 2 (1982), 9-10; NMM, GSN 7/2, meetings of 26 May 1836, 28 February 1837 and 27 August 1839; and GSN 7/3, 26 August 1851; Palmer, "Shipping Industry;" and AUL, MS 2479/3, ASN, Directors' minute books, 7 May 1844.

[19]AUL, MS 2479/3, ASN, Directors' minute books, 11 June 1845; MS 2479/9, 30 December 1862; and NMM, GSN 7/3, meeting of 26 August 1845.

[20]A.M. Northway, "The Tyne Steam Shipping Co.: A Late Nineteenth Century Shipping Line," *Maritime History,* II, No. 1 (1972), 75-76.

freight rates and fares and probably sailings were arranged to avoid competition and provide a regular supply of shipping space to consignors.

If, then, it has been demonstrated that on both the east and west coasts in the early days of steam shipping conferences were not unknown and could last for some time, with the relative sophistication of "joint purse," shared earnings or even shared profits, there is another element of Dyos and Aldcroft's treatment of coastal conferences that needs to be revised. Implicit in their statement that "early Glasgow steamship owners fixed freights by conference" is that the practice had fallen out of use in later periods. This is reinforced by their explicit suggestion that by 1913 conferences were in widespread use "except for the British coasting trade."[21] In fact, there is sufficient evidence, admittedly fragmentary and scattered, to suggest that conferences were still quite common among coastal liner companies in the late nineteenth and early twentieth centuries.

In 1885, two independent companies running steamships between Liverpool and London and calling at many West Country and south coast ports *en route* arranged different sailing days for their ships to avoid duplication and ensure a regular service departing three times per week. They also had the same passenger fares for the run: £2 for a saloon return, twenty-five shillings one-way and fifteen shillings one-way for deck passengers. The two companies traded as the Liverpool, Bristol and London Steamship Co. and the London and Liverpool Steamship Co. but were effectively the Powell and Hough Lines, which were later to merge to form the nucleus of Coast Lines.[22] There was no attempt to hide the collaboration, for public handbills stated explicitly that the two companies ran the eleven steamers on this route. It seems likely that as well as common passenger fares and joint sailing schedules, the two companies also agreed freight charges for the various ports. This was in existence by 1893, as a copy has survived, and it is highly likely that earlier examples were drawn up at the outset of the collaboration. This agreement was intended to benefit both shippers and shipowners in the same way as the earlier conferences: reduced competition, higher freights and ships carrying closer to capacity cargoes leading to better profits for the shipowner and more regular sailings and stability of transport prices for the shipper. There were also two elements of cost reduction for the shipowners. Because by the early twentieth century coastal steam liners were moderately large ships (600 register tons was

---

[21]Dyos and Aldcroft, *British Transport,* 269.

[22]Handbill, Coast Lines, Liverpool; E.R. Reader, "World's Largest Coaster Fleet," *Sea Breezes,* New ser., VII (1949), 88-93; George Chandler, *Liverpool Shipping: A Short History* (London, 1960), chap. 2; and Peter Mathias and Alan W.H. Pearsall (eds.), *Shipping: A Survey of Historical Records* (Newton Abbot, 1971), 36-39.

not unusual and some reached 900 register tons),[23] they needed appropriate terminal facilities to ensure rapid loading and unloading and hence fast turn-around. In some ports, such as Liverpool, they used the quays provided by the dock company, but in others they used their own wharves; for example, the Aberdeen Steam Navigation Co. owned a wharf and warehouse in London from 1877 when it purchased one and a quarter acres at Emmett Street, near West India Dock, because its lease at Wapping had expired and could not be renewed. The new wharf cost around £50,000 and was equipped with hydraulic machinery for loading and unloading cargo. The Tyne-Tees Steam Ship Co. owned Free Trade Wharf in London.[24] These were expensive pieces of capital equipment which might be under-utilized by one company. An advantage of a conference was that the two shipping companies shared the cost since they scheduled sailing and arrivals to ensure that only one ship needed to use the facilities at a given time. It was also quite common for conference shipowners to use the same agent in the less important ports; for example, the Powell and Hough Lines used the same agent in Southampton, which was of minor importance to them, but retained separate agents in London, which was their main market. Inasmuch as there were any fixed costs of paying agents, these could be shared to reduce unit costs. These were not likely to be great since most agents worked on a commission basis, and thus their cost to the shipowner was directly proportional to the freights earned.

The co-operation between the Powell and Hough Lines was long-lasting. Although only fragmentary evidence has survived, it is clear that it continued from before 1885 to 1913, when the two companies formally merged. In addition, at some point along the way at least one other Liverpool-based coastal liner company, the John Bacon Line, was brought into the conference before it was amalgamated with Powells in 1910[25] and in turn subsumed under the Coast Lines banner.

There was also a conference at least from 1893 on the Liverpool-to-South Wales route. The parties to this were the South Wales and Liverpool Steamship Co., based in Liverpool and managed by Robert Gilchrist, and the Volona Shipping Co., also of Liverpool, managed by Messrs. Roger and Bright and Messrs. Michael Murphy of Dublin.[26] The last-named firm was

---

[23]NMM, CST/39/15.

[24]AUL, MS 2479/17, ASN, Alex Hogarth, Chairman ASN, to shareholders, 2 November 1874; and Northway, "Tyne Steam Shipping Co.," 85.

[25]NMM, CST/39/16; Reader, "World's Largest Coaster Fleet," 92; and Mathias and Pearsall (eds.), *Shipping,* 37.

[26]Coast Lines, Liverpool, Freight and Charges, Liverpool to Cardiff, February 1914.

entirely excluded from the trade except for carrying returned empties from
Cardiff to Liverpool. The agreement of 1893 was renewed and revised in April
1900 and again in February 1914, so the conference lasted for more than
twenty years. It conformed broadly to the model that has already been developed. An agreed schedule of freight rates was based on the Railway Clearing
House (RCH) classification into eight categories, and since "exceptional" rates
were more normally quoted by the railway companies than the standard class
rates, the Mersey-South Wales conference also drew up a schedule of "exceptional" rates for the commodities which it commonly carried. These extend to
fourteen foolscap pages. The parties agreed "to adhere to these rates and give
no rebates or alter the rates without mutual consultation." To make the agreed
freight rates adhere there was also a large number of clauses laying down the
details of pricing policy: charges were sea freight only, i.e., quay-to-quay,
terminal, wharfage and cartage were specified; the conditions under which the
shippers' own wharf could be used and the rebate this would attract; the discount given if the shipper used his own labour to discharge; the pricing of consignments for interior towns; and conditions for exceptionally heavy, long or
awkward pieces of freight. It is a very detailed document meant to cover contentious points discovered in operating previous versions. There is no evidence
of any revenue or profit-sharing scheme of the "joint-purse" type, but the various parties did share the same berth at Cardiff in the Bute Docks and so cut
their overhead costs *pro rata*. This agreement was renewed in 1914 for a further seven years, but the First World War put an end to its importance.[27]

There was no reason why one shipping firm should not participate in
more than one conference. This was true of the Powell Line in the early twentieth century. As well as the agreement on the Liverpool-to-London trade already discussed, it was involved with the Bacon Line and Messrs. Hosken,
Trevithick, Polkinhorn and Co. of Hayle in a conference on the route between
Liverpool and Bristol.[28] The documentary evidence is for 1908, but again it is
quite conceivable that this was a renewal or revision of an existing agreement.
The ingredients were similar to previous recipes. The signatories set agreed
freight rates based on the RCH classification and then listed dozens of exceptional rates for the items they normally carried. There was to be no deviation
from or alteration of these rates except by mutual agreement. Like the Liverpool and South Wales conference, the various shipping lines economized on
some of their operational and organizational overheads by sharing wharves and
offices and using the same small coasting firms to forward their goods to ports
at which they did not call direct, such as Bideford and Bridgewater. In order to
eliminate the possibility of competition over terminal charges they also agreed

---

[27]NMM, CST/1/1, minutes, 31 March 1914.

[28]NMM, CST/39/18.

common rates for port dues and landing charges at Bristol, cartage and lighterage in Liverpool, and cartage in Bristol and to nearby suburbs. This was another detailed agreement intended to cover all exigencies and so ensure that there were no loopholes through which price competition could occur.

Thus, the evidence suggests that conferences were fairly widespread among coastal liner companies until the First World War. Given the potential of inter-modal competition from the railway network which was substantially complete by the 1870s, it may seem surprising that more explicit competition did not occur. However, the railway companies reduced competition among themselves via pools and traffic agreements. The earliest pools appeared after the railway mania of the mid-1840s (the Sextuple and Octuple Agreements of 1851, the English and Scotch Traffic Agreement of 1856 and the Humber conference of 1855), and many were in force for long periods: the Humber conference continued until 1904, the Anglo-Scotch lasted until 1869. Thus, the railways endeavoured to eliminate price competition and, through the mechanism of the RCH, ensured that each line received a share of the available traffic.[29] This collaboration included setting common rates for freight transport. Hence, the railways attracted traffic onto their lines not by highly competitive pricing but rather confined their rivalry to the quality of the service they offered. Thus, the market for long-distance, reliable and rapid freight transport was not perfectly competitive but rather an oligopoly between two sets of collaborating companies: railway pools and coastal conferences. The two groupings reached an unwritten and largely informal *modus vivendi,* although on some occasions coastal liner companies were included in railways pools (for example, the Dundee Perth and London Shipping Co. was a party to the English and Scotch Traffic conference from 1856 to at least 1880), and the West Riding London Traffic conference included several coastal steamship companies.[30] This lack of price aggression from the railways created an ethos compatible to the continuation of coastal conferences.

Railway price competition for freight transport was not only remarkably lacking but even the level of service was not threatening to coastal liner companies. Wagons remained small, were of a bewildering variety and lacked continuous brakes and hence were moved only slowly. Because of the small size of consignments, there was also much shunting and marshalling, further slowing the speed of transit. The conference steamships running to a regular

---

[29]Geoffrey Channon, "Railway Pooling in Britain before 1900: The Anglo-Scottish Traffic," *Business History Review,* LXII, No. 1 (1988), 78-79; and Philip S. Bagwell, *The Railway Clearing House in the British Economy, 1842-1922* (London, 1968), chap. XI.

[30]Channon, "Railway Pooling," 87; and Great Britain, National Archives (TNA/PRO), RAIL 1080/465.

timetable and offering relatively rapid speeds were well able to compete with the indifferent service offered by many railway companies for freight.[31]

This article has dealt only with coastal conferences in existence up to the First World War. The war changed railway organization dramatically and permanently with the grouping into four regional monopolies. This largely eliminated the need for railway pools and conferences. The structure of British coastal shipping moved in the same direction, towards concentration, but through market forces rather than government edict. During the war Coast Lines, a subsidiary of the Royal Mail Steam Packet Co. until its disgrace and break-up in the 1930s, acquired a number of coastal lines, and this amalgamation policy was continued throughout the 1920s and early 1930s so that by 1937 over two dozen coastal firms sailed under the Coast Lines flag. This reduced the need for conferences and sailing agreements between coastal shippers. With the advent of road transport both railway and ship were put under new competitive pressure, and they continued a common approach through rates conferences both for specific routes, such as London to the Humber, and on more general issues.[32]

Some qualifications need to be made to this study. It does not claim to be a comprehensive guide to all coastal liner conferences in the nineteenth century. There may well have been many more, and perhaps this article will stimulate others to bring them to light. What has been demonstrated is that conferences existed on both the east and west coast routes and both at the beginning of the steamboat era and towards its end, just before the First World War. Thus, Dyos and Aldcroft's statements about coastal conferences need serious revision. Again, although the evidence is fragmentary it is clear that most of the classic ingredients of a conference appeared among the coastal versions: an agreed schedule of freight rates and passenger fares; a common *rota* of regular sailings; freight or profit-sharing agreements of the "joint purse" variety; the sharing of common facilities and services, such as agents, wharves, offices and through carriers. The one thing that was missing from the coastal conferences which was widespread in the overseas versions was the deferred rebate. There is no evidence of its use in any of these sources, and two witnesses before the Royal Commission on Shipping Rings of 1909 confirmed that deferred rebates were not used in the coasting trade of the United

---

[31]John Armstrong and Philip S. Bagwell, "Coastal Shipping's Relationship to Railways and Canals," *Journal of the Railway and Canal Historical Society*, XXIX, No. 5 (1988), 218.

[32]Philip S. Bagwell, *The Transport Revolution, 1770-1985* (London, 1988), 234-240; Mathias and Pearsall (eds.), *Shipping*, 37-39; Edwin Green and Michael S. Moss, *A Business of National Importance: The Royal Mail Shipping Group, 1902-1937* (London, 1982) 36, 44-45 and 183; and TNA/PRO, RAIL 108/470-471 and 1080/467.

Kingdom.[33] Some coastal liner companies used an agreement drawn up between the shipper and the shipping company whereby the former agreed to send all of a particular type of commodity by the conference ships for a minimum period, usually a year, and in return the latter quoted a particularly low rate, usually even below the "exceptional" rate set in the freight books.[34] This was a combination of a reward for loyalty and a discount for bulk consignments. The rebate was not held back and paid later but offered immediately as an inducement to sign a long-term agreement. In both cases this guaranteed a large trade to the shipowner and a discount to the shipper.

Two features of the coastal conferences did not occur in overseas trade conferences. Both were more prominent in the early nineteenth century than later. One was the attempt to standardize the wages of crews, especially the engine room staff who were in relatively short supply and could command high wages. The other was the concern demonstrated in the early Clyde-Mersey agreement to ensure that there was adequate capacity rather than trying to restrict it. Both were concerns peculiar to their time: the former because of the short-run imbalance in the supply of and demand for experienced steam engineers inherent in a new and rapidly expanding technology; the latter because steamship services were relatively new, demand was growing rapidly and steamship owners were concerned to offer an improved means of transport for their own goods and those of their colleagues rather than maximize their direct return from providing shipping services.

This study sheds no light on the effectiveness of these coastal conferences. It might be deduced that since some were renewed, were drawn up for long periods and appear to have run the whole term, and were still in existence in the early twentieth century, they were not a complete failure as otherwise the shipping firms would have gone out of business or ceased to pursue an ineffective method. Ideally, this could be tested by seeing if freight rates were raised when coastal conferences were inaugurated (there is some evidence that when firms collaborated they did raise freights and fares)[35] and whether loading factors, revenues and profits rose for those firms in the agreement. There may have been factors external to the conference which pushed results in the same direction and would need to be identified and, ideally, disentangled. However, this has not proved possible because few records have survived for coastal liner companies, and certainly no run of accounts, minute books or statistics which would allow this sort of investigation. Until these are discov-

---

[33]*BPP*, 1909, XLVII, qq. 8713 and 9645.

[34]NMM, CST/38/19, indicates at least twenty-six such contracts in 1912 on the Powell and Hough Lines between Liverpool and London. The ASN material at AUL, MS 2479/8 and 9, shows that long-term contracts were used from the 1860s.

[35]NMM, GSN/7/4, meeting of 28 February 1860.

ered, it must remain a matter of reasoned conjecture. The other unexplored area is the degree of non-price competition between firms nominally collaborating in a conference. For the railways it has been suggested that stable freight charges led to improved services, such as greater speed or frequency.[36] In overseas conferences faster, larger and more luxurious ships were built. Certainly ship sizes rose on coastal routes and journey times were reduced, while newspapers claimed that passenger facilities were yet more impressive so that an element of "service competition" may well have replaced price competition. For instance, as early as 1837 the General Steam Navigation Co. believed that its "new and superior ships" would "obtain a much larger portion of the trade" than its partner on the Edinburgh-to-London run.[37]

It needs to be stressed that although the emphasis of this paper has been the collaboration between coastal liner companies, it is not intended to suggest that competition ceased to exist. There was often intense rivalry between firms on the same route; for example, across the Irish sea in the early days of the steamboat.[38] The advent of the railway may have forced a more conciliatory attitude on erstwhile rivals. However, throughout the nineteenth century if a coasting company appeared to be making reasonable profits, an actual or putative competitor would threaten to enter the trade. Given that barriers to entry in the coastal trade were lower than in overseas trade (ships were smaller and thus cheaper, the lag between investment and return was shorter and a less extensive agency system was required), this acted as a brake on unrestrained income maximization on the part of coastal shipping companies. This article suggests that among coastal shipowners there was a tendency to move away from Adam Smith's invisible hand of competition towards Alfred Chandler's visible hand of management whenever conditions allowed.

---

[36]Channon, "Railway Pooling," 85-87.

[37]NMM, GSN/7/2, meeting of 28 February 1837.

[38]Bagwell, *Transport Revolution*, 57-59.

Chapter 6
Coastal Shipping: The Neglected Sector of Nineteenth-Century British Transport History[1]

The popular view of internal British nineteenth-century transport history in undergraduate texts may be characterized as the triumph of railways over older modes of travel, such as stagecoaches, horse-drawn wagons and canal boats. This is sometimes used as evidence of the supremacy of modern over pre-industrial technology and as such contains more than a smidgen of Whig history. The railway is portrayed in Darwinian terms as the more advanced species that killed off other forms of transport by offering superior service, speed and lower prices. Thus, long-distance coaches and wagons ceased operations when a rail line was completed and were then relegated either to areas that lacked railways or to service as intra-urban, short-haul carriers between railway stations and factories or workshops. Canals entered an era of long-term, genteel decline, helped by the fact that some railways absorbed canal companies and at times even laid their tracks in drained canal beds. This can also be seen as progress, since new technology superseded the outdated.

Even a cursory glance at the tables of contents of established transport history texts, such as Dyos and Aldcroft, Barker and Savage or Bagwell reveals that substantial portions are devoted to the advent of the railway, its impact on other forms of transport and its effects on the society and economy.[2] Moreover, there are a plethora of enthusiast books on the history of particular rail lines, classes of engine, towns, engineers and types of carriage. The sheer size of Ottley's massive bibliography, plus its supplement, gives some idea of the amount of information available on various aspects of railway history.[3] In this obsession with the new technology of the iron rail, coastal shipping has

---

[1]This essay appeared originally in the *International Journal of Maritime History*, VI, No. 1 (1994), 175-188.

[2]H.J. Dyos and Derek H. Aldcroft, *British Transport: An Economic Survey from the Seventeenth Century to the Twentieth* (Leicester, 1969; reprint, Harmondsworth, 1974); Christopher I. Savage, *An Economic History of Transport* (London, 1959); T.C. Barker and Christopher I. Savage, *An Economic History of Transport in Britain* (London, 1974); and Philip S. Bagwell, *The Transport Revolution from 1770* (1974; new ed., London, 1988).

[3]George Ottley (comp.), *A Bibliography of British Railway History* (London, 1965; 2nd ed., London, 1983) and *Supplement* (London, 1988).

been relatively ignored. The tacit assumption seems to have been that, like the horse-drawn wagon, coach or barge, coasters were obsolete and hence "naturally" superseded by the railway. An excursion through the pages of the standard texts shows that the space devoted to coastal shipping is minuscule.[4] In comparison to the nearly 13,000 items in Ottley's compilation, a bibliography on coastal shipping and trade from the seventeenth to the twentieth centuries would contain only about 250 entries, even including fleet lists barren of all but the most basic factual information. Nor can it be claimed that such neglect was a phenomenon only of the "first generation" of transport history texts. While Simon Ville's recent book on European transport from 1750 to the First World War devotes a chapter apiece to railways and shipping, the impression of equality is shattered because most of the chapter on sea transport deals with overseas trade and shipping.[5] Although Ville briefly discusses the role of the coaster, it is treated as only a tiny corner of shipping and is dealt with in little depth compared to the railway. An even more recent volume on transport developments in France and England during the industrial revolution devotes parts of four pages (of a total of 238) to coasting, while the rest of the work is concerned with roads and canals. Given that the author set out to explain why Britain industrialized earlier than France, and that he saw England's superior transport network as the key, it is obvious that the huge coastal trade deserved greater attention.[6] In sum, the opinion in the standard works – and it must be impressionistic given the small space allocated to the topic – is that the coaster was unimportant, unworthy of serious study and technologically obsolete.

If we examine the massive volumes on nineteenth-century shipping history, the result is equally disappointing. Although much has been published on Britain's overseas trade and the shipping firms that facilitated it, the focus is on foreign trade and overseas routes, with their large liners, exotic destinations and important products. Relative to such treatment, only a tiny part has been devoted to the less glamorous coastal trades. The implicit suggestions are that coasters are unsuitable for scholarly scrutiny and that their nineteenth-century role was marginal.

This essay argues that such impressions are false. The coaster continued as a crucial component of British internal transport until at least the First World War. Indeed, its role expanded between 1830 and 1914; by the terminal

---

[4]There is one honourable exception: Bagwell, *Transport Revolution,* devotes an entire chapter to nineteenth-century coastal trade and looks at it again, albeit more briefly, in the twentieth century.

[5]Simon P. Ville, *Transport and the Development of the European Economy, 1750-1918* (London, 1990).

[6]Rick Szostak, *The Role of Transportation in the Industrial Revolution: A Comparison of England and France* (Montréal, 1991).

year it was performing as much work as rail, albeit in a different fashion. Rather than being rendered obsolete, the coaster was able to compete successfully in certain markets and to reach agreements with railways to eliminate competition in others. The ability to compete (or collude) rested partly on technological innovations in coastal ships (and ancillary shore-based berthing and handling facilities) and partly in the structure and organization of the coastal trade.

Despite the paucity of surviving records of firms and individuals in coastal shipping, there is much evidence that the sector did not decline during the railway age. The government, keen to monitor Britain's naval potential, recorded activity in both overseas and coastal shipping. Throughout the period it published returns of the number and tonnage of ships entering and leaving each port of the kingdom, disaggregated by mode of propulsion, whether carrying cargo or ballast, and whether in overseas or "coastwise" trade. Printed regularly in the Trade and Navigation Returns in the *Parliamentary Papers,* these data shed much light on the significance of coasting.[7] Unfortunately, the basis on which information for coasters was collected changed during the period. Before 1873 vessels carrying certain low-value cargoes, such as chalk, flints and manure, were omitted; thereafter, they were included. In 1898, trade within the Thames estuary, which had previously been included as "coastal," was redefined as "estuarial" and so ceased to be recorded.[8]

Nonetheless, when these shifts are controlled for, it is possible to estimate that coastal tonnage entering British ports with cargo grew by about 1.5 percent per annum between 1830 and 1914,[9] a statistic which does not support the contention that the coaster was of diminishing importance. Although available figures only report the number and registered tonnage of *ships* rather than the quantity of cargo carried, given that those in ballast were excluded and that the relationship between registered tonnage and carrying capacity (deadweight tonnage) was improving, the evidence points to a steady increase in the quantity of goods carried by coastal shipping.

The government also collected some statistics on coastal trade. From 1833 it printed returns of the quantity of coal shipped coastwise from each port and hence the aggregate tonnage of coal moved by coaster for the whole coun-

---

[7]It would be tedious to list all of the *Parliamentary Papers;* the author can provide a list to anyone preternaturally interested. For part of the period, readers may consult David J. Starkey, *et al.* (eds.), *Shipping Movements in the Ports of the United Kingdom, 1871-1913: A Statistical Profile* (Exeter, 1999).

[8]Michael J. Freeman and Derek H. Aldcroft (eds.), *Transport in Victorian Britain* (Manchester, 1988), 172.

[9]This, and the growth rate for coal carried coastwise in the next paragraph, was calculated from end-point ratios using the formula $r = (m \sqrt{x_n/x_1}]-1) \times 100$.

try.[10] In the eighty years to 1913 there was an annual compound growth rate of nearly two percent, with volumes rising from about 5.8 million to 22.9 million tons. Given that coal was the most important commodity, by both volume and weight, carried by coasters, this also contradicts the argument about declining coastal trade.

The third piece of evidence pointing to a flourishing coastal trade throughout the period is the survival of an extensive network of coastal liner services. Liners operated scheduled services on particular routes and were usually the fastest, best-appointed and largest ships in the trade. A glance at the newspapers of any large port, or at local trade directories for coastal towns, will demonstrate that there were frequent coastal services. The Aberdeen Steam Navigation Co., for example, ran between Aberdeen and London; the Tyne Tees Steam Shipping Co. connected these two rivers to London; and the Carron Company connected Edinburgh (or at least Boness and Grangemouth) with London.[11] Nor were liners confined to the east coast, which was most favourable for their operations because the coastline allowed a fairly direct route. Although the sea journey from London to Liverpool or Glasgow was several times longer than the land route, there were regular liner services. Between Liverpool and London the Powell, Bacon and Hough Lines called at Bristol, Plymouth and Southampton.[12] Glasgow and Liverpool were connected by ships managed by J. and G. Burns;[13] the south coast was served by the Red Funnel and Little Western Lines; and Langlands linked the northeast and northwest coasts via the Clyde and Forth Canal. This list is by no means exhaustive, but the point is that there was a comprehensive network of liner services connecting all major and many minor UK ports. Since the carrying capacity of the vessels is known – as is their frequency – some calculations could

---

[10]The same comment applies as to note 5.

[11]Geoffrey Channon, "The Aberdeenshire Beef Trade with London: A Study in Steamship and Railway Competition," *Transport History*, II, No. 1 (1969); Clive H. Lee, "Some Aspects of the Coastal Shipping Trade: The Aberdeen Steam Navigation Company, 1835-80," *Journal of Transport History*, New ser., III, No. 2 (1975), 90-103, reprinted in John Armstrong (ed.), *Coastal and Short Sea Shipping: Studies in Transport History* (Aldershot, 1996), 90-103; and A.M. Northway, "The Tyne Steam Shipping Company: A Late Nineteenth-Century Shipping Line," *Maritime History*, II, No. 1 (1972), 69-88.

[12]George Chandler, *Liverpool Shipping: A Short History* (London, 1960), chapter 1.

[13]Anthony Slaven, "John Burns," in Slaven and Sidney G. Checkland (eds.), *Dictionary of Scottish Business Biography, 1860-1960* (Aberdeen, 1990), 11.

be made of the total capacity of this network.[14] But regardless of the precise levels, it is clear that it was large. Such an extensive network was inconceivable if in fact trade were declining.

If we accept that there was an extensive role for coasters, then their owners might have survived by restricting operations to tiny niches in which it was not worthwhile for the railway to compete. This in fact did happen: some communities were too remote, isolated or small to merit railway connections and hence continued to provide markets for small coasters, which often ran directly onto a shelving beach if there was no proper port.[15] But there is also evidence that coasters could be competitive with railways in certain trades.

The coal trade to London is one example. Before the completion of long-distance rail lines, the vast majority of coal was brought to the capital from the North East by collier brigs.[16] Although by 1867 there were sufficient through routes to enable the railways to capture the lion's share of this trade, coastal shipowners did not accept this as a *fait accompli* and invested in improved services, such as larger ships and faster methods of unloading.[17] The industry was also aided by the construction of waterside power stations and gasworks, which consumed coal in large quantities. As a result, by 1898 the colliers, now mostly steel-hulled and screw-propelled, brought more coal to London than did the railways, *a status quo* which was maintained until the outbreak of war.[18] This suggests that the simple view that coasters were superseded by railways as conveyors of bulk commodities is incorrect. Competition in fact spurred the owners of coasters to seek greater productivity, which then led to increased market shares. The effect of iron, steel and steam was to per-

---

[14]This work could mirror the methods used by Chartres, Turnbull and Gerhold to calculate service capacities in the road haulage business of industrializing Britain; see John A. Chartres, "Road Carrying in England in the Seventeenth Century: Myth and Reality," *Economic History Review*, 2nd ser., XXX, No. 1 (1977), 73-94; Gerald L. Turnbull, "Provincial Road Carrying in England in the Eighteenth Century," *Journal of Transport History*, New ser., IV, No. 1 (1977), 17-39; and Dorian Gerhold, "The Growth of the London Carrying Trade, 1681-1838," *Economic History Review*, 2nd ser., LXI, No. 3 (1988), 392-410.

[15]Basil Greenhill, *The Life and Death of the Merchant Sailing Ship, 1815-1965* (London, 1980), 15.

[16]Lionel Willis and Basil Greenhill, *The Coastal Trade: Sailing Craft of British Waters, 900-1900* (London, 1975), 22.

[17]Freeman and Aldcroft (eds.), *Transport in Victorian Britain*, 185.

[18]Peter J. Cain, "Private Enterprise or Public Utility? Output, Pricing and Investment on English and Welsh Railways, 1870-1914," *Journal of Transport History*, 3rd ser., I, No. 1 (1980), 15 and 25-26.

mit increases in cargo capacity. As well, they enabled reductions in the time required per voyage, thus increasing the number of trips that could be made in a given period. In 1840 the average collier brig carried about 220 tons of coal from Newcastle to London.[19] By 1880 the average load on this route, mostly carried by screw colliers, had increased to 700 tons and by 1900 to over 1100 tons.[20] Where in the early nineteenth century the collier brig completed perhaps twenty voyages in a good year, by 1900 seventy was not impossible for the hardest-driven ships.[21]

Further evidence that the coaster could compete comes from the liner section of the trade. Liners did not carry the bulkiest, lowest-value cargoes, such as coal, ore and china clay, but rather mixed freights, often including manufactures, foodstuffs and small quantities of high-value raw materials, as well as passengers and sometimes livestock. The mix of these cargoes negates the hypothesis that railways engrossed all of the high-value goods, leaving coasters only high-bulk, low-value freights. Coasting liners quite normally carried Bovril, Vaseline, soap, cigars and pianos.[22] On some routes they were able to compete by providing a scheduled, speedy service for high-value goods. That rail magnates took this competition seriously is indicated by their readiness to negotiate with shipowners over pricing and frequency of services, and that they even agreed to a sort of revenue-sharing. This sort of collaborative arrangement was more common for long-distance routes; for instance, the Dundee Perth and London Shipping Company was included in the English and Scotch Traffic Agreement, organized by the Railway Clearing House initially for railway companies running between London and Scotland, from at least 1855 to 1880. This included collaboration on price-fixing, pooled receipts and allocation on a prearranged basis. On similar lines, Messrs. G. and J. Burns and Messrs. Langlands, who ran steamboats to a regular schedule between the Clyde and the Mersey, cooperated with the West Coast Conference of four railway companies from at least 1858 to 1878. They agreed freight rates, based

---

[19]E.E. Allen, "On the Comparative Cost of Transit by Steam and Sailing Colliers, and On the Different Modes of Ballasting," Institute of Civil Engineers *Proceedings,* XIV (1855), 318.

[20]Raymond Smith, *Sea Coal for London: History of the Coal Factors in the London Market* (London, 1961), 324.

[21]Simon P. Ville, "Total Factor Productivity in the English Shipping Industry: The North-east Coal Trade, 1700-1850," *Economic History Review,* 2nd ser., XXXIX, No. 3 (1986), 359; and *Nautical Magazine,* LXX, No. 3 (1901), 181.

[22]John Armstrong, "Freight Pricing Policy in Coastal Liner Companies before the First World War," *Journal of Transport History,* 3rd ser., X, No. 2 (1989) 188-191, reprinted in Armstrong (ed.), *Coastal and Short Sea Shipping,* 112-129.

on Railway Clearing House categories, and only altered them after mutual consideration. The extent and details of these collaborative arrangements have been published elsewhere, but the point needs to be emphasized that the very existence of such schemes indicates that railway owners perceived coasters as serious rivals.[23] If it had been otherwise, presumably railwaymen would not have bothered to include various coastal liner companies in "conferences" and "pools."

A final indication that the coaster cannot be dismissed easily as a provider of transport before 1914 may be found in an article published a few years ago which estimated the amount of work performed by the coaster compared to the railways and the canals.[24] Using the Trade and Navigation Returns, a Royal Commission report on canals, railway returns and Bills of Entry, an estimate was made of total ton-miles of productive work in each of the three modes in 1910. Because of the methodology, the figures must he regarded as tentative and should be taken only as indicative of a rough order of magnitude. What the analysis showed was that coasters provided as much (if not more) transport as the railways and that both furnished much more than canals. Yet the way that each supplied their respective services was rather different. The average distance that any one cargo was moved by rail was about forty miles, while mean hauls by coasters were in the range of 250 miles. Although railways moved much more cargo than coasters, because the haul was so much shorter total ton-miles were in fact likely slightly less. Even allowing for substantial margins of error in these calculations, it is clear that coasters carried a significant share of inland trade in 1910.

How were coasters able to compete with the railways in some markets, to expand the amount of work performed and to attract a large share of internal trade? The answer is that the owners of coasters managed to segment the market and to offer a unique service for each. The vessels used were heterogeneous: there were a variety of types, each suited to a particular kind of work. For simplicity they can be subdivided into three broad groupings, although the reality was even more complex.

The first of these was the liners. This was the most "up market" segment, offering frequent services which departed and arrived according to pre-

---

[23]John Armstrong, "Railways and Coastal Shipping in Britain in the Later Nineteenth Century: Cooperation and Competition," in Chris Wrigley and John Shepherd (eds.), *On the Move: Essays in Labour and Transport History Presented to Philip Bagwell* (London, 1991), 76-103.

[24]John Armstrong, "The Role of Coastal Shipping in UK Transport: An Estimate of Comparative Traffic Movements in 1910," *Journal of Transport History*, 3rd ser., VIII, No. 2 (1987), 164-178, reprinted in Armstrong (ed.), *Coastal and Short Sea Shipping*, 148-162.

determined schedules.[25] The ships employed were large and fast, leading to service comparable to that provided by railway goods trains. Liners operated from major ports, often from dedicated wharves equipped with the latest cargo-handling equipment to ensure rapid turnarounds. There was usually no minimum size for consignments; coasters accepted virtually all cargoes, the only exceptions being dangerous, anti-social or unpleasant commodities, such as dynamite, coal or hides, which were inappropriate for mixed-cargo, passenger-carrying vessels. Because service was superior and costs were greater, coastal liner firms charged premiums. Nonetheless, the price per ton-mile for anything other than short journeys was usually less than the corresponding charge levied by the railways. We should also avoid being seduced by stories of crack passenger trains travelling at speeds of seventy or eighty miles per hour into assuming that railways provided fast transport for *freight*. Passenger trains were given priority, while freights were frequently shunted onto sidings. Moreover, freight trains required much marshalling, and wagons spent significant periods in shunting yards. The large number of private wagons, many lacking continuous brakes, further slowed freight trains which had to stop before descending a slope to allow the guard to run alongside applying separate brakes on each car.[26] Indeed, as late as the 1960s the Beeching Report described freight trains as "slow and unpredictable."[27] Against this poor service the coastal liner offered a relatively rapid and reliable form of transport.

The steam tramp was the second category of coaster. These ships did not run on any set schedule, going wherever cargoes were to be had whenever they were on offer. Part of the art of managing tramps was to minimize the number and distance of ballast passages. Because it was obviously more economical to operate with a full load, there were often minimum consignments established. Tramps could be reasonably fast: by the late nineteenth century nine knots (nearly 10.5 miles per hour) was probably the most economical speed.[28] Yet unless they were operated on time charters for firms requiring regular deliveries, they were less reliable than liners, since service depended upon whether a suitable vessel was available at the right time and place.

---

[25]E.R. Reader, "World's Largest Coaster Fleet," *Sea Breezes*, VII (1949), 91-92.

[26]Dyos and Aldcroft, *British Transport*, 334; and Bagwell, *Transport Revolution*, 219-220.

[27]Dyos and Aldcroft, *British Transport*, 333-334.

[28]Robin Craig, "Aspects of Tramp Shipping and Ownership," in Keith Matthews and Gerald Panting (eds.), *Ships and Shipbuilding in the North Atlantic Region* (St. John's, 1978), 209-228, reprinted in Craig, *British Tramp Shipping, 1750-1914* (St. John's, 2003), 15-39.

Agents who acted for tramp firms had to know when and where a tramp was likely to arrive and to have a cargo available to minimize unproductive time. Because service was slower and less reliable, steam tramps charged lower rates than liners for any given distance or commodity. Tramps were particularly important in bulk trades, such as coal, grain, iron and china clay.[29]

The cheapest form of coastal vessel in this tripartite typology was the sailing ship. Although its importance in coasting peaked as early as 1845 (measured by the number and tonnage entering ports coastwise), the decline was long and slow; even in 1900 nearly four million tons of sail entered British ports in the coastal trade, comprising more than 12.5 percent of total tonnage.[30] Because such vessels were subject to the vagaries of wind and tide, they were the least reliable form of transport. Vessels could, and were, delayed for days awaiting favourable winds. Once underway their arrivals were unpredictable; fast passages were possible but could not be guaranteed. Like steam tramps, sailing vessels preferred full cargoes to maximize revenue. Their lesser reliability and need for at least minimum cargoes kept sail in the bulk trades. By the late nineteenth century sailing vessels carried the least valuable freights, such as sand, stone, bricks and coal. Because their service was relatively unpredictable, rates were lower than for the other two categories. Yet sail was able to serve some markets for which liners and steam tramps were unsuitable, such as small communities lacking rail links and sometimes even formal harbours. Small, flat-bottomed coasters were able to run up to a shelving beach and unload directly to carts at low tide or over the side into shallow water if the cargo was unaffected by salt water.

It is thus clear that different routes, cargoes and customer needs were met by distinct types of service. Indeed, the patterns were even more complex than this simple tripartite division suggests. Services available actually constituted a continuum, with the fast and reliable (but expensive) steam liner at one end and the cheaper, slower and less dependable sailing vessel at the other. In between were a wide variety of services. Some steam tramps, such as east coast screw colliers, specialized in certain routes or cargoes and offered services very much like liners. Similarly, some sailing vessels offered service superior to the more obsolete steamers. A variety of services was available to meet the differing needs of various market segments.

In addition, it is of course unrealistic to portray even the sailing coasters as homogenous. There was a variety of hull, rig and tonnage choices to suit

---

[29]Tom Coppack, *A Lifetime with Ships: The Autobiography of a Coasting Shipowner* (Prescot, 1973), 24-68, gives an idea of the variety of cargoes carried by coastal steamers.

[30]Freeman and Aldcroft (eds.), *Transport in Victorian Britain*, 172.

different cargoes, routes and sea conditions.[31] The rantipike, for example, was a particularly strong schooner or brigantine that in the mid-nineteenth century carried pig iron and heavy castings from the Clyde to Liverpool. The Thames sailing barge was built for the narrow channels and congested waters of the Thames and its estuaries, although it could also be used on longer voyages. The Kentish hoy, Humber keel and Severn trow were local variations on a basic design.[32] There were also numerous rigs, ranging from two-masted, square-sailed brigs to fore-and-aft schooners or ketches, and even to brigantines, which carried square sails on one mast and fore-and-aft canvas on the other. Each type produced slightly different sailing characteristics, and craft were chosen for their suitability in prevailing water depths, winds and tides in specific trades. As a result, coasters offered a wide range of slightly different services to cater for a variety of conditions and needs.

The other factor which explains the ability of coasters to maintain their share of internal trade was that they were not technologically stagnant but rather underwent significant technical modifications during the century. Most changes cut costs, increased speed or in some way improved the quality of service. It should be remembered that the use of steam to propel vessels predated a similar application to land transport by more than a decade: the first commercial steamboat service began in 1812 on the Clyde,[33] while the first steam railway was either the Stockton-to-Darlington (1825) or the Liverpool-Manchester (1830), depending on definitions.[34] Early steamboats consumed large amounts of coal and hence were suitable only for short-hauls, which explains why they were initially confined to river, estuarial and coastal routes. But continuing improvements eventually removed such constraints; some of the most important entailed modifications to engines and boilers to reduce coal consumption. Speed, economy and manoeuvrability were aided by the shift from paddles to screw propellers. Iron, and later steel, added strength and durability. Watertight bulkheads provided greater security; increased hatch sizes eased loading and unloading; and derricks and winches speeded cargo-handling. The use of double bottoms and water ballast obviated the need to buy and discharge solid ballast, thereby reducing turnaround times. Turbines, like

---

[31]An idea of the wide variety of types of rigs and vessels can be gained from Willis and Greenhill, *Coastal Trade*.

[32]Frank G.G. Carr, *Sailing Barges of Britain* (London, 1931; reprint, Lavenham, 1989).

[33]Hereward P. Spratt, *The Birth of the Steamboat* (London, 1958), chapter 3.

[34]Maurice W. Kirby, *The Origins of Railway Enterprise: The Stockton and Darlington Railway, 1821-1863* (Cambridge, 1993); and Thomas J. Donaghy, *Liverpool and Manchester Railway Operations, 1831-1845* (Newton Abbot, 1972).

those installed in *King Edward* (1901),[35] led to unprecedented speeds, especially for passenger ships.[36] Such improvements were matched by new shore-based facilities, as cranes, grabs, suction and other methods combined with dockside railways to speed unloading.

Not all changes were the exclusive province of steamers; sail also benefited. The shift from square to fore-and-aft rigs improved sailing characteristics while reducing the number of crew (and hence costs). Winches and self-furling gear lessened the tasks that had to be performed aloft, further reducing labour requirements, while winches and derricks were installed to speed loading and unloading. By the late nineteenth century schooners which could carry 200 tons of cargo with a crew of two men and a boy were cheap and efficient conveyors of low-value, non-perishable goods.[37]

The point that must be emphasized is that coasting technology was no more static than in railways or deep-sea shipping. Indeed, many of the improvements applied to deep-water shipping were pioneered in the coastal trades.[38] The net effect of these technological developments was to increase the speed and size of liners; to reduce the coal consumption of steamers; to lower operating costs for both steam and sail; and to allow for the more intensive use of capital.

The final feature that aided the survival of coasters in the so-called "railway age" involved their cost structures. Unlike the railways, the owner of a coaster did not need to purchase land or lay tracks, sidings, points or stations. Nor were people employed to operate, maintain and guard a permanent right-of-way. Most coastal shipping firms did not own any land at all, although some liner companies bought or rented quays and wharves. The fixed capital required by coastal shipping companies thus was a much smaller component of total costs than was the case with railways. Since pilotage, wharfage, loading and unloading, trimming and other port dues incurred by coasters were levied irrespective of voyage length, costs per mile declined as distance increased. Hence, it was more economical for coasters to undertake long than short voyages. These cost characteristics gave coasters distinct advantages over the

---

[35]J. Graeme Bruce, "The Contribution of Cross-Channel and Coastal Vessels to Developments in Marine Practice," *Journal of Transport History*, 1st ser., IV, No. 2 (1959), 65-80, reprinted in Armstrong (ed.), *Coastal and Short Sea Shipping*, 57-72.

[36]Andrew McQueen, *Clyde River Steamers, 1872-1922* (Stevenage, 1990), 107-108.

[37]Basil Greenhill, *The Merchant Schooners* (2 vols., London, 1951-1957; 4th rev. ed., London, 1988).

[38]Bruce, "Contribution," 65-80.

railways. As a recent author put it, "shipping with the highest terminal but lowest journey cost is most effective over long distances."[39]

In conclusion, I have endeavoured in this brief essay to marshal a range of evidence to show that rather than being eliminated by railways, coasters survived and prospered. In part, the two transport modes were complementary rather than competitive. For a firm situated close to a port which shipped products to a destination that was similarly sited, it made sense to employ coasters. On the other hand, if both the point of despatch and receipt were well inland, railways faced little competition. It is too simple to suggest that railways carried high-value goods while leaving low-value commodities for coasters; indeed, coastal liners carried a fair share of manufactured goods while railways often hauled coal. Where both modes were available, the nature of the commodity in terms of value, bulk, need for rapid transit, fragility and perishability all became part of the decision calculus, as did distance, location and the range of services available.

As well as being complementary, there frequently was outright collusion between railway companies and coastal liner firms to fix freights, prices, frequencies and terminal charges, especially in long-distance trades. In addition, growth in one mode often benefited both, since the aggregate tonnage of goods and overall economic activity both increased as transport costs fell. Similarly, the rapid growth of Britain's overseas trade benefited the coaster, for it was used to marshal exports and distribute imports from the major to the smaller ports.

Nineteenth-century transport history thus needs to be revised to ensure that the humble coaster is given its rightful – and much more prominent – place. Coastal vessels were neither stagnant nor obsolete, but rather changed and developed to ensure improved efficiency. As a result, they continued to play important roles in the transport service of the UK at least until the outbreak of the Great War.

---

[39]Ville, *Transport*, 7.

Chapter 7
Railways and Coastal Shipping in
Britain in the Later Nineteenth Century:
Cooperation and Competition[1]

Among Philip Bagwell's many publications, one of the earliest was on the Railway Clearing House (RCH).[2] This was the first definitive history of the establishment, functions and mechanisms of the RCH. One of the functions which Philip highlighted was its role as an impartial administrator of the various pooling agreements, conferences and grouping arrangements that the independent railway companies concluded to ensure through working and to reduce inter-company competition, especially on long-distance hauls. This work predated Philip's interest in coastal shipping and, as befits a book on a railway institution, there is relatively little about competing modes of transport. Yet the railway companies did not confine their attempts to regulate long-distance traffic to their own transport mode. For long hauls their chief rival was the coastal steamboat, which remained surprisingly competitive until the Great War. Hence, many railway conferences were only too keen to bring their seaborne rivals into an agreement in order to restrict competition, raise freight rates and allocate traffic on a "reasonable" basis. Bagwell noted that the Dundee, Perth and London Shipping Company was brought into the English and Scotch Traffic Agreement in 1856, as did Channon in his thesis,[3] and that in 1867 "a similar kind of agreement" was concluded on the Clyde-Mersey route by railways and steamboat firms,[4] but neither he nor Channon pursued the analysis to show what these agreements implied about coaster-railway competition.

The purpose of this essay is to demonstrate that there was a range of methods of restricting inter-modal competition which have previously been

---

[1]This essay appeared originally in Chris Wrigley and John Shepherd (eds.), *On the Move: Essays in Labour and Transport History Presented to Philip Bagwell* (London, 1991), 76-103.

[2]Philip S. Bagwell, *The Railway Clearing House in the British Economy, 1842-1922* (London, 1968).

[3]*Ibid.*, 254; and Geoffrey Channon, "Pooling Agreements between the Railway Companies Involved in Anglo-Scottish Traffic, 1851-59" (Unpublished PhD thesis, University of London, 1975), 363-364.

[4]Bagwell, *Railway Clearing House*, 255.

ignored. The use of pools and conferences by the railway companies is now well known, as is the adoption in the later nineteenth century of conferences among shipping companies in some foreign liner trades.[5] Recently it has been shown that agreements, analogous to conferences in all but name, existed from a much earlier date among firms plying the coastal trade using steam liners and that these were widespread geographically and continued in existence at least until the First World War.[6] In other words, the coastal liner companies endeavoured to minimize competition and regulate trade among themselves just as the railways did. This essay will show that there was also inter-modal collaboration over freight rates and levels of service, both formally through written agreements between railway conferences and coastal liner companies and by more informal understandings.

In that it draws heavily on the records of the RCH it follows in Philip Bagwell's footsteps, but inasmuch as he did not pursue the steamboat theme it builds on that work in trying to elucidate the nature of railway-steamboat competition and collaboration, drawing essentially on railway-generated sources because of the paucity of coastal shipping records that have survived.

It has been established elsewhere that an east coast steamboat pool came into existence in 1839, was undoubtedly in operation in 1845 and was probably still active in the 1860s.[7] One of the lines which made up this pool was the Aberdeen Steam Navigation Company (ASN). It is clear that as well as cooperating with other steam coaster firms, this company was also collaborating with the long-distance railways at least from the later 1850s. In March 1856 the goods managers of the railway companies which made up the English and Scotch Traffic Committee worked out with the ASN revised rates for goods traffic between London and Aberdeen.[8] This deal was then put to the Octuple Agreement for its approval.[9] Although two of the railways – the Aber-

---

[5]See, for instance, Geoffrey Channon, "Railway Pooling in Britain before 1900: The Anglo-Scottish Traffic," *Business History Review*, LXII, No. 1 (1988), 74-92; and H.J. Dyos and Derek H. Aldcroft, *British Transport: An Economic Survey from the Seventeenth Century to the Twentieth* (Leicester, 1969; reprint, Harmondsworth, 1974), 269-272.

[6]John Armstrong, "Conferences in British Nineteenth-Century Coastal Shipping," *Mariner's Mirror*, LXXVII, No. 1 (1991), 55-65.

[7]*Ibid.*, 58-59.

[8]Great Britain, National Archives (TNA/PRO), RAIL 1080/508, minute 18.

[9]The Octuple Agreement commenced in 1851. It derived its name from the fact that it involved eight railway companies which operated through routes from London to Edinburgh and Glasgow. Effectively, the Octuple controlled all Anglo-Scottish long-distance traffic. See Bagwell, *Railway Clearing House*, 251-253.

deen and the Scottish Midland Junction – were unhappy with the details, the Octuple insisted on the increases as the previous rates were not remunerative.

The terms of the agreement with the shipping company can be pieced together. Freight and passenger rates were set mutually, and there was a differential in favour of the steamboat company of roughly ten shillings per ton. Cattle traffic was to be divided proportionately between the two modes, with the railways taking two-thirds and the steamers one-third. There is no detail of the workings of this agreement in its early years, but in the 1860s the ASN was in frequent correspondence with Mr. H. Ormond, the Liverpool cattle traffic manager of the London North Western Railway Company (LNWR) about alterations in freight rates. For instance, on 26 September 1861 Ormond wrote requesting that "the Steam Packet companies...agree to advance the present ruinously low rate [on cattle] from 15s to 20s per head, as in that case the Railway companies will be spared the necessity of making the reduction proposed."[10] This suggests continuity in preferring mutually beneficial pricing policies rather than cutthroat competition. At the time of this request the ASN felt unable to agree to the rise because it was facing severe competition from another shipping company, the Northern Steam Company, whose cattle rates had been reduced below fifteen shillings. However, once the two competing steamboat companies agreed to amalgamate they wasted little time in writing to Ormond on 25 October to assure him they "were in a position to give effect to" his proposed increase in cattle rates,[11] and on 19 November the secretary of the ASN confirmed that the rate on cattle was being advanced to twenty shillings per head "after Saturday next."[12]

That this sort of collaboration was not unusual but rather part of the normal methods adopted to regulate competition is shown by a further incident in the mid-1860s. In the spring of 1864 a deputation from the English and Scotch Traffic Sub-Committee called upon the ASN to try to agree rates for livestock carriage between Aberdeen and London and possibly north of Aberdeen as well.[13] The railway companies hoped to conclude a five-year binding agreement which would cover not merely the freight rates to be charged but also how the total traffic was to be divided between the two modes of transport.[14] Although there was some disagreement between the various railway

---

[10]Aberdeen University Library (AUL), MS 2479/8, Aberdeen Steam Navigation Co. (ASN), minute, 26 September 1861.

[11]*Ibid.*, minute, 25 October 1861.

[12]*Ibid.*, minute, 19 November 1861.

[13]TNA/PRO, RAIL 1080/509, minute 1108.

[14]*Ibid.*, minute 1210.

companies who were parties to the conference as to the best course of action, there was no fuss about the principle of talking with the steam packet companies. This suggests that it was perceived as a normal course of action which had been used before to prevent "ruinous" competition.

This view is supported by evidence of events two years later when the railway companies decided to increase their rates on cattle and meat traffic from Aberdeen to London. They wrote to the ASN, *inter alia,* informing the coastal company of the new rates and subsequently called upon its manager "with a view to induce that company to advance their rates for meat and cattle from Aberdeen to London."[15] Initially, they met with little success, for the steam packet company replied that "while not averse to entertain the consideration of an increase in the rates on cattle from Aberdeen, [we] cannot in the meantime and without further information on the effects of the through rates proposed by the railway companies, agree to the terms of rates stated in their letters."[16] What is evident from the tone of this letter is that the coastal firm was not surprised to be approached in such a manner by the railway companies. Quite the opposite; it was obviously a normal part of business. The only question was whether to accede to the request or how much resistance to put up.

The reason for the railway companies' repeated approaches to the shipping company is straightforward. In the eyes of the railway firms the steamers were proving annoyingly effective in retaining a significant share of the trade in cattle and meat. This is surprising, for it is in just such perishable or damageable high-value commodities that the railway should have had a significant advantage over the coaster and hence have been able to charge a premium price to compensate for their rapid speed of transit. Yet throughout the 1860s and 1870s there were complaints voiced in the railway conference of the coasters carrying too large a share of cattle and meat. For instance, in 1861 the railways needed to reduce their rates on both cattle and sheep "with the view of getting back the Traffic at present diverted to the sea route."[17] In 1868 it was suggested that for the railways to compete with the steamers the cattle rate needed to be the same on both modes.[18] This implied that the railway service was perceived as comparable to the steamer. An investigation held in 1869 revealed that while in 1867 the railways carried sixty-four percent of total cattle, their share of the extensive sheep traffic was only 3.5 percent and they carried no pigs. In 1868 their market share was even worse, being only forty-

---

[15]*Ibid.*, minute 1498 and appendix.

[16]AUL, MS 2479/10, ASN, minute, 20 November 1866.

[17]TNA/PRO, RAIL 1080/509, minute 857.

[18]*Ibid.*, RAIL 727/1, minute 508.

three percent of cattle and the same low shares for sheep and pigs as the previous year.[19]

In September 1872 the railway companies again approached the coastal firm to inform it of a contemplated rise in freight rates. The steamboat company raised its rates in sympathy.[20] Not content with this rise, in March 1873 the railways were keen to raise their rates yet again, believing that the coastal company would agree to an advance of about eight percent. A railway delegation called upon the coastal company.[21] The directors of the ASN then agreed to consider it at their next board meeting in April 1873.[22] In April 1875 the railways were horrified to learn that the steamers were charging only fifty shillings, delivered in London, for meat from Aberdeen, whereas the equivalent charge by railway was 79s 2d, a premium of nearly sixty percent.[23] Not surprisingly, much of the trade was now going by sea, so much so that in March 1876 the railway companies considered reducing their rates because their current charges "had the effect of causing a larger proportion of meat to be sent by sea."[24] As it transpired, no reduction was made because the railway companies thought a lower rate unremunerative.

As a result, the coasters continued to carry a "large proportion" of meat traffic, so that in May 1878 goods managers were again considering a reduction in rates "to secure a better share of the traffic."[25] By February 1879 they had decided against any such cut, as the return on the traffic did not justify any reduction.[26] As late as July 1889 the railways were still concerned over "the decreased tonnage [of dead meat traffic] carried by rail and the large weight taken by sea" and were considering "a more convenient and accelerated train service for dead meat and other perishable traffic from Aberdeen to London."[27] They had obviously abandoned any idea of trying to compete merely

---

[19]*Ibid.*, RAIL 1080/510, minute 1905. The trade in sheep and pigs was extensive, amounting to nearly 28,000 sheep and 4500 pigs in 1867 compared to 7200 cattle.

[20]*Ibid.*, RAIL 1080/511, minute 2634.5.

[21]*Ibid.*, minute 2725.

[22]*Ibid.*, minute 2749.

[23]*Ibid.*, RAIL 727/1, minute 1379.

[24]*Ibid.*, minute 1420.

[25]*Ibid.*, minute 1646.2.

[26]*Ibid.*, minute 1673.

[27]*Ibid.*, RAIL 727/2, minute 2420.

on price because they could not bring their costs close to those of the coaster and instead were considering competition via the quality of their service to justify a premium price. In October 1887 the railways expressed concern over the "large number of passengers travelling by sea" between Aberdeen and London and contemplated running excursion trains at special low fares.[28] Given the railway's superior speed of travel for passenger traffic it is surprising to find the steamers competing successfully in this area. They did so through the appeal of more luxuriously-appointed new ships, greater speed brought about by improved engine technology and the invigorating air enjoyed on a sea cruise.

   Two important points come out of these minutes. Firstly that on this long-haul route the railways and steamboat companies quite normally consulted and informed each other of changes in freight rates and tried to keep some sort of balance between the two, though the coaster rate was always below that of the railway. Secondly, that even in the sort of commodities which should have been the most ready to switch to the train, the railway companies had no easy or lasting victory. Through at least until the 1890s the steamer continued to carry a significant share of such traffic. This certainly confirms Channon's view, confined to the cattle and meat trade, that competition between rail and steamboat "was very real and very effective."[29] When the sheep and pig traffic is included, the steamboat carried the lion's share in some years.

   Collaboration between coasters and railway companies on the very long-haul Aberdeen-to-London route seems to have been informal, sporadic and uncertain. However, there is one example of long-lasting formal collaboration between the railways and a coastal company. As Philip Bagwell pointed out in his work on the RCH, from about 1855 the Dundee, Perth and London Shipping Company (DPLS) was brought into the English and Scotch Traffic Agreement. This included all railway companies in the cross-border trade which pooled receipts and then divided them among the various railways.[30] The DPLS ran steamers and sailing smacks between London and the two Scottish towns. The railways and steamboat company had a formal agreement in which they collaborated on fixing rates for bale traffic and also agreed to pool the receipts from this traffic and divide it between the two modes in pre-arranged proportions. Philip Bagwell was mainly concerned to show the role of the RCH in collecting statistics, calculating apportionments and acting as an objective arbitrator to promote through traffic. He did not deal with the significance

---

[28]PRO, RAIL 1080/514, minute 6291.

[29]Geoffrey Channon, "The Aberdeenshire Beef Trade with London: A Study in Steamship and Railway Competition, 1850-69," *Transport History,* II, No. 1 (1969), 18.

[30]Bagwell, *Railway Clearing House,* 254-255.

of this agreement for the coaster or the light it threw on inter-modal collaboration and competition.

There were in fact three separate but contiguous agreements between the DPLS and the railways which gave a continuous period of twenty-four years of collaboration from 1856 to 1879. There were some common features in all three. Each ran for a set period: fourteen years for the first, five years for both the second and third. Freight rates were determined mutually for each mode and could not be altered without joint consultation. However, the rates fixed were not the same for both forms of transport; the steamboat charge was always considerably less than the railway's rate. This policy of rate fixing was not so rigid that it prevented a rapid response when sea competition occurred. On several occasions a rival shipping firm entered the trade, offering lower rates than the DPLS, but each time this was short-lived for the sea freights were cut and the opposition soon withdrew. For instance, in May 1876 *Brigadier* was put on the route charging only fourteen shillings per ton, one shilling less than the DPLS.[31] The latter immediately cut its rate to 12s 6d and then, as the owners of *Brigadier* went to ten shillings, matched that rate. By mid-August the intruding steamship had been withdrawn, and the sea freight returned to its normal fifteen shillings per ton.[32] All three agreements contained "joint-purse" pooling arrangements. At the end of each half year the RCH presented statistics of the tonnage carried by each mode and worked out how much was due to whom.

The changes that were implemented in successive agreements were partly a matter of detail. For instance, the freight rate by ship in the first term was fifteen shillings whereas in the second and third it was twenty-five shillings. Similarly, the rail rate rose from 37s 6d to 41s 8d at the same time.[33] While the first agreement was current, the rate for calculating the transfer payment from the transport mode which carried more than its agreed share to that which was under-subscribed was set at 1s 6d per barrel bulk, roughly 16s 6d per ton, minus twenty percent for working expenses.[34] In the second and third agreements the transfer price was set at fifteen shillings per ton less twenty percent to cover working expenses.[35] Of more significance perhaps was the changing share of the total traffic allocated to each form of transport. In the first agreement the split was twenty-five percent to the railway, seventy-five

---

[31]TNA/PRO, RAIL 1080/510, minute 1616.

[32]*Ibid.*, minute 1713.

[33]*Ibid.*, minutes 2154; and RAIL 1080/511, minute 3155.

[34]*Ibid.*, RAIL 1080/510, minute 243.

[35]*Ibid.*, RAIL 1080/511, minute 2378.

percent to the shipping company.[36] In the second agreement, from 1870, the railways' share was reduced to ten percent with the coaster receiving ninety percent.[37] The third agreement, in force from 1875 to 1879, saw the railway's share increased to 12.5 percent.[38]

The earliest agreement, as well as fixing freight rates jointly, also set passenger fares in collaboration. This practice ceased with the first agreement and was never renewed.[39] Similarly, during the first period, the railway companies guaranteed a minimum sum for cattle traffic receipts to the shipping company. That too was not incorporated into the second agreement. Admittedly, the sum guaranteed was quite small, £500 each half year, equal to £1000 per annum. However, on the two occasions when the committee received information on the cattle trade, in 1856 and 1857, the railways paid £315 and £491, respectively, to the coastal company. It seems likely that similar payments were made in other six-month periods, though the minutes are silent on the topic. The logic of the railway companies in guaranteeing a minimum income for cattle traffic to the shipping firm seems obscure, until two clauses in the third agreement are added into the conundrum. These read, "The shipping company to discourage as far as possible any Meat or Fish traffic being sent by their steamers"[40] and, "The Shipping Coy to discourage cattle traffic being sent by their route."[41] The £1000 per annum was compensation paid by the railways to the steamers for allowing the meat, fish and cattle traffic to go predominantly by rail. The justification for this may have been speed, since if the railway ran special fast ice-cooled trains, rather than using the regular freight trains, they could cover the distance much faster than the coastal ship. This also allowed the railway to cream off the higher-rated traffic, leaving the bulkier, lower-rated goods, such as bales of jute, pieces of cloth and hanks of yarn, to the steamers.

To those who espouse the conventional wisdom that railways superseded coasters because of their lower costs and charges, the proportions officially allocated to the coaster may come as a surprise, for it was expected to carry the vast majority of the cargo. Even more surprising from this viewpoint, as shown in table 1, is that for those years where the tonnages are re-

---

[36]*Ibid.*, RAIL 1080/510, minute 1713.

[37]*Ibid.*, minute 2298.

[38]*Ibid.*, RAIL 1080/511, minute 3945.

[39]*Ibid.*, RAIL 1080/510, minute 2226.

[40]*Ibid.*, minute 2154.

[41]*Ibid.*, RAIL 1080/511, minutes 3155 and 3189.

corded, the coaster consistently carried more than its official share and ended each half year owing money to the railways. Thus, the vast majority of customers preferred to send their goods by the cheaper sea route rather than by the dearer railway, even if the train was faster.

Table 1
Freight Carried from Dundee to London, 1856-1879

Six Months Ending	Tonnage Carried by Sea	Percentage of Total Traffic
31 August 1856	7041	78
31 March 1857	6250	84
30 Juned 1870	7572	
31 December 1870	8367	92
31 December 1872	11,716	92
30 June 1873	10,478	92
31 December 1873	12,543	93
30 June 1874	12,215	93
31 December 1874	10,723	93
30 June 1875	5075	87
31 December 1875	5825	88
30 June 1876	5175	89
31 December 1876	4945	88
30 June 1877	4346	87
31 December 1877	4354	90
30 June 1878	4870	87
31 December 1878	5758	92
30 June 1879	6299	90
31 December 1879	5715	90

Source: Great Britain, National Archives (TNA/PRO), RAIL 1080/509-512.

The agreement involving the DPLS was the most formal sort of collaboration between the railway companies and a coastal liner business. There was also mutuality of price fixing between the railway conference and two other companies, both sailing between the Firth of Forth and London. These were the General Steam Navigation Co. (GSN) and the London and Edinburgh Shipping Co. (LES). As early as 1866 the rail companies were writing to the owners of the London and Edinburgh steamers (as well as those plying from Dundee and Aberdeen) to inform them that the railway rate on cattle was to be advanced from 1 December in the hope that the steamer companies would follow suit.[42] Similarly, in April 1872 the railway companies set up a subcommittee to call upon the two steamer firms plying between Edinburgh and

---

[42]*Ibid.*, RAIL 1080/509, minute 1498.

London to try to arrange for the freight rates by both modes to be advanced.[43] After detailed negotiations it was agreed by the railway companies, the GSN and the LES to raise rates in tandem from 1 November 1872.[44] Appended to the agreement is a long list of rates for a range of commodities which indicates that the steamboat price was always lower than that by railway. Although the differential varies from commodity to commodity, the premium charged by the railway lay in the range of twenty to thirty-five percent. The variety of products for which steamer rates were quoted, including manufactured goods, such as biscuits, books, pianos and drapery, perishables, such as oranges, lemons and meat, and frangibles, such as glass chimneys and globes, indicates that the coastal liners were not confined to bulky, low-value commodities but were carrying a wide range of high-value goods as well. In 1873 the two steamboat companies and the railway companies that were party to the conference again made minor adjustments to the agreed rates.[45] Throughout 1875 and 1876 tripartite talks took place between the two coastal liner firms and the railways making detailed alterations mutually to freight rates. Sometimes this was done at meetings, at other times by an exchange of letters.[46]

The agreement between the two transport modes was sufficiently flexible to allow the steamer companies to respond to threatened competition. For instance, in April 1875 the London Leith and Glasgow Steam Shipping Company threatened a twice-weekly steam service between Leith and London in direct opposition to the GSN and LES. As a result, the coastal companies wrote to the railway conference requesting a suspension of their agreement. In this the railways concurred.[47] This opposition by Messrs. Burrell and Son proved long-lived, and it was not until January 1877 that the coastal firms could report that the competition had ceased and they were willing to revert to the rates agreed in October 1875.[48] Only a few months later, in May 1878, opposition was threatened again, and in response the rail and shipping companies cut the rates both by rail and sea by about ten percent to maintain their share of the traffic.[49] Again in April 1879 a new shipping company was offer-

---

[43]*Ibid.*, RAIL 1080/511, minute 2577.

[44]*Ibid.*, minute 2634.

[45]*Ibid.*, minutes 2708 and 2709.

[46]*Ibid.*, RAIL 1080/512, minutes 3135, 3168 and 3198.

[47]*Ibid.*, minutes 3194 and 3211.

[48]*Ibid.*, minute 3573.

[49]*Ibid.*, minute 3895.

ing to carry paper from Edinburgh to London at a lower rate than that in force, so the railway companies agreed to the steamer firms reducing their rate by one shilling per ton.[50]

This agreement between the two shipping firms and the railway conference was still in existence in September 1886 when it was cited by the railways as an exemplar of its kind.[51] By then it also included the Carron Company, which ran a liner trade between the Firth of Forth and London.[52] There is also evidence of this agreement continuing, essentially unchanged, until at least the spring of 1889,[53] by which time it had been superseded by the agreement discussed below.

The east coast agreement between railway and steamboat was to serve as a model for the west coast and was to usher in an even larger grouping; in September 1886, the railway companies in the Anglo-Scotch Conference were negotiating with firms operating on the Glasgow-to-London route to bring them into an agreement.[54] The Clyde Shipping Company joined in talks with the railway companies, the GSN, LES, and Carron Company. The aim of these discussions was to agree a set of freight rates for both the east and west coast routes between London and Scotland by both rail and sea. After some months of meetings and negotiations an agreement was drawn up with mutually agreed freight rates to be effective from 1 March 1887.[55] The agreement was subject to a three-month notice of intention to withdraw. The terms were similar to those which had applied to the east coast trade. A schedule of freight rates was drawn up collaboratively by the various parties and no unilateral alterations could be made in them. The rates by the various routes and types of transport were not identical. The railway rate was the most expensive, with the east coast sea route being on average between twenty-five and fifty percent less than the railway rate. Generally the steamboat enjoyed a larger differential on the more expensive rates. The freight charges by the west coast steamboats were in all cases slightly less than by the east coast sea route. The discount enjoyed by the west coast over the east was quite small, varying from four to eleven percent, with the larger discount on the higher-rated commodities. The

---

[50] *Ibid.*, RAIL 1080/513, minute 4117.

[51] *Ibid.*, minute 4276.

[52] Ian Bowman, "The Carron Line," *Transport History*, X, Nos. 2-3 (1979), 143-170 and 195-213.

[53] TNA/PRO, RAIL 1080/514, minute 6693.

[54] *Ibid.*, minute 5969.

[55] *Ibid.*, minutes, 21 December 1886.

reason given for this additional discount was that the west coast was the "long sea route" which would mean greater time taken in transit; hence, the lower freight rate was to compensate for the slower speed of delivery.[56]

This agreement was still in force in June 1888 when Paisley and Greenock were brought into the group of stations and ports which were included in the definition of Glasgow.[57] The concord was also sufficiently alive and healthy for it to be worthwhile in October 1893 to reprint the lists of rates applying to the various routes, as there had been numerous changes in individual charges.[58]

Collaboration between the two modes was still going strong in January 1895 when the railway companies agreed to the shipping firms temporarily reducing their rates on some products because of "competition by outside steamers," that is, non-members of the conference.[59] This opposition was long-lasting, for these exceptional rates stayed in force until December 1898. In the autumn of 1899 and spring of 1900 the conference agreed to rises in freight rates of between two and seven percent on both modes.[60] Various references in the minutes indicate that this concord between railways and coastal liners continued at least until the spring of 1911,[61] and that in 1908 there was an attempt to implement identical terminal rebates:[62] that is, where customers were given discounts from published rates because they did their own collection or delivery or unloaded directly, not using the transport mode's carts or labour. This was to remove hidden price competition and so eliminate another element of discretion in rate making. By 1911 the concord between the coastal liner companies had become further formalized, for there was a body in Glasgow known as the London Shipping Conference which acted for those companies trading between the Clyde and the Thames in negotiations with the railway conference.[63] Although it is not absolutely clear, it seems likely that this cooperation between railways and coasters on both the east and west coast routes between Scotland and London continued until at least the First World War.

---

[56] *Ibid.*, minute 6179.

[57] *Ibid.*, minute 6494.

[58] *Ibid.*, minute 7713.

[59] *Ibid.*, RAIL 1080/516, minutes 8010 and 8055.

[60] *Ibid.*, minutes 9569 and 9656.

[61] *Ibid.*, RAIL 1080/518, minute 12891.

[62] *Ibid.*, minute 12103.

[63] *Ibid.*, minutes 12891, 12948 and 13130.

Railway-coaster cooperation was a normal part of business life on the long-distance east coast route between Scotland and London from the mid-nineteenth century. From at least the 1880s this was extended to embrace liner companies plying the west coast route to the Thames. The pattern was similar in all cases, collaboration to fix freight rates and then alteration only by mutual agreement. Competition was restrained essentially to non-price competition via speed and frequency. This did not deny choice to the intending shipper, for the freight rates were not identical. The merchant could decide between the cheaper sea route, which was a little riskier and might be marginally slower and less frequent, and the dearer railway rate which might be faster, more frequent and more reliable. The difference in price probably reflected the difference in costs of the two modes.

There is plenty of evidence of railway-steamboat collaboration on the west coast. Specifically, the trade between the Clyde and the Mersey was the subject of formal cooperation between the two transport modes for at least twenty years from 1858 to 1878. In the former year the railway companies met with Messrs. G. and J. Burns and Messrs. Langlands to negotiate over the trade.[64] Few details seem to have survived of their agreement, but it is highly likely that it included the mutual fixing of freight rates with a differential, giving the railways a premium price for what was perceived as a superior service: in 1862 the two steamboat companies wrote to the West Coast Conference about the "differential rates for traffic between Glasgow and Liverpool and Manchester by railway and steam packet,"[65] suggesting that the 1858 agreement was still in force and was concerned, *inter alia*, with such matters. What is certain is that the railway monitored the traffic between Liverpool and Glasgow from 1860 to 1863 very closely, comparing the number of passengers by rail and steamboat.[66] The latter carried twice as many as the former. These comparative figures are likely to have come from the coastal companies themselves, further supporting the idea of collaboration.

In February 1867 a formal agreement was drawn up between the four railway companies (LNWR, Caledonian, North British and Glasgow and South Western) and Messrs. Burns and Langlands and Son.[67] By then the steamer companies were operating a daily service each way, and in addition there was

---

[64]*Ibid.*, RAIL 1080/511, minute 2616. On Messrs. G. and J. Burns, see Anthony Slaven and Sidney Checkland (eds.), *Dictionary of Scottish Business Biography, 1860-1960, Volume 2: Processing Distribution and Services* (Aberdeen, 1990), 266-267.

[65]TNA/PRO, RAIL 727/1, minute 185.

[66]*Ibid.*, minutes 152, 172 and 256.

[67]*Ibid.*, RAIL 1080/511, minute 2616.

a weekly service by the steamer *Jacinth* operated by McArthur Brothers and a regular service by sailing schooners operated by Lewis Potter and Co. The 1867 agreement is quite explicit. The parties drew up a list of differential freight rates for a whole range of goods. The basis of these rates was the five classes of the RCH classification. However, the differential between the railway and steamer rates was much greater than on the Edinburgh-to-London route, for on the west coast the steamer rates lay normally in the range forty to fifty-six percent of the railway price, giving the railway a premium of forty-five to sixty percent for its service. It was also agreed that these rates should apply to Manchester-Glasgow traffic, giving both sides through rates to the inland town, the steamer traffic using the canal to maintain an all-water route. The only goods for which a special rate was quoted were box and bale goods – essentially textiles – where the steamer rate was set at two-thirds of the railway rate. The agreement was not for any particular fixed period but simply terminable on one-month's notice by any of the parties. Here again the pattern was confirmed: railway and coastal liners collaborating to regulate traffic by fixing mutually agreed freight rates with a premium differential in favour of the railways.

These agreements were not set in stone, and when one side felt aggrieved it was likely to request reconsideration. This occurred in April 1870 when the railways considered "the present arrangement with the Glasgow and Liverpool Steam Boats [has] been found unsatisfactory, especially with respect to the down traffic from Liverpool to Glasgow."[68] They consequently approached the coastal liner companies to discuss the matter. The basis of the railway's complaint was that the shipping companies had extended their operations "by quoting through rates to the interior of England."[69] The steamers were now using various canal carriers such as on the Duke of Bridgewater's canal to enable them to quote all-water through rates to inland towns in Staffordshire and the Midlands at much lower prices than the through railway rate.[70] This had diverted a significant amount of the traffic, hence the railway businesses' unease. Eventually, in January 1871 the steamboat proprietors agreed to a meeting[71] and were willing to negotiate on the through rates.[72] These negotiations dragged on through 1871 and well into 1872. In September 1872 a new agreement was drawn up whereby both railway and steamboat

---

[68] *Ibid.*, RAIL 727/1, minute 804.

[69] *Ibid.*, minute 928.

[70] *Ibid.*, RAIL 1080/510, minute 2228.

[71] *Ibid.*, minute 2296.

[72] *Ibid.*, minute 2372.

rates were raised "in consequence of the serious increase in working expenses" from 2 September.[73] At this time the class rates were not altered but rather the special rates, running to fourteen printed pages, were altered individually and the bale-and-box rate was increased by 2s 6d per ton by both means of transport. As in the case of the DPLS, the variety of goods for which steamer rates were quoted supports the idea that the coastal liner was carrying not only bulky, low-value goods but a wide range of commodities including perishables, such as apples and damsons; manufactured goods, such as bottled lemonade and ketchup; fragile products, such as window glass and empty jars; and high-value goods, such as sewing machines, macaroni and lithographic stones.

In late 1872 negotiations resumed over the knotty problem of the shipping firms quoting through rates to the Midlands.[74] For example, Burns was carrying sugar from Greenock to Birmingham and Wolverhampton.[75] The railway companies wished the ships to confine their through rates to Liverpool and Manchester at the one end and Glasgow and Greenock in Scotland in order that all traffic bound for inland destinations should travel by rail.[76] The shipping companies desired some latitude as to which final destinations were covered by a through rate, suggesting that any place within a radius of twenty-five miles of Liverpool and Glasgow should be "grouped" as within the scope of these rates. They declined to cease offering through rates to inland towns, preferring to set up a separate conference to include the Bridgewater Navigation Co. as well as the railway and steamboat firms, and suggesting a general rise in rates, which they considered "unremunerative."[77] Negotiations were terribly slow, many meetings being postponed or inconclusive.[78] The railway and steamer companies continued to collaborate in setting freight rates for the Liverpool-Glasgow run but could not agree which towns should be included in each grouping or on the scale of steamer through rates to the inland towns of the Midlands.[79] The cause of this was that the railway which worked the area around the Scottish terminus, the Glasgow and South Western Railway, re-

---

[73] *Ibid.*, RAIL 1080/511, minute 2616.

[74] *Ibid.*, minute 2705.

[75] *Ibid.*, minute 2941.

[76] *Ibid.*, minute 2705.

[77] *Ibid.*, minute 2744.

[78] See, for example, *ibid.*, minutes 2792, 2818 and 2845.

[79] *Ibid.*, RAIL 1080/512, minute 3247.

fused to bring nearby stations into the grouping, presumably because it would lose the extra freight revenue for the local carrying.

The railway companies still felt that the agreement was not working in their favour. In autumn 1876 they were so concerned by the cheap passenger fares offered by the steamboat companies during the summer that they contemplated running special excursion trains or offering cheap fares themselves. In the end they decided against any reduction because of the effect this would have on other through passenger fares which included the Liverpool-Glasgow leg.[80] In June 1878 the railway companies made a last effort to resolve the problem. They proposed a general rise in rates, a strict definition of Liverpool and Glasgow so that nearby towns were not included, and that coastal liners should not charge less for through traffic on the Liverpool-to-Glasgow part than the agreed rates. The latter would have ruled out the steamers offering any special rates for long-distance traffic. Since they already had arrangements with the Bridgewater Navigation Company for through rates to inland cities, such as Manchester, Warrington, Birmingham and Wolverhampton, the coastal firms refused this suggestion. Upon inspecting the new rates proposed by the railway companies the shipping lines also discovered that the differential between the railway and steamer rates had been much narrowed.[81] The railways felt that they had been receiving too small a share of the traffic and that a larger proportionate rise in the steamer freight rates would reduce their attractiveness to shippers, thus diverting traffic onto the railways. It might be seen as an admission by the railways that the service they now offered was proportionately not as superior to the steamboat as it had been when the agreement was drawn up in 1867. Certainly the average speed of coastal liners was increasing as improvements in engine technology were introduced. From the 1870s the railways' determination to carry even small consignments may have slowed down the speed of freight trains, as more time was spent shunting and marshalling wagon loads rather than train loads.

The result was stalemate between the two sides: the steamers were unwilling to cease their through rates to the interior or to accept the drastic narrowing of the differential, while the railways insisted on both and that the agreed rates should apply to only Glasgow, Liverpool and Birkenhead. As a result, on 25 October 1878 the railway companies gave notice of termination of the agreement as of 25 November. Attempts were made to find an "amicable arrangement," and there was an extensive exchange of letters and many meetings at which disputes arose over the share of total traffic each mode had enjoyed, whether or not shippers insured, and precisely which goods attracted

---

[80]*Ibid.*, RAIL 727/1, minutes 1445, 1470, 1479 and 1492.

[81]*Ibid.*, RAIL 1080/512, minutes 3956 and 3989.

Town and Dock dues.[82] No resolution was found, and from 1 April 1880 the railways introduced new, lower rates in consequence of the breakdown of these talks.[83] These seemed to be of little short-term advantage to the railways: in June 1881 the conference was complaining of the falling off in the weight of goods carried by rail between stations in the Manchester and Glasgow groups because of the "low rates charged by the shipping company throughout, using the canal from Manchester to Liverpool and then the steamer to Glasgow."[84]

The conclusion to be drawn from this incident is that the railway companies preferred to fix prices with the steamboat companies rather than engage in price competition. When this failed, because the railways felt they were not getting their fair share and that "the Water Route practically enjoys a monopoly of the traffic to the exclusion of the Railway Route,"[85] their initial reaction was to attempt a reduction in the differential between the railway and water rates to attract customers to the railways. After the steamboat owners refused this, the railways reluctantly accepted that they must act separately and reduced their rates. This was to little avail, as about a year later they still felt that they were losing traffic to the steamers.

On routes such as Glasgow to Liverpool, or the east coast, the distance by sea was little different from that by rail, since both modes had a fairly direct route. This was patently not true for traffic moving from one coast to another. Here the inland route had a direct line whereas the coaster had to go around the coast, travelling a much greater mileage than the railway route. Thus, for the journey from Liverpool to Aberdeen or Dundee, the coaster had a distinct disadvantage, the distance by land being, respectively, 330 and 270 miles, whereas by sea it was 648 and 714 miles: the coaster had twice as far to go. While it might be thought that the railway had nothing to fear from the coaster on this sort of route, this was not the case.

As early as October 1858 the English and Scotch Traffic Committee was considering concluding long-term agreements with merchants because of a threatened service by steamer between Dundee and Liverpool.[86] By January 1859 this threat had partly materialized.[87] From at least 1863 the railways

---

[82]*Ibid.*, RAIL 1080/513, minutes 4041, 4060, 4097 and 4116.

[83]*Ibid.*, minute 4334.

[84]*Ibid.*, minute 4593.

[85]*Ibid.*, E.G. Rider, Secretary of English and Scotch Traffic Rates Conference, to Messrs. Burns and Messrs Langlands and Sons, 5 May 1879.

[86]*Ibid.*, RAIL 1080/509, minute 549.

[87]*Ibid.*, minute 584.

found their through traffic affected by the North British Railway Company (NBR) collaborating with a steamboat service between Liverpool and Silloth to offer cheap rates, not just between Cumberland and Dumfrieshire but also for traffic from Dundee to Liverpool. The railway/steamboat rate was less than the through rail rate, and the conference felt compelled to match this price.[88] To try to eliminate this competition, the West Coast Conference requested the NBR in 1866 to discontinue the steamer service and send all its traffic by rail. The rail rates would then be raised to remunerative levels, and the NBR would be given facilities via the LNWR's tracks through Lancaster and Preston.[89] No changes had occurred by February 1867,[90] but in June 1868 the rates between Silloth and Liverpool were raised mutually by the railway companies and the steamboat, giving a differential in favour of the coastal ship. Here again is an example of the two transport modes collaborating to fix freight rates to provide reasonable returns and eliminate competition. The railway conference antici- pated that the steamboat would soon be withdrawn,[91] but this hope was not realized, for in October 1869 the steamboat was still plying and the West Coast Conference noted "the very great falling off in jute between Liverpool and Dundee and in other traffic between Liverpool and Manchester and Dundee."[92] The railways faced not merely steamboat competition on the west coast but also on the east, so that goods could travel from Liverpool to Dundee in three stages: by steamboat from Liverpool to Silloth; then by rail to Newcastle; and from there to Dundee by steamer, hence maximizing the cheaper water rate.[93] Additionally, there was now a new threat: steamers plying direct from east to west coast via the north of Scotland.

In 1870 the railways noted that traffic between Liverpool and the east coast Scottish ports was "still unsatisfactory."[94] This was because the NBR was now running a daily steamer service between Liverpool and Glasgow whereas in 1867, when the agreement on differential rates had been concluded, the coastal service had been thrice weekly[95] and because of the direct steamer

---

[88] *Ibid.*, RAIL 727/1, minute 319.

[89] *Ibid.*, minute 372.

[90] *Ibid.*, minutes, 16 February 1867.

[91] *Ibid.*, minute 444.

[92] *Ibid.*, minute 691.

[93] *Ibid.*, minute 692.

[94] *Ibid.*, minute 904.

[95] *Ibid.*, RAIL 1080/510, minute 2066.

which had been recently put on the route.[96] By the autumn of 1871 the direct steamer was doing so well that in a four-month period it carried nearly 7000 tons of goods between Liverpool and Dundee and Aberdeen, equivalent to more than 20,000 tons per annum.[97] Throughout 1872 Liverpool-to-Aberdeen and Dundee traffic by rail continued "to show a decrease" because of "the competition by direct steamer,"[98] so that in the autumn the railway companies commenced negotiations with Messrs. Langlands and Sons, who ran the direct steamer, to try to get the freight rates by both modes of transport raised.[99]

On this route the railway companies faced a double difficulty for they had competition from one of their number – the NBR operating in conjunction with the steamboat between Liverpool and Silloth – as well as the direct steamer operated by Langlands. Hence, the all-rail share of the traffic was negligible. By the autumn of 1874 the railway conference was reporting "a large falling off in the goods traffic carried by the rail through route," blaming it on "the increased number of steamers between Liverpool and Aberdeen and Dundee."[100] In December of that year Langlands stole a march on the railways by having a steamer built especially for this trade that could travel by the Caledonian Canal, cutting the mileage by sea drastically. As a result, in the autumn of 1874 the coaster was carrying over 1700 tons per month between Liverpool, Aberdeen and Dundee whereas the direct rail route was carrying about 200 tons per month.[101] The coaster was not confined to low-value, high-bulk commodities, carrying syrup, oils and rice as well as soda and jute from Liverpool and mostly "bale goods" – manufactured textiles – back.

In order to compete, the railway conference recommended a reduction in their freight rates from January 1875 to a flat rate of £1 per ton, station to station, for a whole range of commodities, as "it would result in Messrs Langlands seeking an interview with the companies to arrange differential rates on a fair basis as between sea and railways."[102] This proved of no avail, and in January 1875 the flat rate was reduced to fifteen shillings per ton.[103] Even this

---

[96] *Ibid.*, RAIL 727/1, minute 999.

[97] *Ibid.*, minute 1050.

[98] *Ibid.*, minute 1125.

[99] *Ibid.*, RAIL 1080/511, minute 2659.

[100] *Ibid.*, RAIL 1080/551, minute 2659.

[101] *Ibid.*, minute 1339; and RAIL 1080/511, minute 3095.

[102] *Ibid.*, RAIL 1080/511, minute 3095.

[103] *Ibid.*, RAIL 1080/512, minute 3136.

drastic reduction did not benefit the railways, for they calculated that in the first quarter of 1875 the direct steamer increased the tonnage it carried from Dundee to Liverpool by about 400 tons per month, whereas the direct railway route carried 700 tons *less* per month.[104] Since outright price competition was not working in the railways' favour, and since Langlands was now running its direct steamer on to Leith as well as Aberdeen and Dundee, the railways felt their only option was to come to an arrangement with Langlands. Consequently, in July 1875 a formal agreement was drawn up between Langlands and the English and Scotch Conference whereby the former agreed virtually to abandon the trade with Leith, except where the quantities offered were in excess of fifty tons, and the coaster agreed to charge the same rate as the direct rail route. For lots of fifty tons or more of a single article the steamboat charged a price lower than the direct rail route by 6s 8d per ton. For the Liverpool-to-Dundee and the Liverpool-to-Aberdeen routes the parties agreed a series of rates based on the RCH classification in which the steamers enjoyed a differential of between ten and thirty-two percent below the railway rate for the class rates. The differential in favour of the coastal ship was rather greater for some "exceptional rates." The differential was greatest on the lower rates and narrowed as the freight rate rose for higher-value commodities.[105] The railways then cancelled their special fifteen shilling rate and the parallel £1 to Inverness.[106]

Even this agreement seems to have benefited the railway companies very little, for in 1881 the conference was complaining that the direct steamers were carrying about 2000 tons of bale goods per month from Scotland to Liverpool whereas the direct railway line was carrying only about 650 tons, a quarter of the total traffic. In part, they blamed this on the increased frequency of the steamboat service, now running twice a week.[107] Having reached an agreement with Langlands on freight rates, they could not reduce these and considered improving the service. At the same time, the English and Scotch Conference still had the problem of the rail-steamer route offered by the NBR via the steamboats from Liverpool to Silloth. Although a formal agreement had been reached in 1868 between the NBR and the conference, in 1877 the latter complained that the NBR was not sticking to the agreed rates and hence was carrying too large a proportion of the traffic. Its daily steamboat service was

---

[104]*Ibid.*, RAIL 727/1, minute 1374.

[105]*Ibid.*, RAIL 1080/512, minute 3242.

[106]*Ibid.*, minute 3297.

[107]*Ibid.*, RAIL 727/1, minute 1938.

seen as too frequent.[108] Not merely this, but the NBR was quoting lower through rates to a large number of destinations in Scotland, such as Edinburgh and Leith, as well as those cities further north.[109] In addition, there was obviously non-price competition between the through railway and steamer-rail routes in terms of the terminal services they offered. In 1878 the Liverpool agents for the Silloth steamers, Messrs Johnson Grainger and Co., were charging only 1s 3d per ton for cartage in Liverpool, whereas the railway conference charged 1s 8d. Also, the steamboat allowed its men to stillage ale in the consignee's cellars, whereas the railway normally charged for this another five pence per ton.[110] The conference thus urged the NBR to bring its rates into line with those of the conference. However, even if these methods of non-price competition were eliminated, the railway conference remained unhappy about the quantity of freight going via the Silloth Steam Packet Co. rather than using the all-rail route.[111] In 1884 it tried to get the NBR to cancel the price differential in favour of the steamboat which had been agreed in 1868.[112]

There is one further set of RCH records which shed light on the nature of railway and coaster cooperation. In 1902 a conference existed to deal with goods traffic between London and the West Riding of Yorkshire; the parties to this conference included ten railway companies, seven coastal shipping firms and two canal businesses.[113] It is clear that this organization had been in existence at least since 1897 and that it continued until at least mid-1904.[114] However, because only one volume of minutes survives, the date of establishment is unknown, and its history before 1902 remains a mystery. A number of significant points germane to the theme of this essay emerge from the one surviving volume of the West Riding, London Shipping Traffic Conference (WRLSTC).

Firstly, it is obvious that collaboration between the two transport modes involved a larger number of separate firms than any of the previously-cited agreements. Although there was some variation at the margins, there was a hard core of ten railway companies (the Great Central, Great Eastern, Great

---

[108]*Ibid.*, RAIL 1080/512, minute 3707.

[109]*Ibid.*, minute 3820.

[110]*Ibid.*, minute 3878.

[111]*Ibid.*, RAIL 727/2, minutes 1949 and 1966.

[112]*Ibid.*, minute 2088.

[113]*Ibid.*, RAIL 1080/465.

[114]*Ibid.*, meeting, 17 June 1903.

Western, Hull and Barnsley, Great Northern, Midland, Lancashire and York-shire, Cheshire, London and North Western, and North Eastern); seven firms involved in the coastal liner trade (the GSN, Jescott Steamers, George R. Hal-ler, France Fenwick and Co., Sollas and Sons, the London Yorkshire and Lancashire Carrying Co., and Fisher, Renwick and Co.), and two inland wa-terway businesses (the Rochdale Canal Co. and Aire and Calder Navigation) in the conference.[115] There were also additional companies for some of the time; for example, the Crofton Shipping Co. joined in June 1902, while the Mer-chants Carrying Co., another coastal shipper, was in negotiation to join in No-vember 1903.[116]

The agreement was a formal, written document with a code of confer-ence regulations stating its objectives, membership and procedural rules. Its main aim was to facilitate long-distance traffic between London and the West Riding of Yorkshire whether via the Humber or Manchester by sea and rail, rail direct, sea and canal, or any combination of the three transport modes. The freight rates were set by the West Riding Conference and were mutually agreed and binding on conference members once approved. As in other agree-ments that we have examined, the rates by each method of transport were not identical, but rather the all-water route was cheaper than the direct rail rate. The discount in favour of the shipping companies varied within the range four-teen to twenty-one percent, with the average being about seventeen percent.[117] Again, this was no doubt meant to reflect both the lower cost structure of the coaster in long-distance freight haulage compared to the train and the superior service which the railways claimed to offer in terms of speed and frequency.

Mutually enforced freight rates were not enough. A whole host of an-cillary services and drawbacks had to be mutually priced and agreed by all parties to ensure that these did not become back-door methods of giving dis-counts on agreed freight rates since, as we saw above, the railway companies had claimed the Silloth steamer was doing in its charges for cartage in Liver-pool and its free stillage of ale. The WRLSTC's solution to this was to draw up a long appendix to the code listing the agreed rates of as many of these types of ancillary charges as possible.[118] For instance, the scale of drawbacks to be paid when the public did their own carting or barging in London but where the agreed rates included collection or delivery were laid down; the normal practice on the cost of customs entries and clearances, cost of insur-ance on cargoes, cost of marking and carding imported colonial wool in Lon-

---

[115]*Ibid.*, various minutes.

[116]*Ibid.*, minutes 630 and 746.

[117]*Ibid.*, minutes 591 and 686.

[118]*Ibid.*, minute 582, appendix.

don and where separate deliveries were needed because consignments were sent in separate batches were all codified. By drawing up such a detailed set of regulations the conference hoped to eliminate all forms of price competition, not merely the main freight rate.

Further to reduce competition between the various transport modes, the WRLSTC agreed to divide up the traffic in one major commodity. This was for raw wool brought into London from the colonies or foreign parts and then moved to the West Riding as a raw material for the woollen and worsted industries. There were two categories involved – "overside" and "sales" wool. The coaster's allocated share of each in December 1901 was seventy-five percent and thirty-eight percent, respectively, giving it a slightly larger slice of the traffic than the railways.[119] If either mode carried more than its share in any given quarter then it had to pay a levy per ton on the "excess carryings" to the other route. In the summer of 1903 there was some discussion of this split between modes, but after some bargaining, there was no real change, the coaster retaining seventy-five and 37.5 percent, respectively.[120]

It is also clear that the coaster was not pushed out of carriage of the high-value commodities. The conference appointed a number of inspectors to check that the goods at the quays were as described in the manifests and that higher-rated goods were not being passed off as commodities which incurred a lower freight rate. Thanks to these returns we get quarterly snapshots of some of the goods which were being carried by the coasters between London and Humber ports such as Hull, Grimsby and Goole, and also Manchester. These commodities included manufactured goods such as Bovril, Vaseline, Spratts dog food and sewing machines, and frangible products such as lamps and glass tumblers.[121] Thus, the coastal liners were carrying a wide variety of products, including the more valuable and higher-rated goods on this route, just as we have discovered on other routes.

Although neither aggregate figures nor quantities of specific cargoes carried by the two transport methods are given, there is one exception which indicates that the coaster was not necessarily relegated to carrying a tiny proportion of total traffic while the railway moved the lion's share. This is suggested by the split agreed by the conference, referred to above, where the coaster was assigned slightly more than fifty percent of wool traffic. This allocation was borne out in practice, for in 1901-1903 the coaster increased consistently the proportion of "overside" wool which it carried from seventy-four to ninety-two percent and kept its proportion of "sales" wool between forty-five and forty-nine percent. Overall, the coaster carried forty-seven percent of all

---

[119]*Ibid.*, minutes, 17 December 1901.

[120]*Ibid.*, minutes of sub-committee appointed by minute 704, 12 June 1903.

[121]*Ibid.*, minutes 592 and 664.

wool traffic in 1901 and fifty-five percent in 1902 and 1903.[122] This is not consistent with the hypothesis of a transport method beaten into submission by the technologically superior railway. Rather, it suggests a rough parity between the two types of transport with a margin of superiority for the coaster.

This conclusion is borne out by the only other direct observation in this file of relative traffic carried by direct rail compared to the coasters. For the first half of 1903 the WRLSTC monitored the quantities of rags, hides and kips carried between London and the West Riding. It discovered that the coasters carried fifty-three percent of all rags and fifty-eight percent of all kips and hides.[123] Admittedly, these were not high-value goods, and kips and hides could be smelly, unpleasant cargoes and hence the railway may not have been sorry to see them go by sea, but certainly this reinforces the view that the coaster was able to hold its own against the railways and indeed carry slightly over half the total traffic. The WRLSTC bears out the findings of the other examples examined earlier that collaboration between railway companies and coastal liners to eliminate price competition was normal and formal. This could also be extended to include inland navigation firms. The essence of the agreement was mutually approved freight rates with a small premium in favour of the railways. The coaster did carry high-value commodities, such as manufactured goods, and was able to attract a significant share of total traffic on long-distance routes, perhaps slightly more than half.

A number of important points emerge from this story that are of relevance to railway-coaster competition. Firstly, there is no question but that the steamboats were able to compete successfully with the railways long after the latter had through rail routes. Secondly, the railways did not cream off all the high-value freight, relegating the coastal liner to low-value bulk goods. The coaster continued to carry high-value commodities, such as manufactured goods and exotic imports. Thirdly, the railway conferences considered it quite reasonable to try to eliminate competition by fixing freight prices with rival steamboat companies so that the rates were mutually remunerative. This is quite contrary to the view propounded recently by one transport historian who believed that "coasting was exempted [from conference agreements] because railway competition made rate-fixing impossible."[124] These findings suggest that the railways were no hindrance to rate-fixing but rather that they encouraged collaboration with coastal liner companies to fix mutually agreed rates. These agreements took the form of either formal written documents, some highly codified, or less formal *ad hoc* cooperation through an exchange of let-

---

[122]*Ibid.*, minutes 662 and 771.

[123]*Ibid.*, minute 764.

[124]Simon P. Ville, *Transport and the Development of the European Economy, 1750-1918* (London, 1990), 95.

ters or meetings as and when necessary. This cooperation was not restricted to rate conferences but in some cases involved pooling of traffic receipts and division among the parties on a prearranged basis. Fourthly, the steamboat was perceived as having lower costs than the railways, which was reflected in its lower freight rate compared to the railway when the agreed schedules were drawn up. The railway was accepted as having higher costs and thus needed a higher freight rate to remunerate it. In theory, such a premium price should have been reflected in a higher quality service but, on their own evidence, the railways seemed to be suggesting that many shippers did not see it this way for long-distance traffic and preferred to send their goods via the coaster. Finally, when agreement had been reached on freight rates so that price competition had been eliminated, the two transport modes continued to try to outdo each other via service competition. This could take the form of frequency of sailings, differences in terminal charges or the level of service offered within such charges, and of specialized vessels allowing shorter and therefore faster journeys, such as via the Caledonian Canal.

There are many lacunae in this essay, and some of the issues which remain unresolved are of major importance. Perhaps the most significant is to assess the outcome of the cooperation and competition between the two modes of transport. Did collaboration pay off for each form of transport or did one benefit at the expense of the other? Did the steamboat make higher profits when collaborating with the railway companies or when competing? It would be a little easier to attempt an answer to some of these questions if we had runs of figures on how much of what goods were carried between which destinations. However, there is a dearth of precisely this material. The railway records do not contain such data for their own network let alone the potential competitors. The few coastal companies for which records survive do not normally include even the aggregate tonnage carried in any given year, let alone any breakdown of commodities and routes. These are difficult to get around and must mean that our picture of nineteenth-century transport remains broad brush and rather impressionistic.

It should also be stressed that although much of this essay has been devoted to discussing the extent of collaboration between the coaster and the railway, this only applies to coastal liner services. There is no evidence of any such activity between the tramping coaster and the railway company, and indeed none is to be expected, since the tramp was offering a totally different service to the liner or railway. In terms of the quantity of cargoes carried, the tramp sector of the coastal trade was almost certainly the largest. The single largest commodity conveyed was coal, and this was entirely inappropriate to liners. Hence, the liner trade was the smaller in quantitative terms but carried the higher-value commodities and could command a premium price over the tramp ship because of the higher quality of the service it provided. Thus, it was the smaller but up-market segment of the coastal sector which collaborated with the railways.

The RCH records also support some of the generalizations made elsewhere about the economic characteristics of the coastal trade. They reinforce the view that the coaster was particularly able to compete in long-distance goods transport and that the average haul of the coastal ship was likely to be well in excess of 100 miles. This suggests that the idea of a freight market segmented in part by distance with the coaster filling the long-haul niche while the railway's average haul was much shorter, is not unreasonable. It also confirms that there was normally a significant differential between the price charged by the coasting liner and the railway train for carrying an identical commodity between the same two points, and that the discount offered by the coaster was normally in the range from twenty-five to fifty percent. Finally, it also reinforces the idea that coastal liner companies were likely to cooperate with each other in setting freight rates, arranging timetables mutually and even agreeing to some prearranged division of the traffic and revenue. All of these elements were present in at least one of the examples of railway-coaster collaboration, and cooperation with other lines was obviously a normal part of coastal liner operations.

Such inter-modal collaboration is in line with previous thinking on railway policy in the last quarter of the nineteenth century. A number of writers, including Philip Bagwell, have shown that the railways virtually eliminated competition in freight rates in this period[125] and that service competition, in terms of facilities, frequencies, speed and willingness to move small consignments was the extent of inter-company rivalry.[126] If the railways were to eliminate price competition completely, they needed to include any competing mode which offered a similar service. The coastal liners did just this, hence the railway companies' eagerness to bring them into agreements to fix freight rates and to cooperate.

---

[125]Bagwell, *Railway Clearing House*, 263.

[126]Peter J. Cain, "Private Enterprise or Public Utility? Output, Pricing and Investment on English and Welsh Railways, 1870-1914," *Journal of Transport History*, 3rd ser., I, No. 1 (1980), 18-19.

Chapter 8
The Crewing of British Coastal Colliers, 1870-1914[1]

I

There is a large and diverse literature on the employment conditions and shore-based activities of the merchant seaman. One view, now rather out of favour, saw the mariner ashore as dissolute, easily led astray and a breed apart from civilized society.[2] This view has been challenged, and seamen have been seen simply as "working men who got wet" with similar problems and experiences as any other group of wage labourers. Marxist and other writers have debated degrees of exploitation and how changes in technology, organization and capi-

---

[1]This essay appeared originally in *The Great Circle*, XX, No. 2 (1998), 73-89.

[2]An early exponent of this traditional view was Hadley, who talks of streets crowded with sailor boys spending their time in open violation of "decency, good order and morality and at leisure to exercise their dissolute manners:" George Hadley, *A New and Complete History of the Town and County of the Town of Kingston-upon-Hull* (Kingston-upon-Hull, 1788), 424. Runciman, writing more than a century later, suggested that it was "the popular idea" that old time sailors were "a hard-drinking race:" Walter Runciman, *Collier Brigs and Their Sailors* (London, 1926; reprint, London, 1971), 135. Frederick William Wallace, that pioneer of Canadian maritime history, also stressed cursing, belligerent crewmen who were shipped aboard "drugged and drunk" by unscrupulous crimps: Wallace, *Wooden Ships and Iron Men* (London, 1924; reprint, Belleville, ON, 1976), 165-166, 176, 180 and 187. This view has been perpetuated in more modern writing: Ralph Davis wrote that "drunkenness and wildness characterised the merchant seaman" and "some went [to sea] willy-nilly, drunk or unconscious as the crimp made up the required crew as best he could:" Davis, *The Rise of the English Shipping Industry in the Seventeenth and Eighteenth Centuries* (London, 1962; reprint, Newton Abbot, 1972), 122 and 153. Alfred G. Course, in his classic work *The Merchant Navy: A Social History* (London, 1963), 241-243, emphasized the dangers of shore life for the naive sailor beset by crimps, prostitutes, publicans and boarding-house keepers. Perhaps the zenith of this view is Stan Hugill, *Sailortown* (London, 1967), who devoted a whole large book to mapping and telling stories of the areas in port towns which served sailors with "sordid pleasure, unlimited vice and lashings of booze" (xviii). For an appraisal of this view and a statement of the counter case, see the essays in Rosemary Ommer and Gerald Panting (eds.), *Working Men Who Got Wet* (St. John's, 1980).

tal intensity affected maritime labour.[3] Recently there have been discussions about wage rates and the efficiency of labour markets,[4] and Williams has shown the high level of concern, in government and outside, with the quality, skills and supply of merchant seamen in the mid-Victorian period.[5] To review and criticize this body of literature adequately would require a whole essay in its own right, so this article cannot attempt that.

Two points emerge from a perusal of the extant literature. Firstly, the vast majority of it deals with maritime labour conditions on sailing ships, and very little is devoted to exploring this topic on steamships. Even Sager, who sees the iron steamship as equivalent to an industrial factory bringing de-skilling and alienation, devotes only one brief final chapter to employment circumstances in steam.[6] The second point to emerge from the literature is that virtually all of it is about sailors in deep-water trades and almost none of it on coastal employment. Kaukiainen devotes some space to "peasant shipping," by which he means essentially small sailing ships in the coastal trade.[7] In this segment he believes working conditions were more egalitarian and democratic

---

[3]Marcus Rediker, *Between the Devil and the Deep Blue Sea: Merchant Seamen, Pirates and the Anglo-American Marine World, 1700-1750* (Cambridge, 1987), and the "Roundtable" review of Rediker's book in *International Journal of Maritime History*, I, No. 2 (1989), 311-357; and Eric W. Sager, *Seafaring Labour: The Merchant Marine of Atlantic Canada, 1820-1914* (Montréal, 1989), and the "Roundtable" review of Sager's book in *International Journal of Maritime History*, II, No. 1 (1990), 227-274.

[4]Lewis R. Fischer, "International Maritime Labour, 1863-1900: World Wages and Trends," *The Great Circle*, X, No. 1 (1988), 1-21; Fischer, "A Dereliction of Duty: The Problem of Desertion on Nineteenth-Century Sailing Vessels," in Ommer and Panting (eds.), *Working Men Who Got Wet*, 51-70; Fischer (ed.), *The Market for Seamen in the Age of Sail* (St. John's, 1994); and Charles P. Kindleberger, *Mariners and Markets* (New York, 1992).

[5]David M Williams, "Mid-Victorian Attitudes to Seamen and Maritime Reform: The Society for Improving the Condition of Merchant Seamen," *International Journal of Maritime History*, III, No. 1 (1991), 101-126; and Williams, "The Quality, Skill and Supply of Maritime Labour: Causes of Concern in Britain, 1800-1914," in Lewis R. Fischer, *et al.* (eds.), *The North Sea. Twelve Essays on Social History of Maritime Labour* (Stavanger, 1992), 41-58.

[6]Sager, *Seafaring Labour*.

[7]Yrjö Kaukiainen, "Five Years before the Mast: Observations on the Conditions of Maritime Labour in Finland and Elsewhere," in Lewis R. Fischer and Walter Minchinton (eds.), *People of the Northern Seas* (St. John's, 1992), 47-62, reprinted in Lars U. Scholl and Merja-Liisa Hinkkanen (comps.), *Sail and Steam: Selected Maritime Writings of Yrjö Kaukiainen* (St. John's, 2004), 31-44.

than in long, deep-water voyages because the crews were small, often related to one another or from the same village. Sager, too, believes small coastal craft were run on pre-industrial lines which minimized social distinctions and ensured close, informal working relationships. He, like Kaukiainen, is talking about small sailing craft in the coastal trade. No study has been made of the coastal steam sector of the market. Given this gap, this article examines some aspects of employment conditions on British coastal ships between 1870 and 1914, and particularly on what might be seen as one of the most "modern" and "industrial" segments of the British coastal trade, namely the steam-engined, screw-propelled, steel-hulled collier fleet plying the east coast of England carrying coal from ports such as Newcastle, Shields and Sunderland to London and the South East of England.[8]

Two primary sources have been used in this investigation. Firstly, there are the series of censuses of seamen taken in the late nineteenth and early twentieth centuries by the British government.[9] These disaggregated the population of mariners by the types of trade in which they were employed, by age and by grade. Valerie Burton has pioneered work on these censuses and the earlier attempts to construct a register of seamen.[10] She concludes that by the end of the nineteenth century they were "a valuable source for the study of seafaring employment,"[11] which provides reassurance on their suitability as one of the bases for this study. As useful as Burton's analysis is, she never deals with the coasting crews alone, always including coastal mariners under the home trade. Thus, there remains some mileage in examining these censuses to see what they say about the crewing of coastal ships.

The other primary source which has been employed are the Crew Agreements held partially in the National Archives in Kew and the National Maritime Museum in Greenwich, but mainly in the Maritime History Archive at Memorial University of Newfoundland in Canada.[12] Two methods were used to sample the vast number of crew agreements. The names of screw col-

---

[8]J.A. MacRae and Charles V. Waine, *The Steam Collier Fleets* (Wolverhampton, 1990); and John Armstrong, "Late-Nineteenth-Century Freight Rates Revisited: Some Evidence from the British Coastal Coal Trade," *International Journal of Maritime History*, VI, No. 2 (1994), 76-78.

[9]Great Britain, Parliament, Parliamentary Papers (*BPP*), 1902, XCII; 1908, XCVI; and 1912-1913, LXXVI.

[10]V.C. Burton, "Counting Seafarers: The Published Records of the Registry of Merchant Seamen," *Mariner's Mirror,* LXXI, No.3 (1985), 305-320.

[11]*Ibid.*, 305.

[12]Keith Matthews, "Crew Lists, Agreements and Official Logs of the British Empire, 1863-1913," *Business History,* XVI, No. 1 (1974), 78-80.

liers were identified from various secondary sources,[13] and the crew agreements were sought for these ships. In addition, random searches were made of boxes looking for ships whose voyage patterns looked like those of steam colliers and whose ownership confirmed that they were indeed colliers, as they were owned by well-known names in the coal trade such as Stobart, Lambert, Cory and Pelly.[14] Eighty crew agreements were examined in all. By the standards of the Atlantic Canada Shipping Project, this is a very small number, but two things provided some reassurance. On some characteristics, especially pay rates, there was very close similarity between agreements. This suggests that there was very little dispersion around the mean and that those agreements accessed were typical of the population of collier crews as a whole. The other reassuring feature was that each crew agreement contained details of a number of mariners, and hence the number of seamen analyzed (1220) was reasonable. That said, the following findings must be seen as tentative until more research is carried out on the topic. Also, the findings are relevant only for screw colliers. Whether they are typical of any other part of the coastal trade is totally unknown.

II

One of the key features of any employment is the payment of wages. The amount, method of calculation, frequency and regularity of payment, seasonality and degree of casualness of employment are all important considerations when assessing conditions of employment. This section will investigate such basic aspects of working on a coastal screw collier.

There appear to have been three methods of payment which were used in the British coastal trade in the nineteenth century. Wages could be based on a simple time rate, namely by the week or the month for the duration of the voyage. Alternatively, the mariner could receive a piece rate for the job, that is, a fixed sum was offered for a particular round trip irrespective of the time taken to complete it. Finally, the sailor might receive a form of performance related pay, that is, a proportion of the revenue earned, namely a fixed percentage of the freight rate.

Payment on piece rates appears to have been a normal method on sailing ships engaged in the coal trade in the eighteenth and nineteenth centuries. Willan, writing about the eighteenth century, said that "wages in the coasting trade were generally paid by the Voyages and not by the week, month or year

---

[13]For instance, MacRae and Waine, *Steam Collier Fleets*.

[14]*Ibid.*, 47-59.

and therefore varied according to the anticipated length of the voyage."[15] His examples of *Hopewell* in 1724 on a trip from Sunderland to Leigh and other examples from 1725, 1729 and 1740 were of colliers.[16] Hughes confirmed that piece rates were not confined to the east coast, suggesting that a return voyage with coal from Cumberland to Dublin in 1770 earned a seaman two guineas,[17] and that earnings were normally assessed per trip. Simon Ville, making an in-depth examination of the Henley vessel *Freedom* between 1791 and 1816, showed that for all of the coastal voyages carrying coal from the North East, mostly to London and occasionally to Portsmouth, the seafarers were paid by the voyage.[18] He draws a distinction between this form of payment and that preferred when the ship was deployed in foreign voyages. On the latter the crew were paid "by the month rather than by the voyage."[19] Hodgson confirms that piece rates continued to be normal in the east coast coal trade, claiming that in the 1830s on collier brigs from the Tyne to London "the men's wages" were "reckoned by the trip,"[20] as did Moffat, speaking from experience of the 1860s, who stated that "men were engaged by the voyage in the coal trade."[21] This is confirmed by an anonymous contemporary reporter who suggested that the practice was institutionalized to the extent that the seamen "issued notice" of the amount to be paid per London voyage.[22] Thus, in the era of sail, seafarers in the coastal coal trade were quite normally paid by the voyage rather than by time rates.

---

[15]T.S. Willan, *The English Coasting Trade, 1600-1750* (Manchester, 1938; reprint, Manchester, 1967), 17-18.

[16]*Ibid.*

[17]Edward Hughes, *North Country Life in the Eighteenth Century. Vol. II: Cumberland and Westmorland, 1700-1850* (Oxford, 1965), 196.

[18]Simon P. Ville, "Wages, Prices and Profitability in the Shipping Industry during the Napoleonic Wars: A Case Study," *Journal of Transport History*, 3rd ser., II, No.1 (1981), 48-51, reprinted in John Armstrong (ed.), *Coastal and Short Sea Shipping: Studies in Transport History* (Aldershot, 1996), 21-34.

[19]*Ibid.*, 43.

[20]George B. Hodgson, "The Genesis of the Screw Collier," *Nautical Magazine*, LXX, No. 3 (1901), 174.

[21]Henry Y. Moffat, *From Ships-Boy to Skipper, with Variations* (London, 1911), 44.

[22]*North and South Shields Gazette,* 3 October 1860.

The logic of piece rates in the coastal coal trade under sail is clear. Piece rates were used by the owners to contain the costs of running the ship if it was delayed by adverse weather conditions or crew laxity. The sailing ship was only too susceptible to weather and tide. If the wind was excessive, in the wrong direction or too light to counteract a tide the sailing ship stayed in harbour. Similarly, as the crew was outside of effective supervision of the owners, any lack of motivation or cause to dawdle reduced the freight-earning power of the coaster. Hence, knowing that the earnings would be the same, however long the return journey took, acted as a motivator to the crew. "No one got paid overtime in these ships and however long the voyage took the money was that agreed upon when he first signed on."[23] Thus, whether as a result of crew procrastination or natural conditions, if the ship was delayed the owners limited their expenditure on wages to a known amount per voyage and provided the crew with an incentive to speed up loading and discharging and to crack on as much sail as was safe. At the extreme this could cause crews to take risks, perhaps carrying too much canvas, leaving harbour when conditions were barely safe, overloading or taking other chances.

The screw collier was much more predictable in its journey times. By the 1880s it was usually also much larger than the collier brigs with consequent economies of scale. The method of payment in these ships appears universally to have been a weekly rate.[24] The owner could predict the time taken on the voyage to fine tolerances and with this uncertainty removed was willing to commit himself to fixed wages. In addition, the crew were not routinely discharged at the end of a voyage in order to save costs when the ship was idle in port, for the essence of profitability for these screw colliers was a fast turn-around at both ends of the trip so as to ensure a large number of revenue-earning voyages and hence to maximize income. Thus, to hire the crew only for the duration of the journey might be counter-productive if as a result the collier had problems recruiting a fresh crew and its departure was in any way delayed. To obviate this possibility, the crew was treated like other members of the industrial workforce and paid a regular weekly wage.

The other benefit to the owners of paying time rather than piece rates was that any gains in productivity achieved by reducing the period the screw collier spent in port or on the voyage would accrue to the shipowner. The labour cost per time period was now fixed, and if the collier performed an extra voyage or two as a result of improved loading or discharging methods, then the additional revenue was not matched by extra costs, and the owners profited

---

[23]Susan Campbell-Jones, *Welsh Sail. A Pictorial History* (Llandysul, 1976), 96.

[24]This statement is based on an examination of approximately eighty crew agreements for steam-powered colliers covering over twenty different ships for four benchmark years in the period 1875-1912.

disproportionately. This was important, since there is evidence that by the late nineteenth century the average number of trips performed by the steam collier was increasing as mechanical methods for unloading coal were introduced and improved.[25] The payment of time rates allowed the shipowners to benefit from this improvement, whereas payment by the trip would have allowed the crew members to participate, and hence total costs would have advanced more in line with revenue.

III

The mode of payment is not the only feature of the screw colliers that is of interest. Equally important is the level of wages. We shall take able seamen and firemen as being the "working class" on board the steam colliers, for they were the lowest-paid grades, apart from "boys," who were probably apprentices as well, and the grade "ordinary seaman" appears not to have been used. These two grades almost invariably were paid the same weekly rate. Over the period 1870 to 1914 and between ships the variations were minimal, the range being £1.40 to £1.63 and the average being £1.48. Nor is there any obvious trend over time in these earnings; it was not the case that there was either a rise or fall over the forty-year period. It is interesting to compare these rates of remuneration with those earned by the deep-water sailor and with shore-based employment, and to see whether the conditions of employment of labour deteriorated on steamers compared to sail. Fischer calculated the average monthly earnings of able seamen in deep-water sailing ships who were recruited in England in the period 1860 to 1900 as between £2.87 and £3.26.[26] Given that the AB on a coastal collier was earning about £6 per month, this seems an enormous difference, and the idea that the switch to steam brought a deterioration in working conditions, at least in this regard, seems not to be borne out. Hence, for able seamen and firemen employed on the screw colliers the level of wages shows a marked improvement over both deep-water sailing crews and those on coastal sailing ships. When compared to shore-based workers, the complication arises as to how to judge the level of skill of the able seaman. He obviously was a skilled worker who had transferable skills within the industry, and the fireman at least could also find similar work ashore charging gas retorts, on the footplate of locomotives or stoking stationary boilers. Whatever the precise comparison, the coastal crew member was not doing badly when compared to the shore-based, semi-skilled worker given that unskilled labour-

---

[25]Armstrong, "Late Nineteenth-Century Freight Rates Revisited," 45-81.

[26]Fischer, "International Maritime Labour."

ers working for large industrial companies on shore could earn £1 to £1.42 per week, and unskilled labourers in London about £1 per week.[27]

One difference between the deep-water sailor and those on the coastal screw colliers was that the former usually received food on board as part of the total remuneration package whereas the latter was often expected to supply his own. Nearly all their agreements carry the phrase "crew to find their own provisions." This could account for some of the difference in the wages of the coastal and blue-water mariner. Feinstein calculates that forty-eight percent of total household income was spent on food in the period 1870 to 1914,[28] and thus the deep-water mariner's wages were worth nearly double if the provisions he received were equivalent to those on shore. Even allowing for shipowners' parsimony by estimating their inferior provisions as worth only one-third of wages, the deep-water sailor's monthly salary range becomes £3.91 to £5.40 with a mean of £4.53, closer to but still inferior to the wage of the able seaman on the coastal screw collier.

It might be argued that this comparison is still rather unfair since Feinstein calculated that another 13.3 percent of working class household incomes were spent on rent, rates, fuel and light,[29] which the sailor did not need to incur as these were provided by the ship. However, this is as true for the coastal as for the ocean sailor. The only point of difference might be if deep-water sailors were more likely to be single than coastal crewmen, and the coastal crews were more likely to have families and hence had to keep a home financed, whereas the single sailor saved all expenses of home maintenance when at sea. Then the deep-water sailor's wage would need to be further enhanced. This difference in marriage rates is supported by Burton's findings. She discovered that in 1891, fifty-four percent of home trade seafarers (which included the coastal trade) were married but only thirty-eight percent of foreign-going men.[30]

The crewing of coastal ships has given rise to several generalizations which can be investigated by examining the data in the crew lists. It has been claimed that the coastal trade was "the nursery of seamen" in the sense that mariners first learned their seamanship in the coastal ships and then when they

---

[27]The title of Maud Pember Reeves, *Round about a Pound a Week* (London, 1913), indicates the remuneration level of the unskilled in London just before the Great War.

[28]Charles Feinstein, "A New Look at the Cost of Living, 1870-1914," in James Foreman-Peck (ed.), *New Perspectives on the Late Victorian Economy: Essays in Quantitative Economic History, 1860-1914* (Cambridge, 1991), 159.

[29]*Ibid.*, 170-171.

[30]Burton, "Counting Seafarers," 316.

had graduated they moved into the deep-sea trades.[31] The other way in which the coastal trade acted as a nursery was for the Royal Navy.[32] In this sense all sailors in the merchant navy were potential crews for warships in wartime. Perhaps coastal ships, by virtue of touching a port more frequently, were easier prey for the imprest service and hence were a more immediate source of fighting sailors. A closely related theory is that the coastal trade was crewed largely by boys and older men. The former were learning their trade, while the latter had been in the deep-water trades but had married, settled down and wished a form of employment which would allow them more regular nights in their own bed rather than in a bunk.

If these descriptions of the crew characteristics are accurate, they can be tested against the ages of the crews in the coastal trade and in particular in the screw colliers. This seems particularly apposite since the collier fleet plying the east coast was seen as the nursery of seamen *par excellence.* The most famous Englishman who learned his trade as a seaman on colliers was Captain James Cook, and so enamoured was he of these early experiences that the ship in which he explored, *Endeavour,* was a converted collier barque.[33] Of course, the screw-propelled, steam-powered colliers were generations apart from a collier barque, but they should demonstrate whether the "nursery effect" persisted into the late nineteenth century.

The implications of the theories are that there will be a large number of young mariners in the coastal trade, but as they finish apprenticeships and gain a few years experience they will move on to deep-water berths. Thus, one would not expect to find many crewmen in their late twenties or thirties. The age at which "settling down" might commence is less clearly defined. Given that the average age of spinsters at marriage between 1871 and 1900 was about twenty-four to twenty-six,[34] and allowing for the bachelor to be two or three years older than his bride, this suggests the average male married in his late twenties. This might be seen as one measure of when "settling down" might occur. This suggests there should be a surge of mariners in their early thirties, as they quit deep-water trades and take up coasting. These generalizations can be tested against the age distribution of coastal crews in the late nineteenth and

---

[31]Kindleberger, *Mariners and Markets,* 13-16.

[32]For a contrary view see David J. Starkey, "War and the Market for Seafarers in Britain, 1736-1792," in Lewis R. Fischer and Helge W. Nordvik (eds.), *Shipping and Trade, 1750-1950: Essays in International Maritime Economic History* (Pontefract, 1990), 25-42.

[33]Ronald Hope, *A New History of British Shipping* (London, 1990), 222.

[34]H.J. Habakkuk, *Population Growth and Economic Development since 1750* (Leicester, 1971), 56.

early twentieth centuries thanks to the Censuses of Seamen conducted by the government. The results for 1896 and 1911 are shown in table 1.

Table 1
Numbers and Percentages of Seamen Employed in the
Coastal Trade Broken Down by Age, 1896-1911

1896

Age of	Sail		Steam		All Vessels	
Seamen	Number	%	Number	%	Number	%
<20	2373	17.2	661	4.4	3034	10.6
20s	4789	34.7	3891	26.0	8680	30.2
30s	2380	17.3	4447	29.8	6827	23.8
40s	2107	15.3	3486	23.3	5593	19.5
50s	1508	10.9	1979	13.2	3487	12.1
60+	628	4.6	477	3.2	1105	3.8
	13,785		14,941		28,726	

1911

Age of	Sail		Steam		All Vessels	
Seamen	Number	%	Number	%	Number	%
<20	860	13.9	847	4.9	1707	7.3
20s	1853	30.0	4145	24.2	5998	25.7
30s	1233	20.0	4926	28.8	6159	26.4
40s	961	15.6	4048	23.6	5009	21.5
50s	837	13.6	2454	14.3	3291	14.1
60+	427	6.9	704	4.1	1131	4.8
	6171		17,124		23,295	

Source: Great Britain, Parliament, *Parliamentary Papers (BPP)*, 1897 LXXVIII, 409; and 1912-1913, LXXVI, 88-89.

There is a little evidence to support the nursery hypothesis in the figures for sailing coasters in either year and in all coasters in 1896, inasmuch as the largest group of seamen were in their twenties and the cohorts then declined in significance. However, in steam coasters the modal group was in its thirties, as was the case for all coasters in 1911, which gives sonic support to the idea. What is clear is that there are distinct differences in the age characteristics of crews in sail and steam. Sailing coaster crews were on average younger than steam coaster crews, with much larger proportions of sailing crews in their teens and twenties than was the case for steamers. This hints at easier work schedules on steamships than on sail, since seamen were able to continue employment to a later date on steamers rather than being too old, perhaps worn out by hard physical work or more aware of the greater dangers of the job. Burton suggested there were "increased opportunities for older men in the stokeholds" in the later nineteenth century, though explaining that "coal trimmers and firemen were young men's occupations."[35] These figures give little credence to the notion either of only young men working in coasters and moving on once qualified or of there being an influx later in life. The distribution is consistent with a large entry of young people in their teens and twenties, peaking in their thirties and then gradually drifting away from the sea as they reached their forties and fifties.

If we turn to the coastal steam colliers, we can examine the validity of these conclusions for this particular subset of the coastal trade. These generalizations are less likely to apply to the officers as they have had to gain experience and qualifications. The bulk of seamen in the colliers were employed either as "able seamen" or "firemen." If we look at the distribution of these grades in table 2 it does not seem to conform to the pattern which would be anticipated.

Table 2
Age Distribution of Crewmen on Coastal Screw Colliers, 1870-1905

	Able Seamen		Firemen	
	No.	%	No.	%
1-19	2	0.7	0	0
20-29	76	26.3	55	28.9
30-39	92	31.8	71	37.4
40-49	77	26.6	51	26.8
50-59	38	13.1	12	6.3
60+	4	1.4	1	0.5
	289	99.9	190	99.9

Source: Great Britain, National Archives (TNA/PRO), various crew agreements.

The largest numbers of such mariners were in their thirties with an approximately equal number in both their twenties and forties. There is no preponderance of crewmen in their twenties, no "gap" in the thirties and very few teenagers. In fact, it looks like a normal distribution around a mean of about thirty-five. In addition, if the number of sailors in their twenties is broken down, it is not the case that there were any more in their early than in their late twenties. In fact, the opposite is true: a higher number of both able seamen and firemen were in their late twenties than in their early twenties. Thus, there seems no evidence from the age distribution of the "working class" on board these ships that coastal colliers acted as a nursery for young, inexperienced mariners who when qualified and skilled moved on to blue-water employment.

---

[35]Burton, "Counting Seafarers," 316.

There is further evidence that supports the view that the coastal trade was perceived as a career in its own right and not simply as a route to entering the deep-water trades. This is the average ages of the officers on the coastal colliers. It was these grades who needed skill and experience in terms of learning where there were hidden perils such as submerged rocks, reefs or sandbanks, which routes were too shallow at certain states of the tide and where there were shifting channels and fast tides and currents which could affect the ship's course. These were different skills to those of the deep-water officer, but just as vital, needing experience and judgement. If the coasting trade was perceived as a necessary step in the progress to deep-water, but only that, not a career in its own right, we might expect the officers to be relatively young, building up experience before graduating to the deep-water routes. This is not borne out by the evidence in table 3.

Table 3
Average Age of Officers on Coastal Screw Colliers

Captain	46.2
Mate	45.3
Second Mate	41.5
First Engineer	44.2
Second Engineer	36.0
Donkeyman	35.2

Source: See table 2.

The average age of all three deck officers was in the forties, showing that they were not young men soon to move on but were career officers. The degree of progression in their ages also suggests that there was a promotion ladder whereby with sufficient suitable experience and qualifications they could move upward in status and salary. Similarly, the range of ages of the engineering officers suggests that these too were not all "learners" but included men of mature experience and that progression was normal.

Associated with the concept of the coastal trade as a nursery for seamen was the underlying fear that the Royal Navy would be unable to recruit sufficient sailors in time of an emergency. This was part of the logic which had reserved the coastal trade to British ships and seamen until 1849 and 1853, respectively.[36] The removal of the manning clause raised the spectre of British ships crewed by foreigners of dubious dependability in time of war, and a Royal Navy made impotent by a lack of experienced British sailors. There

---

[36]Hope, *New History*, 282; Adam W. Kirkaldy, *British Shipping: Its History, Organisation and Importance* (London, 1914; reprint, Newton Abbot, 1970), 27; and R.H. Thornton, *British Shipping* (Cambridge, 1939; 2nd ed., Cambridge, 1959), 42-43.

were objective grounds for this fear, inasmuch as the number of foreign seamen in the British fleet did rise in the late nineteenth and early twentieth centuries.[37] However, there is significant disagreement in the literature over the proportion of the labour force on British ships which was of foreign nationality. Burton, using the censuses of seamen, is the most recent and reliable estimate. She calculates the proportion of Lascar and foreign seamen, expressed as a percentage of "trading vessels" only (that is excluding fishing boats), as in table 4.

Table 4
Number and Proportion of Lascar and Foreign Seamen in the
Total British Fleet

	Lascar		Foreign	
	No. (K)	Percent	No. (K)	Percent
1891	21	12.3	24	13.8
1896	28	15.5	27	15.2
1901	34	18.0	33	17.5
1906	38	19.0	35	17.3
1911	43	20.6	29	13.8

Source: Burton, "Counting Seafarers," 318.

These figures make the estimates of Thornton, Hope and Prescott seem exaggerated.[38] Nor should the category "Lascar" be added to "foreign" to arrive at the total of foreign crewmen, for the majority were likely to be colonial since they comprised "Asiatics and East Africans employed on vessels trading either from India to this country or entirely in Asiatic or Australian waters."[39] It should be stressed that table 4 is not the total population of seamen but those actually employed at the date of the census. Hence, to arrive at the total number of seamen available it would be necessary to add in the figure of those between voyages, resting, ill and in temporary shore jobs, which Burton estimates to be an additional one-third of the number counted.[40] However, we are interested in the proportion and not the number of the maritime labour

---

[37]Kindleberger, *Mariners and Markets*, 70; and Burton, "Counting Seafarers," 318.

[38]Thornton, *British Shipping*, 82; Hope, *New History*, 342; R.G.W. Prescott, "Lascar Seamen on the Clyde," in T.C. Smout (ed.), *Scotland and the Sea* (Edinburgh, 1992), 199-212; and Conrad Dixon, "Lascars: The Forgotten Seamen," in Ommer and Panting (eds.), *Working Men Who Got Wet*, 265-281.

[39]*BPP*, 1902, XCII, 281.

[40]Burton, "Counting Seafarers," 313.

force which was foreign, and the former is not affected by the inadequacies of the census. Table 4 shows that the proportion of foreign crews was between 13.8 and 17.5 percent but with no overall trend. The proportion of Lascars in the British maritime labour force rose steadily from just over twelve percent to over twenty in the same period. However, as explained earlier, it would be wrong to perceive these as foreign, since the majority were likely to be of colonial origin.

The Censuses of Seamen also break down the aggregate numbers into the overseas and coasting trades. The proportion of crews made up of foreigners in the coasting trade was as in table 5. This demonstrates that the coasting fleet had a very small proportion of foreigners compared to the overseas fleet, just as Burton noted for the home trade, and that it peaked at just over four percent in 1901 and then fell away a little.

Table 5
Coastal Trade: Proportion of Foreign-born in the Crews

	Sail	Steam	All Ships
1896	2.1	2.6	2.4
1901	4.7	3.7	4.1
1906	3.1	3.7	3.5
1911	2.7	3.4	3.2

Sources: BPP, 1897, LXXVIII, 406; 1902, XCII, 296-297; 1908, XCII, 116-117; and 1912-1913, LXXVI, 86-87.

Table 6
Countries of Origin of Foreign Seamen, 1891-1911 (%)

	1891	1896	1901	1906	1911
Sweden	19.5	19.0	18.4	14.9	12.5
Germany	17.8	18.8	15.7	14.7	17.0
Norway	14.2	13.4	11.9	9.9	7.4
USA	8.2	8.1	8.4	6.6	5.4
Russia	6.8	7.1	6.2	5.2	6.6
Denmark	6.2	5.5	4.9	4.2	4.5
Italy	2.9	3.2	4.8	6.1	4.8
Holland	3.4	3.9	3.7	3.3	4.6

Source: BPP, 1912-1913, LXXVI, 74.

The censuses also identified the countries which contributed the largest numbers of foreign seamen. Table 6 gives the percentage of all foreign seamen from the top eight countries represented. North European countries were the most important, with the Scandinavian countries contributing the largest proportion, though declining from four-tenths to one-quarter of all foreign seamen between 1891 and 1911. The single most important country of origin was Sweden contributing between one-fifth and one-eighth of overseas mariners.

We now have a broad picture from the census figures into which the coastal screw colliers can be contextualized. About one-fifth of all seamen on British ships were foreigners, with Lascars making up an equal proportion. However, the proportion on coastal ships was much smaller, rising from two and a half percent in 1896 to four percent in 1901 and then falling to just over three percent in 1911. How well do these generalizations fit the case of our sample of colliers? Table 7 gives the bare bones of the analysis.

Table 7
Foreign Crewmen on Coastal Colliers

	1870-1875	1882-1887	1892-1898	1905-1906
No. of ship/years	7	5	8	9
Crews	401	124	222	474
Foreign	12	6	19	115
% Foreign	3	5	9	24
Min. Foreign %	0	0	0	8
Max. Foreign %	10	11	21	48
Most Frequent Nationalities	Swedes	Danes	Norway	Swedes
	-	-	Danes	Norway
	-	-	Swedes	German
Scandinavians: No.	7	3	14	70
Scandinavians: %	58	50	74	61

Source: See table 6.

It should be stressed that this is based on a very small random sample of steam colliers, and thus the results are very tentative. However, a number of trends emerge clearly. Firstly, the proportion of the total labour force made up by foreigners rose over time from a few percent in the 1870s to nearly a quarter in the early years of the twentieth century. In this respect the colliers were typical of the British fleet as a whole, although the proportion of foreigners on the colliers was less than in the aggregate fleet. However, the proportion of foreign seamen on these colliers was much higher than on the coastal fleet in the censuses whether we take the aggregate coastal fleet or steamers alone. In addition, looking at the position on individual ships, the range of foreign participation widens over time. The increase in foreign participation on the colliers was not a steady process. There seems to have been a slow rise in the last thirty years of the nineteenth century but a very rapid influx of foreign mariners between the 1890s and 1905-1906. Explaining this is more difficult. Given that the wages on the coastal colliers appear to have been superior to those earned in deep-water sailing ships, it could be that as sail declined abroad those sailors displaced sought employment on the more remunerative steam coasters. This aspect requires more exploration. The other characteristic which is clear from table 8 is the unimportance of Lascar seamen among the foreign crew members on coastal colliers.

Table 8
Foreign Crewmen by Nationality on Coastal Screw Colliers

	No.	%
Swedes	54	35.5
Norwegians	24	15.5
Germans	14	9.2
Danes	13	8.6
Australians	6	3.9
USA	6	3.9
Canadians	4	2.6
Greeks	4	2.6
Finns	2	1.3
Spaniards	2	1.3
Jamaicans	2	1.3
Sierra Leoneans	2	1.3
West Indians	2	1.3
Others	17	11.2
	152	99.5

Source: Derived from crew agreements in TNA/PRO.

Strictly speaking of course, no Lascar seamen could be employed in the coastal trade, since definition required them to be employed on ships trading wholly or partly with the Far East. However, if we dispense with this strict interpretation and use the term to mean mariners from the Far East we find only one Indian able seaman among these crews. If we take Lascar in its wider meaning of non-white mariners, the number increases a little by the addition of six West Indians and two West Africans. However, even then this makes only nine "Lascar" crewmen comprising about six percent of the total number of foreign crewmen and less than one percent of the total crews. So if Lascar seamen had made large inroads into the British merchant marine it was not in the coastal collier segment. This is not surprising. Those ships and lines which plied to the parts of the globe where the Lascar was native were likely to be the largest employers of such individuals. Hence, those companies which sailed to the Far East or to East Africa, such as P&O, BI or Blue Funnel were likely to have had large numbers of Lascars.

The other important finding from the coastal colliers is the nationalities which predominated among the foreign crewmen. There were relatively few from the British Empire (about fourteen percent), so it was not a case of Britain predominantly drawing on her possessions to staff her fleet. Above all, the foreign mariners came from other European countries (about eighty percent), and among these countries the Scandinavian states predominated. Sixty-one percent of all foreign crewmen came from just four Scandinavian countries, with Sweden the largest contributor by far with over thirty-five percent of all foreign crew members. In both of these respects the experience of the coastal colliers is in line with that of the aggregate fleet. This common feature needs explaining. In part the answer may lie in the trade links between the ports of the North East and the Baltic and Scandinavia. Thus, foreign seamen arriving in the overseas trade from one of the Scandinavian countries may have seen the relatively high wages offered in the coastal colliers and tried their hand in one. It is also quite conceivable that some arrived in Newcastle or Sunderland on boats from the Baltic but then fancied a spell in deep-water trades from one of the more internationally-orientated ports and so worked their passage from the North East to London to try to find a berth in a blue-water ship going to distant parts of the world. This is borne out inasmuch as some foreign crewmen only stayed on board the collier for one voyage to London.

IV

If the employment conditions on shipping were harsh, one way to reconcile workers was to offer the possibility of escape from the drudgery. This might be via the prospects of promotion from the most physically demanding grades, such as firemen and able seamen, via the petty officer grades, such as bosun, donkeyman or carpenter and perhaps from there to the officer class as second engineer or second mate. Of course, such promotions were the exception, only a minority of the workers could hope to achieve them, and some required qualifications, obtained often by part-time study at sea or a period of full-time study on shore, as well as experience and the appropriate attitudes and values. The coastal colliers were no exception to this generalization. The gifted, determined and obedient could climb the social and financial ladder. Given that any promotion involved some risk, as the ability of an individual adequately to perform his duties was unknown until he was tried in the position, a method of testing out promotees for a specific period was likely to be welcomed. This was the case on the coastal colliers. In a number of cases the incumbent in a post was away temporarily and while absent another member of the crew was promoted in an acting capacity. If it was known by all parties that the normal incumbent was away for a specific period and purpose – to take shore leave or sit an examination – then there was no loss of status when the crewman in the acting capacity reverted to his substantive grade on the planned return of the temporary absentee. It gave the promotee a chance to gain experience in the higher grade and hence credibility if he were to apply for such a job permanently, and the master of the vessel was provided with an opportunity to assess the suitability of the temporary promotee for that role. The other advantage of the use of promotion was that it gave the captain patronage to dispense, as it was his decision as to who might be rewarded. This gave him additional power over the crew and provided the crew with an additional incentive to hard work and compliance.

There are numerous examples of temporary promotion on the coastal colliers. For example, on *F. Stobart* in October 1905 the second mate was absent for about a week and an AB was temporarily promoted to act for him.[41] Similarly, the mate of *Ocean* was absent for three weeks in July/August 1906 and so the second mate was promoted in an acting capacity until the original mate returned.[42] Similar examples occurred on *Battersea* in 1906 and *Caenwood* in 1905 where the first engineer stayed on shore "for a few voyages" and the second engineer was temporarily promoted.[43] It seemed quite normal for officers to take temporary shore leave and expect to be given their post back when they returned.

In addition, there are many examples of permanent promotions. For example, J. Keys, aged thirty-three, was the mate on *Charlaw* in February 1882, By 1 June of the same year he was captain, which post he retained at least until June 1885.[44] In a similar vein, L. Hansen, a thirty-eight year old Norwegian, was the mate on *Ocean* in January 1897.[45] In February of that year he was promoted to captain, and he stayed in that post on that ship until at least December 1906.[46] However, promotions were not restricted to the highest grades. In 1905 on *Airedale* the steward left the ship and the donkeyman was promoted to take his place.[47] This seems an odd route, as the logical progression for a donkeyman would be to second engineer. However, the steward's job attracted 1s 6d per week more than the donkeyman's post and hence it might be perceived as a gain in both status and salary. As a result of the donkeyman's move, one of the firemen was promoted into that job. On *Swiftsure* at the end of February 1905 the post of bosun and lampkeeper became va-

---

[41]Great Britain, National Archives (TNA/PRO), Board of Trade (BT) 99/2443.

[42]*Ibid.*, BT 99/2490.

[43]*Ibid.*, BT 99/2460.

[44]*Ibid.*, BT 99/1354 and 1456.

[45]*Ibid.*, BT 99/2002.

[46]*Ibid.*, BT 99/2490.

[47]*Ibid.*, BT 99/2443.

cant.[48] It was left unfilled until early July, when one of the ABs was promoted into it.

Hence, there were two parallel ladders on the steam colliers, the deck route and the engine-room route. Normally promotion was up these ladders with a fireman aspiring to donkeyman and thence to second engineer, while the AB might hope for promotion to bosun or lampkeeper and from there perhaps to second mate. Promotions could and did take place on the same ship, and the promotee was not required to change ships. Among the senior grades there was also a surprising degree of loyalty to particular ships, and brief absences ashore, followed by a return to the same ship and post, were tolerated by the shipowners.

Desertion rates can be seen as the ultimate expression of a breakdown in labour relations. Given that desertion might mean a loss of accumulated pay, possible prosecution and letting down one's crew mates, it was not a step to be taken lightly. Sager saw a rise in the desertion rates in Canadian shipping from the 1860s as an indication of increased class struggle.[49] The small size of the sample of colliers prevents any analysis over time, for out of a total of 814 crew members, only twenty-nine were recorded as "deserted," giving a proportion of 3.6 percent. This is much lower than the desertion rates reported by Sager for British deep-water steamships and sailing ships at nine and sixteen percent, respectively,[50] and by Fischer for sailing ships sailing out of St, John, New Brunswick and Prince Edward Island at twenty-three and twenty-four percent.[51] Of course, desertion might be a totally rational economic course of action. If for instance a seaman was recruited in the UK for a voyage to North America he might well find that the wage offered for a return trip from America to the UK was between one and a half and two times as much as for a return trip originating in the UK.[52] Not surprisingly, the grades of seamen who were most likely to desert were those who were the lowest paid and least powerful, *viz.*, the able seamen and the firemen. Of the twenty-nine "deserters" in the coastal collier sample, sixteen were able seamen and eight firemen. This is in line with Fischer's findings, with the ordinary seamen the most prone to desertion. More surprisingly, the tendency to desert was quite widespread among the petty officer grades, there being one bosun and one lampman in the

---

[48]*Ibid.*, BT 99/2443.

[49]Eric W. Sager with Gerald E. Panting, *Maritime Capital: The Shipping Industry in Atlantic Canada, 1820-1914* (Montréal, 1990), 144.

[50]Sager, *Seafaring Labour*, 254-257.

[51]Fischer, "Dereliction of Duty," 55.

[52]Fischer, "International Maritime Labour," 4 and 18.

sample. These were one step above the basic working grades with the possibility of promotion to officer. It is thus more surprising to see such individuals endangering their future prospects by leaving without permission. It suggests either extreme push or pull factors, and hence should be viewed as exceptional rather than normal. Even more outstanding is that one Second Engineer also deserted his ship. Again, given his high status this suggests a strong incentive, though re-employment might not be too difficult if the high salaries paid to engineers were indicative of the difficulty of recruiting and retaining them. The implication might be that conditions on board the coastal colliers were not as harsh or alienating as in the ocean-going fleet. Alternatively, it could be that the greater frequency of completing a voyage in the coastal trade made desertion less necessary. Either way the seamen had less need to this last resort to express displeasure at working conditions.

V

This study has looked at some aspects of labour conditions in the British coasting trade, and in particular the screw steam colliers plying the east coast. It must be stressed that its findings need to be regarded as tentative until further research is carried out in greater depth, and it has been possible to look at only some aspects. That said, weekly time rates were normal in the collier fleet, and there is some evidence to suggest that changes in technology led to changes in payment systems which gave the shipowner a larger share of the rewards gained from improvements in productivity. There is no evidence to support the hypothesis that the coastal trade was seen solely as a training ground for young men wishing to move on to deep-sea routes or for older men seeking an easy post. It appears to have recruited people because of the advantages of short-duration voyages over long ocean trips.

The colliers were among the most "capitalistic" and "industrialized" of the coasters, but wage rates were better than on deep-sea ships, and employment was regular, with individuals taking time off and returning to the same post on the same ship. Promotion prospects did occur, though for only a minority of men. There was a surprising regularity and continuity of employment, and desertion rates were low by deep-water standards. Although the coastal fleet had a much smaller proportion of foreign crew members than the overseas fleet, the screw colliers had a proportion of foreigners in their crews more in line, both in proportion and nationalities, with the deep-water fleet. This suggests they were an attractive sector of the coastal fleet in terms of working conditions. The coastal steam colliers, by offering less arduous and dangerous employment conditions than coastal sailing ships, and better wages than deep-water sailing ships, provided a regular, reasonably paid occupation that could be pursued as a career into later life.

Chapter 9
Late Nineteenth-Century Freight Rates Revisited: Some Evidence from the British Coastal Coal Trade[1]

The study of maritime freight rates has a substantial and honourable ancestry. Indeed, its honour was recently much enhanced by the award of the Nobel Prize in Economics to one of the subject's progenitors. As long ago as 1914 C.K. Hobson drew up "a freight index number" from 1870 to 1912 to calculate the contribution of shipping earnings to Britain's balance of payments.[2] Isserlis in 1938 developed an index of tramp shipping freights from 1869 to 1914.[3] In the early 1950s Alec Cairncross drew on the work of both Hobson and Isserlis to calculate three indices of freight rates – inward, outward and composite.[4] The early interest in freights arose from a wish to calculate foreign exchange earnings from shipping as one component of Britain's invisibles, but later work has laid emphasis upon the theoretical benefits a reduction in ocean freight rates is likely to have had on international trade and hence regional specialization and economic growth. A lowering of freight rates is a social saving and could be calculated as a proportion of national income, as another Nobel Laureate, Robert Fogel, did with railway rates in America.[5] In addition, assuming reasonably competitive markets, if freight rates fell they would likely have caused a commensurate drop in the delivered price of the commodity carried, which in turn should have spurred demand (assuming a degree of price elasticity). Hence, output would be expanded, perhaps extracting economies of

---

[1]This essay appeared originally in the *International Journal of Maritime History*, V, No. 2 (1994), 45-81.

[2]C.K. Hobson, *The Export of Capital* (London, 1914; reprint, New York, 1983).

[3]L. Isserlis, "Tramp Shipping Cargoes and Freights," *Journal of the Royal Statistical Society*, CI, Part 1 (1938), 53-134.

[4]Alec K. Cairncross, *Home and Foreign Investment 1870-1913: Studies in Capital Accumulation* (Cambridge, 1953; reprint, Clifton, NJ, 1975).

[5]Robert W. Fogel, *Railroads and American Economic Growth: Essays in Econometric History* (Baltimore, 1964; reprint, Baltimore, 1993).

scale, adopting new technology or improving methods of organization – and thus reducing unit cost. This may then have led to another beneficial spiral.

Similarly, a reduction in transport costs may have opened some markets previously closed to a particular good because the final price, including the freight charge, was too high to compete. These new markets would have created additional demand, again perhaps stimulating improved methods of production or extraction and hence lowering unit costs, and thus into the same beneficent loop.[6] In addition, reductions in transport costs might have allowed local monopolies to be breached, bringing competition to rigid markets and hence reducing prices. It has been postulated that one important contributory factor in opening the American west to grain production was the relatively inexpensive transport provided by railways and steamships which enabled Midwest wheat to penetrate European, and especially English, markets.[7] This was a classic effect of a long-term reduction in transport costs, part of which was transatlantic freights.

This is not to suggest, of course, that the price of transport was necessarily its only or indeed its most important feature. There were a whole host of other non-price characteristics which were crucial, such as speed, load sizes, reliability, regularity and quality of service. But for low-value, high-bulk commodities needing no special transport provision – in short, for the greatest volume of goods moving long distances, such as grain, coal, ores, raw cotton or timber – price *was* the most important consideration because low-value goods are less able to stand high transport costs, as the freight rate is a larger proportion of total costs. Any increase in rates thus will have a proportionately larger impact on the final price of low-value goods.

The modern study of freight rates dates from Douglass North's seminal 1958 article in which he discovered two periods of rapid decline in ocean freights, 1815-1851 and 1870-1909.[8] He argued that this was more a function of the increased utilization of shipping – a result of port improvements; the ending of restrictions, such as the Navigation Laws; and improved knowledge of winds and currents, which reduced voyage times – than an outcome of technological change, such as the shift from sail to steam. Thirty years later Knick Harley challenged this view, suggesting that North's pre-1850 rates depended

[6]Rick Szostak, *The Role of Transportation in the Industrial Revolution: A Comparison of England and France* (Montréal, 1991), chapter 1, develops a very full model of the interrelationship between transport improvements and economic growth.

[7]Cormac Ó Gráda, "Agricultural Decline 1860-1914," in Roderick Floud and Donald N. McCloskey (eds.), *The Economic History of Britain since 1700* (2 vols., Cambridge, 1981), II, 196.

[8]Douglass C. North, "Ocean Freight Rates and Economic Development, 1750-1913," *Journal of Economic History*, XVIII, No. 4 (1958), 537-555.

heavily on cotton freights from North America which fell as a result of improved methods of compressing raw cotton so that a given weight occupied less space in a ship's hold. This decline was therefore a function of improved baling technology.[9] Harley also asserted that British freight rates fell more sharply after 1850 and that this was a function of technical change, such as the introduction and improvements in metal hulls and steam propulsion. More recently, Harley has stressed that for the transatlantic trades a single index is inadequate.[10] He argues that patterns of trade were complex and that a range of rates are needed to match this diversity. If this is true of one route, how much more relevant is it when considering a diversity of routes or when using the movement in overseas freight rates as a surrogate for what was happening in coastal waters?

To date, virtually all attempts to build freight rate indices have been for deep-water routes. For instance, North's pioneering work on ocean freight rates used timber from Québec to London, wheat from a number of sources to London and an index of British import freight rates and American exports. Harley, on the other hand, has looked at the rates for wheat and then cotton from the US to Liverpool, while Lewis Fischer and Helge Nordvik have devised an all-inclusive set of North Atlantic freights based upon the actual recorded earnings generated by all transatlantic cargoes carried by the Norwegian merchant marine.[11]

Coastal freight rates can have an impact similar to those for the deep-water fleet, although their impact is likely to be more confined geographically. But inter-modal competition becomes an additional variable in the coastal trades, since there may be more than one form of transport capable of carrying a given commodity. Hence, the coaster may lose or gain traffic to rail, canal or road transport. W.J. Hausman recently used the inverse of the reduction in coal freight rates from Newcastle to London over the long period 1691-1910 as

---

[9]C. Knick Harley, "Ocean Freight Rates and Productivity, 1740-1913: The Primacy of Mechanical Invention Re-affirmed," *Journal of Economic History*, XLVIII, No. 4 (1988), 856-860.

[10]C. Knick Harley, "North Atlantic Shipping in the Late Nineteenth Century: Freight Rates and the Interrelationship of Cargoes," in Lewis R. Fischer and Helge W. Nordvik (eds.), *Shipping and Trade, 1750-1950: Essays in International Maritime Economic History* (Pontefract, 1990), 147-171.

[11]Lewis R. Fischer and Helge W. Nordvik, "Maritime Transport and the Integration of the North Atlantic Economy, 1850-1914," in Wolfram Fischer, R. Marvin McInnis and Jürgen Schneider (eds.), *The Emergence of a World Economy, 1500-1914* (2 vols., Weisbaden, 1986), II, 519-544.

a measure of productivity increases in the coal trade.[12] Unfortunately, his freight rates were not based on contemporary recorded observations but rather calculated from the difference between coal prices in London and Newcastle, the disparity being assumed to be the freight rate. Simon Ville has questioned the theoretical validity of this assumption, and it has also been suggested that Hausman's calculated rates do not conform to recorded freight rates for the last twenty years of his period.[13]

Hausman generously seems to have accepted this and has suggested that his series reflected "shipping costs" rather than freight rates. If this is correct – and given the steeper fall from 1870 in freight rates than in Hausman's series it appears to be – it suggests diminished profitability in the coal trade.[14] This could be investigated empirically to see if firms like Cory or Charrington earned lower profits on their shipping activities. Regardless, it is certainly consistent with what has been dubbed "the great depression," even though that characterization has been much under attack since 1969 when S.B. Saul described it as a "myth."[15] It is also consistent with the huge merger among London's coal merchants and shipowners in 1896 when eight separate firms amalgamated as William Cory and Son. The new company owned thirty steam colliers and handled nearly seventy percent of seaborne coal imports to London.[16] Similarly, the 1901 merger of Fenwick, Stobart with William France and H.C. Pelly to control about sixteen colliers might have been a response to falling profits.[17] An investigation of collier costs might shed light on this.

---

[12]William J. Hausman, "The English Coastal Coal Trade 1691-1910: How Rapid Was Productivity Growth?" *Economic History Review*, 2nd ser., XL, No. 4 (1987), 588-596.

[13]Simon P. Ville, "Defending Productivity Growth in the English Coal Trade during the Eighteenth and Nineteenth Centuries," *Economic History Review*, 2nd ser., XL, No. 4 (1987), 597-602; and John Armstrong, "The English Coastal Coal Trade, 1890-1910: Why Calculate Figures When You Can Collect Them?" *Economic History Review*, 2nd ser., XLVI, No. 3 (1993), 607-609.

[14]William J. Hausman, "Freight Rates and Shipping Costs in the English Coastal Coal Trade: A Reply," *Economic History Review*, 2nd ser., XLVI, No. 3 (1993), 610-611.

[15]S.B. Saul, *The Myth of the Great Depression, 1873-1896* (London, 1969; 2nd ed., London, 1985).

[16]Raymond Smith, *Sea-Coal for London: History of the Coal Factors in the London Market* (London, 1961).

[17]J.A. MacRae and Charles V. Waine, *The Steam Collier Fleets* (Wolverhampton, 1990), 57.

P.J. Cain, trying to explain trends in the output and productivity of British railways in the late nineteenth century, cited increased competition from coastal vessels as a key feature in the environment.[18] He traced the coaster's recapture of the bulk of the coal trade to London from the North East in the late 1890s and explained this largely as a function of falling freight rates, for which he cited the Isserlis index. Yet a careful examination of the more than 300 individual indices for different routes from which Isserlis constructed his overall figures shows that only one was coastal – Tyne-London with coal – and even for this series the figures only commence in 1901. In other words, Isserlis presents coastal freight rates for only just over a decade on a single route.[19] There must therefore be some doubt as to how far his overall index really reflects movements in coastal freights. If coastal rates moved in sympathy with the deep-water, then the Isserlis index is an acceptable measure. But this is an untested assumption, and it would be better to test it against a separately constructed index of coastal rates.

In a recent article Harley calculated a new index of freight rates for the period 1740-1920.[20] One of the sub-series was "freight rates on Tyne coal shipped to London...from 1741 until the First World War." But from his table 9 it is clear that his coal series in fact stops in 1872 and hence ignores the last thirty years of the nineteenth century. His appendix on "sources" confirms that the last date for which he extracted coastal coal rates was 1870, leaving uncertain the provenance of the two figures in table 9 for 1871 and 1872. A check of the source cited for these freights, the *Newcastle Courant,* showed that the series he used petered out in September 1870. Hence, his index of coastal freights in this article stops during 1870 and is not valid for 1871 or 1872.

Harley made this omission good in another article in which he provided a time series of coal freights from Newcastle to London between 1838 and 1913.[21] But it is not clear how large his sample was or the number of observations from which he calculated his yearly averages. Indeed, it appears that he may have based his series on as few as six observations per year. This causes some concern, especially for periods when the market was somewhat volatile, such as during wars or political crises. For this reason I decided to construct an index of coastal coal freights for the period 1870-1913 using a

---

[18]Peter J. Cain, "Private Enterprise or Public Utility? Output, Pricing and Investment on English and Welsh Railways, 1870-1914," *Journal of Transport History,* 3rd ser., I, No. 1 (1980), 13-16.

[19]Isserlis, "Tramp Shipping Cargoes," 106.

[20]Harley, "Ocean Freight Rates," 851-875.

[21]C. Knick Harley, "Coal Exports and British Shipping, 1850-1913," *Explorations in Economic History,* XXVI, No. 3 (1989), 311-338.

larger number of observations and more than one route and to compare this with Harley's series. Moreover, I wanted to examine the relationship between this and the existing deep-water indices with which Harley was not concerned. This would indicate the degree of congruence between deep-water and coastal rates. It was proposed to test the implications of this trend as well as to make some attempt to look at the causes of changes in freight rates.

Methodology

Freight rates were determined not only by the volume of cargo or distance but also by the nature, value and problems of loading and discharging a specific cargo, as well as the cost and difficulty of entering a port or harbour, the propulsion of the vessel, the time of year and other variables. It is thus imperative to ensure that the index is constructed from compatible data, collected for the same commodity on exactly the same route and preferably for the same type of ship and time of year. This made it necessary to search a number of sources.

The single largest commodity in the British coastal trade in the late nineteenth and early twentieth centuries was coal. In 1870 about eleven million tons were moved by ship, a figure that rose to nearly eighteen million tons in 1900 and to just under twenty-three million tons by 1913.[22] The single largest recipient was London, which absorbed about three million tons in 1870, nearly eight million in 1900 and just over nine million in 1913.[23] Most coal sent to London came from two areas, the North East of England and South Wales. Thus, indices for coal from the Tyne and Cardiff to London would encompass two of the largest single trades in coastal shipping at this time. Given the size of the trade there are likely to be large numbers of examples of these rates being fixed. Moreover, because of the volume of the flows and competition, these indices are likely to move in the same direction as other coastal rates and hence to indicate the trend in coastal freights generally. It was decided to construct two indices, one for coal from the North East to the Thames and another from Cardiff to London for the period 1870-1913. The former also has the benefit of complementing both Harley's index and Hausman's longer series of calculated rates.

The simple but tedious methodology was to look for local newspapers containing reports on coal freights on the two target routes. Unfortunately, no single paper was found that covered the entire period. As a result, the index

---

[22]Great Britain, Parliament, House of Commons, *Parliamentary Papers* (*BPP*), "Annual Report on Mines and Quarries," part 3, 1871, LXII, 7; part 2, 1902, CXVI, 377; and 1914-1916, LXXX, 537.

[23]B.R. Mitchell with Phyllis Deane (comps.), *Abstract of British Historical Statistics* (Cambridge, 1962; reprint, Cambridge, 1976), 113; and *Coal Merchant and Shipper,* 1 February 1913 and 3 April 1915.

had to be built up from a number of sources (see appendix 1).[24] Some sources overlapped and hence could be used as checks on other data; there was a high degree of consistency where such audits were possible. For example, for the late 1870s rates were recorded in both the *ICTReview* and the *SMG* for vessels carrying coal from the North East to London. Hence, for 1877-1879 the results from these two sources can be compared (see table 1). It can be seen from this table that the degree of agreement is great and that there was no consistent bias. The difference is less than three percent in most cases, and the greatest disagreement is ten percent. This reinforces the impression that the number of rates reported in the newspapers is sufficiently large to comprise a representative sample of those actually fixed and to boost confidence in the average rates calculated.

Table 1
Comparison of Average Freight Rates for
Coal, Shields-Thames, 1877-1879

	SMG		*ICTR*	
	No. Observations	Rate	No. Observations	Rate
1877				
Winter Steam	5	4s 5½d	31	4s 6½d
Winter Sail	17	6s 2d	12	6s 2d
Summer Steam	3	4s 7½d	10	4s 6¼d
Summer Sail	6	6s	3	6s 8d
1878				
Winter Steam	14	4s 10½d	23	4s 7d
Winter Sail	24	5s 9¼d	21	6s 3¼d
Summer Steam	7	4s 2½d	15	4s 2¾
Summer Sail	11	5s 3½d	19	5s 1½d
1879				
Winter Steam	39	4s 4¾	26	4s 3¼d
Winter Sail	39	5s 10½d	11	6s
Summer Steam	28	4s 1d	8	4s 1¼d
Summer Sail	29	4s 10d		

Sources: Shipping and Mercantile Gazette (SMG), 1877-1879; and Iron and Coal Trades Review (ICTR), 1877-1879.

The results for the two routes, Tyne-Thames and Cardiff-London, are depicted in tables 2 and 3, respectively. Where possible the rates were divided into summer and winter sectors (summer was taken as April-September) and sail and steam. The number of annual observations is in parentheses to provide

---

[24]I gratefully acknowledge the cheerful assistance of Julie Stevenson in the task of extracting the data and thank Robin Craig for suggesting a number of possible sources.

some indication of the robustness of the average rate. There was a great range in the number of observations, but there were rarely fewer than ten for any year. In addition, some of the recorded figures were themselves averages of a range of freights (the precise number now unknown), so the means are more reliable than they at first might appear. Moreover, the number of observations was always greater than what Harley considered adequate in his coal index for the period 1838-1870. Finally, the reliability of the means is enhanced by the low variance in annual rates. For instance, in 1894 the range for coal from Cardiff to London was 3s 9d-4s 6d in winter and 3s 9d-4s in summer. The range rarely exceeded one old shilling.

Table 2
Freight Rates on Coal, Shields-London, 1870-1913 (per ton)

| | Winter | | Summer | | Annual | | |
	Sail	Steam	Sail	Steam	Sail	Steam	All
1870	-	6s 1¼d (7)	6s 3d (8)	5s 3½d (7)	6s 3d (8)	5s 8½d (14)	5s 10d (54)
1871	5s 11½d (64)	4s 11½d (11)	5s 6¾d (4)	5s 1d (7)	5s 9½d (10)	5s (18)	5s 3½d (28)
1872	6s 9½d (5)	5s 2d (4)	8s 3d (2)	5s (1)	7s 2½d (7)	5s 1½d (5)	6s 4d (12)
1873	7s 7½d (12)	-	6s 7½d (5)		7s 4d (17)		
1874		7s 9d (5)		6s 1½d (10)			6s 8d (15)
1875	6s 9¾d (10)	6s 4d (4)					6s 8d (14)
1876	6s 5½d (16)	4s 10d (29)	5s 5d (4)	4s 7½d (26)	6s 3d (20)	4s 8¾d (55)	5s 1½d (75)
1877	6s 2d (30)	4s 6¼d (36)	6s 2¾d (9)	4s 6½d (13)	6s 2¼d (39)	4s 6¼d (49)	5s 3½d (88)
1878	6s 0¼d (45)	4s 8¼d (37)	5s 2¼d (30)	4s 2¾d (22)	5s 8¼d (75)	4s 6¼d (59)	5s 2d (134)
1879	5s 10¾d (50)	4s 4¼d (65)	4s 10d (29)	4s 1d (36)	5s 6d (79)	4s 3d (101)	4s 9½d (180)
1880	5s 6d (22)	4s 3¼d (46)	4s 4¾d (19)	3s 11d (19)	5s (41)	4s 2d (65)	4s 5¾d (106)
1881	6s 4¼d (21)	4s 6½d (53)	4s 10½d (11)	3s 11½d (44)	5s 10d (32)	4s 3¼d (97)	4s 8d (129)
1882	5s 9d (4)	4s 3d (29)	5s 3½d (13)	4s	5s 4¾d (17)	4s 1¾d (52)	4s 5½d (69)
1883	5s 6d (7)	3s 11d (34)	5s 6d (4)	3s 11d (25)	5s 6d (11)	3s 11d (59)	4s 2d (70)
1884	5s 2d (11)	3s 7½d (32)	4s 8d (3)	3s 8½d (22)	5s 0¾d (14)	3s 9d (54)	4s 0¼d (68)
1885	5s (9)	3s 8¾d (34)	4s 6¾d (4)	3s 8d (27)	4s 10½d (13)	3s 8½d (61)	3s 11d (74)
1886	4s 11d (3)	3s 8¼d (14)	-	-	4s 11d (3)	3s 8¼d (14)	3s 10¾d (17)
1887							
1888							

	Winter		Summer		Annual		
	Sail	Steam	Sail	Steam	Sail	Steam	All
1889							3s 9d
1890							3s 9¼d
1891							3s 8d
1892							3s 5¾d
1893							3s 7¾d
1894							3s 3¾d
1895							3s 3¾d
1896							3s 4¾d
1897	3s 4d (48)						3s 4¼d
1898	3s 3d (24)						3s 7d
1899							3s 6½d
1900							4s 1¼d
1901							3s 3d
1902							3s 3d
1903							3s 2d
1904							3s 1d
1905							3s 3d
1906							3s 3d
1907							3s 4d
1908							2s 11d
1909							2s 10d
1910							2s 11d
1911							3s 4d
1912							4s 1½d
1913							3s 4½d

Note: Figures in parenthesis indicate the number of observations

Sources: See appendix 1.

Table 2 also demonstrates that the sailing ship earned consistently higher freights than the steamship. The differential varied by year and season, but for those years where both a sail and steam rate were found it averaged 1s 4½d in winter and 1s 2d in summer. The differential was remarkably stable and showed no sign of diminishing. Given that collier brigs carried the same cargo on the same route as the screw colliers, why did they earn more per ton? At first this seems illogical, for the service they offered was much inferior in that it was less reliable and slower than the steam-propelled ship. Moreover, from the shipowner's perspective sail represented less sunk capital than steam; the capital costs thus should have been lower, and there were no fuel costs. Hence, the sailing collier might be expected to charge a lower price.

Table 3
Freight Rates on Coal, Cardiff-London, 1870-1913 (per ton)

	Winter	Summer	Annual
1870			
1871	9s (9)		
1872			
1873			
1874	6s 8d (3)	6s 6d (2)	6s 7d (5)
1875			
1876			
1877			
1878			
1879			
1880	4s 9d (5)	4s 6½d (6)	4s 7¾d (11)
1881	5s 1½d		
1882			
1883	4s 11d (8)	4s 9¾d (11)	4s 10¼d (19)
1884	4s 5d (56)	4s 2½d (27)	4s 4d (83)
1885	4s 5d (63)	4s 3d (44)	4s 4¼d (107)
1886	4s 2¼d (50)	4s (59)	4s 1d (109)
1887	4s 6¼d (49)	4s 2d (33)	4s 4½d (82)
1888	4s 6d (37)	4s 5¼d (34)	4s 5½d (71)
1889			
1890	4s 11d (14)	4s (1)	4s 10¼d (15)
1891	4s 5½d (50)	4s 1¼d (36)	4s 3¾d (86)
1892	4s 5d (100)	4s (53)	4s 3d (153)
1893	4s 9d (52)	4s 3½d (42)	4s 6½d (94)
1894	4s 1d (73)	3s 10d (26)	4s (99)
1895	4s 1d (37)	3s 10d (47)	3s 11½d (84)
1896	4s 9d (3)		
1897	4s 2¾d (16)	3s 9d (1)	4s 2¼d (17)
1898	3s 11d (3)		
1899	4s 2d (5)		
1900			4s 9d
1901			4s 7d
1902			4s 4d
1903			4s 2d
1904			4s 1d
1905			4s 4d
1906			4s 4d
1907			4s 6d
1908			4s 6d
1909			4s 3d
1910			4s 1d
1911			5s 4d
1912			4s 9d
1913			5s 0d

Note: See table 2.

Source: See table 2.

There are three possible hypotheses which might explain this phenomenon, none of which are mutually exclusive. One is that the extra cost was an implicit demurrage charge. The sailing ship normally had a lower priority than the steamer in docking because the latter represented so much fixed capital. A steamer thus almost always had precedence at a berth; in some cases, even if on a berth and half unloaded, the sailing vessel was moved to allow a steamer access. The second possibility is that the differential compensated for the costs of unloading. There is some evidence that in sailing ships loading and unloading cargoes were part of the crew's duties. Such manpower would have been less necessary in loading, as hoists and drops were available, but mechanical unloading was less likely to be accessible to wooden-hulled, small-hatched sailing ships. Hence, the arduous and time-consuming job of discharging cargo fell to the crew members. This added to the turnaround time, increasing the demurrage costs already discussed, and the extra freight was to compensate the crew for their effort and perhaps to hire a few dockside labourers to assist.

The third possible explanation lies in the superior condition in which the coal arrived at its destination on sailing vessels. It was argued by contemporaries that engine vibrations in mechanically-propelled ships caused the coal to grind and rub, producing a larger proportion of dust and small coal unsuitable for use. In addition, screw propulsion was claimed to aggravate the pitching of the ship, which caused the cargo to shift and hence break up. The first of these features was absent in sail, while the second was less pronounced. It was for similar reasons that sailing ships were preferred, even into the twentieth century, for bricks, slates, tiles and fire clay goods which were liable to chip or crack. It should be stressed that none of these hypotheses has yet been tested empirically. This must await further research.

One drawback, however, is that as yet it has not been possible to construct a complete series for South Wales' freights. Despite an assiduous search of a range of Welsh, London and even Bristol papers, thus far only thirty-two of the forty-four years have yielded enough observations to provide a reasonable level of confidence in the rates. Most of the gaps are for the 1870s. Hopefully, further search will lead to the discovery of an as yet untapped cache of high-quality freight rates.

Results

The money freight rates in table 2 for coal from the North East to London seem to fit reasonably well with those in Harley's two series. Those in the 1988 article overlap by only three years, 1870-1872. In all three cases Harley's figures are above my calculations, albeit by relatively small amounts and in a consistent direction. Turning to the longer series in his 1989 article, one anomaly is obvious. The rate for 1872 is nearly nineteen percent higher than in Harley's earlier article, although no explanation is offered.

Table 4
Average Coastal Freight Rates on Coal,
Tyne-Thames, 1870-1913
(Harley's 1989 Series Compared with Table 2) (Shillings)

	Harley	Table 2
1870-1874	6.5	6.3
1875-1879	5.4	5.4
1880-1884	4.5	4.35
1885-1889	4.1	3.85
1890-1894	3.6	3.6
1895-1899	3.4	3.4
1900-1904	3.5	3.4
1905-1909	3.1	3.1
1910-1913	3.45	3.4

Sources: C. Knick Harley, "Coal Exports and British Shipping, 1850-1913," *Explorations in Economic History,* XXVI (1989), 334-336, table 2.

The concord between Harley's 1989 series and my data in table 2 is very strong. Although there are variations for individual years, for all except three the difference is less than ten percent; the level of agreement is even greater after 1880. The decade with the greatest variation between the two series is the 1870s, but they are more similar thereafter. The average difference for the 1870s is nine percent, while for the 1900s it is just over two percent. The correlation between the two unaveraged series is very strong ($r = .9362$ and $R^2 = 88$). This provides statistical confirmation of the close association between the two series. If the rates are averaged by quinquennia (table 4), they remain close, and the trends are virtually identical. Given that there were slight variations in rates depending upon the actual place of loading and discharge, and that in some years they exhibited particular short-term volatility, the degree of concordance is a ringing endorsement of Harley's work. These two series can be considered a fairly reliable guide to the movement of coastal coal freights for the east coast in the late nineteenth century.

Given the degree of agreement, the series presented here also reinforces the scepticism about Hausman's rates. Since he has accepted this elsewhere, it does not need to he laboured. It is sufficient to say that his figures are definitely not freight rates. Whether they represent "shipping costs," as he claims, remains to be proven.

The Cardiff-London freight rates in table 3 confirm the general trends. The individual rates for the run from South Wales were greater than those from the Tyne in all cases. This is not surprising given that South Wales is nearly twice the distance from London as the Tyne (596 versus 312 miles). When the Cardiff rates are averaged over the same quinquennia as the Tyne series (see table 5), the movements are similar, thus reinforcing confidence in the methodology and results.

Table 5
Coastal Freight Rates on Coal, 1870-1913,
Tyne-Thames Compared with Cardiff-London

	Tyne (shillings)	South Wales (shillings)	Difference (percent)
1870-1874	6.3	-	-
1875-1879	5.4	..	-
1880-1884	4.35	4.6	+6
1885-1889	3.85	4.3	+12
1890-1894	3.6	4.4	+22
1895-1899	3.4	4.2	+23
1900-1904	3.4	4.4	+29
1905-1909	3.1	4.3	+39
1910-1913	3.4	4.8	+41

Source: See table 2.

Two important points arise from a comparison of the two series. First, since the distance from South Wales is roughly twice as far as from Newcastle – but the freight rates for the Welsh run are only slightly higher than those for the east coast route – the freight per ton-mile was much less from Cardiff than from Newcastle. Given the disparity in distance, this suggests that there were significant economies in long journeys. A high proportion of a coaster's costs were incurred at the beginning and end of a voyage in the form of harbour dues, berth fees, charges for loading and unloading, pilotage and the like, all of which were levied irrespective of distance travelled and hence were lower per ton-mile for longer trips. This phenomenon – the economy of the long haul – was well known on the railways and operated to an even greater extent for the coastal ship.

The second point derived from a comparison between the Geordie and Welsh series is that over time the differential increased. As table 5 shows, Welsh rates were on average six percent higher than from Newcastle in the early 1880s but were twenty-three percent greater in the late 1890s and forty percent above in the years just before the Great War. This was a result of Welsh rates falling less rapidly than the English. For example, the reduction in Welsh rates between 1880/1884 and 1905/1909 was only six percent, whereas on the east coast rates declined by twenty-eight percent. I would suggest tentatively that this is both a function of the nature of the product and of railway

competition.[25] Welsh coal was especially prized for its steam-generating prop-
erties and commanded a higher price than that produced in the North East. The
demand for Welsh coal was buoyant before World War I and might explain the
lesser decline in Welsh rates. In addition, the railway companies operating
from the Newcastle region, the NER and GNR, took coal transport more seriously than the GWR, which ran from South Wales to London. As a result, the
east coast rail lines provided better facilities and hence a more realistic alternative to the coaster.

Table 6
Index of Coastal Freight Rates Compared to Deep-water, 1870-1913

	Coastal Rates	USA Export Rates	New York-Liverpool Grain	Cork for Orders	New York-Liverpool Flour
1870-1874	100	100	100	100	100
1875-1879	85.8	92.3	103	85	111
1880-1884	69.2	82.5	59	63	67
1885-1889	61.3	57.4	39	52	45
1890-1894	56.9	56.5	38	45	41
1895-1899	54.6	56.9	37	47	46
1900-1904	53.6	41.8	22	37	34
1905-1909	49.7	38.2	21	34	32
1910-1913	54.6	-	33	-	49

Note: 1870-1874 = 100 for all series.

Sources: Column 1: see appendix 1; Column 2: Douglass C. North, "Ocean Freight
             Rates and Economic Development 1750-1913," *Journal of Economic His-
             tory,* XVIII, No. 4 (1958), 549; Columns 3-5: C. Knick Harley, "North
             Atlantic Shipping in the Late-Nineteenth Century: Freight Rates and the In-
             ter-Relationship of Cargoes," in Lewis R. Fischer and Helge W. Nordvik
             (eds.), *Shipping and Trade, 1750-1950: Essays in International Maritime
             Economic History* (Pontefract, 1990), 167.

        The index created here of coastal coal freight rates can be compared
with previously calculated deep-water freights. Table 6 compares the coastal
index per quinquennium with one of North's series and three of Harley's. The
four overseas series exhibit broadly similar patterns. Overseas rates fell less
steeply than coastal levels in the first quinquennium, but most fell so sharply in
the early 1880s that they dipped below coastal rates. After 1885 ocean freight
rates were consistently lower than coastal (with only one exception), and the

[25]For a more detailed explanation, see John Armstrong, "The Shipping Depression of 1901 to 1911: The Experience of Freight Rates in the British Coastal Coal
Trade," *Maritime Wales,* No. 14 (1991), 99-100.

recovery in the last years before the First World War was less marked. Not surprisingly, the series for American export freights fell least of the oceanic rates, since it covers all exports while the other three are for low-value, bulky, non-perishable commodities and thus similar to coal.

Table 7
Freight Factors on Coal Delivered in London

	1 Coastal Freight Rates	2 Cost of Coal in London	3 Col. 1/ Col. 2 x 100	4 Col. 3 as Index
1870-1874	6s 3½d	24s	26.2	100.0
1875-1879	5s 4¾d	20s 2d	26.8	102.3
1880-1884	4s 4¼d	16s 10d	25.9	98.8
1885-1889	3s 10¼d	16s 6d	23.4	87.8
1890-1894	3s 7d	18s 6d	19.4	74.0
1895-1899	3s 5¼d	16s 2d	21.3	81.3
1900-1904	3s 4½d	18s 11d	17.8	67.9
1905-1909	3s 1½d	17s 5d	17.9	68.3
1910-1913	3s 5¼d	19s 6d	17.6	67.2

Sources: Column 1: see appendix 1; Column 2: B.R. Mitchell with Phyllis Deane (comps.), *Abstract of British Historical Statistics* (Cambridge, 1962; reprint, Cambridge, 1976), 483.

The other tables that North provided are not of crude rates but rather of "freight factors" (freight rates adjusted for changes in the value of the commodity carried and expressed as a proportion of total delivered cost).[26] These provide a measure of the proportion of total delivered cost contributed by transport and hence the potential importance of any reduction. While I am not aware of any freight factors having been calculated for the coastal trade, this ratio can be determined easily since a series of coal prices delivered in London already exists for this period.[27] Table 7 gives the freight factor on coal delivered to London from the North East. When column 3 of table 7 is compared to North's freight factors on wheat delivered in London (see table 8), it is notable that the freight factor for coastal coal was much higher, starting at around twenty-five percent and falling only to about eighteen. On the other hand, the freight factors for wheat range between six and fifteen percent in the opening period and drop to between about three and six and one-half percent by the end of the period. In addition, the index numbers show that there was a much greater proportionate fall in the freight factor on overseas wheat than on coastal coal: the former dropped to between one-third and one-half of the lev-

---

[26]North, "Ocean Freight Rates," 542.

[27]Mitchell with Deane (comps.), *Abstract*, 483.

els of the early 1870s by the first decade of the twentieth century, while the latter fell by less than one-third.

Table 8
Freight Factors on Wheat Delivered in London

	Freight Factors			Index Numbers		
	Baltic	Black Sea	East Coast America	Baltic	Black Sea	East Coast America
1870-1874	6.1	14.3	9.8	100.0	100.0	100.0
1875-1879	5.3	12.5	10.5	86.9	87.4	107.1
1880-1884	4.5	9.5	6.7	73.8	66.4	68.4
1885-1889	4.7	10.7	6.5	77.0	74.8	66.3
1890-1894	3.9	9.8	5.6	63.9	68.5	57.1
1895-1899	4.8	8.9	6.8	78.7	62.2	69.4
1900-1904	4.1	8.3	4.2	67.2	58.0	42.9
1905-1909	3.3	6.2	3.4	54.1	43.4	34.7
1910-1913	3.4	6.4	4.5	55.7	44.8	45.9

Note: 1870-1874 = 100 for all series.

Source: North, "Ocean Freight Rates," 551-552.

In trying to explain the lower freight factors on wheat it is tempting to seize on voyage length as a key variable. The coastal coal voyage from the North East to London was just over 300 nautical miles, while all the deep-water routes were several times longer. It might then be argued that whatever technical innovations, such as larger ships, more efficient engines and navigational aids, were responsible for reducing marine transport costs, they were likely to apply with greater force on longer voyages. This was reinforced by the cost structure of shipping, in which a large portion of operating costs was incurred as terminal charges. But this explanation is less convincing considering that for North's overseas routes, the longest (Black Sea-London) had the highest freight factor while the shortest (Baltic-London) had the lowest.[28]

A more fruitful line of explanation may lie in the nature of competition on the various routes. The coal trade from the North East was increasingly operated neither by tramps nor liners but rather by regular traders, ships that did not run to a regular timetable nor seek cargoes in a range of trades. Many were dedicated to coal and shuttled back and forth along the coast, unloading coal in London as quickly as possible and returning north in water ballast, not even seeking a back cargo.

Evidence for the nature of the trade and the presence of these regular traders can be found in some freight reports in which the name of the ship on

---

[28]North, "Ocean Freight Rates," 551-552.

which the freight rate was fixed is given; the same names reappear regularly.[29] Also, an examination of the "Agreements and Accounts of Crew," lodged in various repositories from Greenwich to Newfoundland, lends support to the existence of regular traders making repeated return trips between the Tyne and the Thames, often running to thirty or forty circuits per year. Some of these colliers were specially constructed with low superstructures and shallow draughts to allow passage under London's bridges.[30] Others had large hatches and holds to allow easy mechanical loading and unloading. In short, these craft were dedicated to a single trade and were ill-suited for flexible deployment, although many could of course carry coal to other ports, both domestic and short-sea. The colliers earned their revenue entirely on the carriage of coal without any return revenues; the overseas traders, on the other hand, also generated revenue from outward cargoes from the UK to, say, grain-despatching ports. The cargoes were usually manufactured or semi-manufactured goods which paid a higher rate per ton than the bulky, low-value inward cargoes. Thus, deep-water vessels earned revenue on both legs and often more on exports from the UK than on imports. Their operating costs could thus be spread over an entire voyage, while a collier had to cover all its costs on a single leg, carrying a low-value, bulky good. It is not surprising that the freight factors for colliers were higher than for imported grain.

The coastal collier was in a peculiarly advantageous position compared to railways. From the 1890s, the latter were reluctant to reduce posted charges because of the Railway and Canal Commission which examined (but seldom approved) all requests for rate increases, and because of the hostility of local traders and merchants, expressed through chambers of commerce, to increased railway charges.[31] Given this environment, railways were unwilling to cut rates even speculatively; since they would be unlikely to be able to reverse them, their charges therefore were static. The coaster's rates were competitive in the 1870s; as rates fell in both current and constant terms, coasters became even more attractive as coal carriers and weaned traffic away from railways.

The issue of freight factors can be pushed a bit further. It can be postulated that they will generally vary inversely with the value of the commodity carried – that is, freight factors will be high on low-value

---

[29]See, for example, *Daily Freight Register* (post-1897); and *Western Mail* (1890s).

[30]Smith, *Sea-Coal*, 335; and D. Ridley Chesterton and Roy S. Fenton, *Gas and Electricity Colliers: The Sea-Going Ships Owned by the British Gas and Electricity Industries* (Kendal, 1984), 6.

[31]Peter J. Cain, "The British Railway Rates Problem, 1894-1913," *Business History*, XX, No. 1 (1978), 87-99.

goods and low on high-value products. For example, the freight factor on culm (a low-quality coal dust) from South Wales to the Lleyn Peninsula in the 1860s was between twenty-eight and thirty-nine percent, while on lime it exceeded fifty percent.[32] By comparison, Gerhold has shown that in the early 1700s freights for serge cloth from Exeter to London by road added only two to four percent to the cost.[33] Despite covering about 150 miles and using one of the most costly forms of transport, the freight factor was low because the commodity's value was high. The implication is that a reduction in freights on low-value goods is likely to have a great impact on final prices and perhaps on demand.

Table 9
Isserlis's Tramp Index Compared to Coastal Coal

	Isserlis	Coastal Coal
1870-1874	100.0	100.0
1875-1879	88.8	85.8
1880-1884	73.9	69.2
1885-1889	63.4	61.3
1890-1894	56.3	56.9
1895-1899	56.5	54.6
1900-1904	52.5	53.6
1905-1909	46.7	49.7
1910-1913	59.6	54.6

Source: L. Isserlis, "Tramp Shipping Cargoes and Freights," *Journal of the Royal Statistical Society,* CI, Part 1 (1938), 53-134.

The other purpose of constructing an index of coastal freight rates was to compare it with Isserlis, which was used by Cain to measure the improved competitiveness of coasters relative to railways.[34] When the two are compared (see table 9) it is clear that although the coastal coal index declined slightly more steeply than Isserlis until the turn-of-the-century, the general trend and velocity of both are broadly similar. Thus, Cain was correct to use Isserlis, since it does reflect with some accuracy the movement of freights on the Tyne-Thames route.

---

[32]R.T. Morris, "David Rice Hughes – A Victorian Entrepreneur," *Maritime Wales,* No. 11 (1987), 136-137.

[33]Dorian Gerhold, *Road Transport before the Railways: Russell's London Flying Waggons* (Cambridge, 1993), 5.

[34]Cain, "Private Enterprise," 13-16.

Given the debate over the performance of the late Victorian economy, often characterized by phrases such as "great depression," "industrial retardation" and "entrepreneurial failure," it is worthwhile to see what an examination of freight rates can reveal. Certainly the trend in coastal rates reinforces the view of a steady downward slope in prices. The timing, however, does not quite fit the portrait of a "great depression" beginning in 1872 or 1873.[35] Coastal rates in fact peaked in 1873 and only began a steep decline in 1876, so any recession was slower to hit coastal trades. The coastal sector also began to recover earlier than the economy in general, for the nadir in coastal rates came in 1894-1895, several years before conventional wisdom posits an upturn. But the most striking feature of coastal rates is that their fall was much steeper than average prices and was more akin to the extremely sharp decline in primary product prices. Charles Feinstein has calculated a cost-of-living index for 1870-1914, using 1900 as the base year (see table 10).[36] In his index the trough came in 1896 when the cost-of-living was about twenty-four percent lower than in 1873. For coastal freights the reduction in the same period was more than fifty percent. Yet given that the nadir for coastal rates occurred in 1894-1895, the decline from the peak was in fact fifty-five percent, an even greater reduction than registered in Feinstein's index. Nor is this disproportionate fall a function of a careful choice of base years or a freak deviation in one or two years. The coastal freight rate index consistently shows greater reductions than Feinstein's index, irrespective of the dates chosen. For instance, if a quinquennial average is calculated, and 1870/1874 compared to 1895/1899, 1900/1904 or indeed 1910/1913, the coastal index shows a greater reduction than Feinstein's series. Similarly, if instead of using a cost-of-living index we use a price index for GDP at factor cost, the coastal series again falls faster and further than the price index (see table 10). If the price index for GDP is employed to calculate a "real" coastal freight index, it can be demonstrated that the real freight rate fell continuously from 1870/1874 to 1905/1909, by which time it was little over half its opening level.

Indeed, the pattern of coastal freight rates is quite different from the cost-of-living index. While the latter declined gently through the late 1890s and thereafter recovered to end up only slightly lower in 1910/1913 than in 1870/1874, coastal rates plunged much more steeply between 1875 and 1899 and, rather than recovering in the twentieth century, continued to fall, so that the lowest rate of the entire 1870-1913 era was recorded in 1909. This explains why James, Sturmey and Aldcroft, among others, have characterized

---

[35]Saul, *Myth*.

[36]Charles H. Feinstein, "A New Look at the Cost of Living, 1870-1914," in James Foreman-Peck (ed.), *New Perspectives on the Late Victorian Economy: Essays in Quantitative Economic History, 1860-1914* (Cambridge, 1991), 170-171.

1901-1911 as an acute shipping depression.[37] Not only did rates fall after the apex coinciding with the Boer War but they also fell below the "great depression." This experience confirms North's statement that "the decline in freight rates was substantially greater than the decline in prices." Although he was discussing deep-water freights for the whole of the nineteenth century, it is no less true for coastal coal rates in the period 1870-1913.[38]

Table 10
Price Indices and Nominal and Real Coastal Freight Rates, 1870-1913

	Cost of Living	Price Index for GDP	Coastal Freight Rates (money)	Real Coastal Freight Rates
1870-1874	100	100	100	100
1875-1879	95.6	93.5	85.8	91.8
1880-1884	90.8	89.7	69.2	77.1
1885-1889	83.5	84.3	61.3	72.7
1890-1894	83.1	85.9	65.9	66.2
1895-1899	80.5	84.2	54.6	64.8
1900-1904	84.7	90.0	53.6	59.5
1905-1909	86.7	90.9	49.7	54.7
1910-1913	89.8	94.4	54.6	57.8

Source: Column 1: Charles H. Feinstein, "A New Look at the Cost of Living, 1870-1914," in James Foreman-Peck (ed.), *New Perspectives on the Late Victorian Economy. Essays in Quantitative Economic History 1860-1914* (Cambridge, 1991), 170-171; Column 2: Charles H. Feinstein, *Statistical Tables of National Income, Expenditure and Output of the UK 1855-1965* (Cambridge, 1976), table 61, T-132.

The fact that coastal coal freight rates fell much faster than prices in general suggests that they were falling in real terms and hence contributed to the decline in prices rather than merely benefitting from the general reduction in the cost of raw materials and other inputs. In addition, coastal coal rates fell much more steeply than the price of coal delivered in London (see table 11). Indeed, comparing tables 10 and 11 shows that the movements of coal prices were more like the overall cost-of-living index than coastal freight rates. Coastal rates fell faster than both the price of the product carried and the over-

---

[37]Frank Cyril James, *Cyclical Fluctuations in the Shipping and Shipbuilding Industries* (Philadelphia, 1927; reprint, New York, 1985), 24; S.G. Sturmey, *British Shipping and World Competition* (London, 1962), 50; Derek H. Aldcroft, "The Depression in British Shipping, 1901-11," *Journal of Transport History*, 1st ser., VII, No. 1 (1965), 14-23, reprinted in Aldcroft, *Studies in British Transport History* (Newton Abbot, 1974).

[38]North, "Ocean Freight Rates," 542.

all cost-of-living. Reductions in coastal freight costs contributed to cheaper coal for London – and presumably all other ports receiving coal by sea – and hence to lower living costs. In this way coastal shipping made a positive contribution to economic growth and welfare. The continued evolution of urbanization and industrialization – dependent almost wholly on coal for heat, light and power – would likely have been retarded if the operating costs of coastal ships had not been reduced so drastically.

Table 11
Price of Coal in London, 1870-1913 (per ton)

	Price	Index
1870-1874	24s	100.0
1875-1879	20s 2d	84.0
1880-1884	16s 10d	70.1
1885-1889	16s 6d	68.7
1890-1894	18s 6d	77.1
1895-1899	16s 2d	67.4
1900-1904	18s 11d	78.8
1905-1909	17s 5d	72.6
1910-1913	19s 6d	81.3

Source: Mitchell with Deane (comps.), *Abstract,* 483.

What is now required is to explain how the coaster was able to increase its productivity and reduce its costs. The nature of the market for coal freights is one important consideration. There were large numbers of ships available, many of which were capable of flexible deployment in terms of ports. Moreover, many operated to the continent, if required. Given the large number of ships and their adaptable deployment, there were no monopoly profits available. Indeed, the market was very competitive, and this intense competition likely kept rates reasonably low, at least in the short term. But if this were true in the long-term, the coaster must have enjoyed some productivity improvements. Part of the explanation must thus rest in the area of cost-reducing procedures, either by the coaster or the shore-based facilities on which it depended.

First, however, several caveats need to be made about the nature of competition. Some large consumers, such as gas works and power stations, negotiated long-term contracts for coal transport; prices for such work were thus likely to vary little. As well, some colliers were constructed for the up-river trade in London which required hinging funnels and masts to sail under some bridges, such as Lambeth and Westminster. Later, special "flat iron" designs were employed; the prototypes were first built in 1878 when T.C. Nichol, a London shipowner, prevailed on Palmers of Jarrow to build *Vaux-*

hall and *Westminster* for the trade.[39] Because these vessels were built for a specific trade they were less likely to he deployed elsewhere. But these caveats notwithstanding, the coal shipping market in the late nineteenth and early twentieth centuries was highly fragmented and competitive.

What then were the responses of the coastal shipping industry to the need to reduce costs to meet increased rail competition? One of the most significant was to increase mean carrying capacity. For example, in 1861 the average coal cargo into London comprised 331 tons, while by 1890 it was over 800 and by 1900 over 1200 tons (see table 12). This lowered unit costs, since construction and operating expenses did not increase commensurately with size. Since the crew required for a large collier was little greater than for a small or medium-sized one, wages per ton were lower on large ships, a classic example of economies of scale. Table 13, which depicts a random selection of colliers operating in 1905-1906 between the North East and the Thames, shows little difference in crew size between the largest, *Ocean,* and *F. Stobart,* little more than half its size. The final column also reveals that man/ton ratios declined as ship size rose: while there was one crew per fifty-three tons on the smallest ship, one crewman could handle eighty-five tons on the largest.

Table 12
Average Cargo of Coal Entering London,
Selected Years, 1861-1910

Year	Coal Arriving (tons)	No. of Ships	Average Cargo
1861	3,567,002	10,765	331
1865	3,161,683	7727	409
1870	2,993,710	6539	458
1875	3,134,846	5497	570
1880	3,723,002	5249	709
1885	4,563,965	5782	789
1889	4,767,876	5789	823
1897			1283
1900			1133
1906			1243
1910			1545

Sources: 1861-1885 and 1900: Raymond Smith, *Sea-Coal for London: History of the Coal Factors in the London Market* (London, 1961), 323-324; 1889: *SMG*; 1897, 1906 and 1910: and *Daily Freight Register.*

Since the crew agreements also provide the wages paid to each seaman it is possible to calculate weekly wage bills per ton (see table 13). Al-

[39]Smith, *Sea-Coal*, 235-238; and Charles V. Waine, *Steam Coasters and Short Sea Traders* (Wolverhampton, 1976; 3rd ed., Wolverhampton, 1994), 106.

though larger colliers usually carried more crew and hence incurred higher labour costs, the increase was not proportional to size, so wage rates per ton fell as size increased. There appear to have been three distinct levels of cost. The most expensive was for ships of around 800 tons. For those around 1000 tons there was a significant reduction in crew costs per ton of about twenty-five percent, while ships of about 1500 tons had wage bills per ton a further fifteen percent lower. Although this is a small sample, there is no reason to doubt its representivity. Similarly, since the cost of fuel and terminal charges was not proportionate to ship size, this gave the large coaster a cost advantage.

Table 13
Tonnage, Crew Size and Wage Bill of Some Colliers, 1905-1906

	Gross Register Tons	Crew Size	Tons per Crew Member	Wage Bill (£ per Week)	Wage Cost Per Ton (shillings per Week)
Ocean	1442	17	85	28.13.6	0.398
Caenwood	1191	17	70	28.13.6	0.481
Airedale	1052	15	70	24.05.6	0.461
Battersea	860	17	51	29.00.0	0.674
F.T. Barry	839	16	52	26.14.0	0.636
Swiftsure	823	15	55	25.17.8	0.629
F. Stobart	800	15	53	25.14.6	0.643

Source: Great Britain, National Archives (TNA/PRO), Board of Trade (BT) 99/2443, 99/2460, 99/2470 and 99/2490.

Table 14
Some Colliers in the London Coal Trade, 1875

	Gross Register Tons	Crew Size	Wage Bill (Per Week)	Tons Per Mariner	Wage Cost Per Ton Shillings Per Week
Fenella	1097	19	£28.07.6	57.7	0.52
Corinna	1103	24	£35.04.4	46.0	0.64
Lord Alfred Paget	866	19	£30.14.0	45.6	0.71
Long Ditton	813	20	£30.11.0	40.6	0.75
Nina	783	19	£31.11.0	41.2	0.81
Berrington	629	23	£32.12.0	27.3	1.04
		18	£24.19.0	34.9	0.79
Fatfield	658	19	£32.06.0	34.6	0.98

Source: TNA/PRO, BT 99/1068 and 99/1078.

There is also evidence in the crew lists that crew per ton fell over time so that labour costs per ton in 1905 were less than in 1875. Table 14 lists a random selection of colliers in the trade from the North East in 1875. Compared with table 13, it is clear that for any given size of collier, fewer crew was hired in 1905-1906. *Fenella* was the only craft in 1875 over 1000 gross tons and had a crew of nineteen. The two ships of similar tonnage in 1905-1906, *Caenwood* and *Airedale*, carried two and four fewer crew, respectively. Similarly, the 866-ton *Lord Alfred Paget* in 1875 had a crew of nineteen, whereas the 860-ton *Battersea* in 1905 carried seventeen. This was also true of smaller colliers.

Table 15
Wage Rates on Some Colliers, 1905 (shillings per week)

	Mate	Second Mate	Steward	ABs	Chief Engineer	Second Engineer	Firemen
Ocean	60	42	34	30	67/6d	47/6d	30
Caenwood	60	42	34	30	67/6d	47/6d	30
Airedale	50	-	35	30	67/6d	47/6d	30
Battersea	50	35	35	30/4d	75	55	30/4d
F.T. Barry	50	40	34	30	65	45	30
Swiftsure	50	37/6d	35	30	70	50	30
F. Stobart	50	36/6d	33	30	67/6d	47/6d	30
Mean	52/10d	38/10d	34/3d	30	68/7d	48/7d	30

Source: See table 13.

Table 16
Wage Rates on Some Colliers, 1875 (shillings per week)

	Mate	Second Mate	Cook/ Steward	ABs	Chief Engineer	Second Engineer	Firemen
Fenella	30	28	24	30	56	36	32
Corinna	30	28	24	30	56	36	30
Lord Alfred Paget	38/6d	35	31/6d	30	70	52/6d	31/6d
Long Ditton	30	28	28	30	57/6d	40	32
Nina	50/6d	36	24	30	87/6d	69	30
					75	55	
Berrington	30	28	24	30	52	30	30
Fatfield	42/6d	37	35	30	75	55	30
Mean	35/11d	31/5d	27/2d	30	66	46/8d	30/9d

Source: See table 14.

Translated into man/ton ratios, a similar trend is apparent. If *Fenella* is compared to *Caenwood* and *Airedale*, the former's man/ton ratio (fifty-eight tons per man) is worse than the seventy tons per man on the latter two. Likewise, the forty-one tons per man of *Long Ditton* and *Nina* in 1875 pale compared to the fifty-plus tons of *F. Stobart* and *Swiftsure*. In the thirty years after 1875 man/ton ratios declined by ten to thirty percent. In monetary terms a similar trend is apparent. In 1875 *Nina* had crew costs of 0.81s per ton-week, while *F. Stobart* incurred crew costs of only 0.64s per ton-week thirty years later. The same trend is visible in the larger ships, although to a lesser extent. *Lord Alfred Paget* cost its owners 0.04s more per ton-week in 1875 than did *Battersea* in 1905. *Fenella* cost about 0.05s per ton-week more in the 1870s than did *Airedale* or *Caenwood* early in the new century.

Given that collier owners were forced to cut freight rates they may also have tried to trim labour bills, not only by reducing numbers but also by slashing wages. In the 1890s this might have been compounded by the success, noted by Frank Broeze, in crushing nascent seamen and dockworkers' unions.[40] Whether this caused wages to decline over the whole period, or merely reversed the gains of the 1880s, remains to be investigated. But a possible pattern can be postulated. Gains in money wages might be expected in the 1880s, when unions were generally successful in improving working conditions. In the 1890s and 1900s, however, it would be reasonable to expect a worsening of pay and conditions as union support crumbled and employers became more militant. This hypothesis can be investigated in a preliminary fashion from the crew agreements. Tables 15 and 16 present actual wages paid to each grade on the same selection of colliers already examined. A number of features deserve comment. First, since wages were not reduced in current terms the notion that collier owners forced rates down can be dismissed. But for able-bodied seaman (ABs) and fireman there were no increases in money wages over the thirty years. Advocates for collier owners might argue that this did not stop real wages from rising, since the cost-of-living fell appreciably in the last quarter of the century. Yet it should also be remembered that Feinstein showed that average money earnings rose by about a quarter between 1880 and 1905.[41] Since the manual labourers on steam colliers received no increase in money wages between 1875 and 1905, their relative position must have worsened. This is reinforced by Feinstein's contention that earnings for transport workers rose about sixteen percent in the same period, which can be compared to

---

[40]Frank Broeze, "Militancy and Pragmatism: An International Perspective on Maritime Labour 1870-1914," *International Review of Social History*, XXXVI, No. 4 (1991), 179-180.

[41]Charles H. Feinstein, "New Estimates of Average Earnings in the United Kingdom, 1880-1913," *Economic History Review*, 2nd ser., XLIII, No. 4 (1990), 608-609.

Lewis Fischer's findings for the wages of able-bodied seamen recruited in England.[42] Actual money wages are not strictly comparable, since his seamen were all employed on sailing vessels, mainly in the overseas trade, but the trends can be examined. While Fischer found an overall wage reduction between the early 1870s and the late 1890s of about twelve percent, coastal ABs had no contraction in money wages. Steam colliers also paid their ABs much more than blue-water seamen in sail. Whereas the coaster paid thirty shillings per week (£6 per month), the sailing ship paid between £2.88 and £3.26 per month. Of course, crew on overseas routes received free provisions while coastal sailors were expected to find their own. Even so, the coastal mariner was better off than his counterpart on sailing vessels in deep-sea trades, even if his money wages were stagnant.

A similar stability of earnings did not characterize men in higher grades. The four officers – first and second mate and first and second engineer – saw their wages rise in current as well as real terms. First mates gained nearly fifty percent and second mates over twenty, while chief and second engineers earned only a few percent extra. The common feature here is that all four were given pay increases, which thus widened the differentials between officers and their less-skilled subordinates. What is also remarkable is the difference in increment between the deck grades and engine room officers. The former increased their earnings much more than the engineering officers. Although the reasons for these relative increases have not yet been investigated, it might be hypothesized that the 1870s was the decade in which steam became entrenched in long-distance shipping and in which the coastal coal trade from the North East shifted substantially from sail. Hence, there was an increase in demand for engine room staff during this decade, which may have pushed their wages up relative to deck officers for whom there was not the same rapid rise in demand. In addition, the 1870s was a decade in which unions of skilled workers, such as the Amalgamated Society of Engineers, were increasing membership and perhaps exerting upward pressure on wages.[43] The high wages of the mid-1870s for engineering officers was thus a temporary situation caused by shortages of supply relative to demand, but this disequilibrium had been alleviated by 1905.

There is some support for this hypothesis in the wage data. If average wages are rank-ordered, in 1875 chief and second engineers both preceded first and second mates. In other words, engine room officers earned more than deck officers. This seems anomalous in that intuitively it might have been expected that at the very least first mates would have earned more than second

---

[42]Lewis R. Fischer, "International Maritime Labour, 1863-1900: World Wages and Trends," *The Great Circle*, X, No. 1 (1988), 1-21.

[43]E.H. Hunt, *British Labour History, 1815-1914* (London, 1981), 251.

engineers, and that second mates and second engineers might have enjoyed roughly comparable wages, albeit perhaps with a small skill premium for the latter. This is borne out by the 1905 data. While in 1875 second engineers earned nearly fifty percent more than second mates, by 1905 this premium had dropped to just under twenty-five percent. The same trend can he seen in the premium earned by the first engineer, which fell from eighty-three percent more than the first officer to thirty percent over a similar period. It seems plausible that temporary market conditions pushed the premium for skill above the customary differential in the 1870s but that this was adjusted by the early twentieth century. Future research should shed more light on this subject.

There is thus no evidence of individual money wages being forced down, although for stokers and ABs there were no increases and a relative decline in earning power. Officers' earnings did rise in money terms, although the increase for engineers was less than the average calculated by Feinstein, while for deck officers it was about average. Collier owners constrained the wages paid to the most numerous grades of mariner. Moreover, by switching to larger vessels with lower crew requirements per ton, and by reducing crew for any given collier, in the long-run they were able to reduce operating costs per ton. This enabled them to remain profitable despite falling freight rates.

Another feature which reduced unit costs was increased capital productivity. Given that a collier earned freight revenue for each voyage to the Thames, if it could make a larger number of trips in any period aggregate income would rise and perhaps enable the vessel to withstand lower freight rates. Indeed, this was what happened. For reasons to be explored later, the average number of journeys a collier could make rose dramatically. One expert calculated that in the early 1860s a collier able to make fifty-seven circuits per year was outstanding. Yet by the late 1860s, sixty-eight trips were considered remarkable and by 1895 seventy voyages were considered average. This increase in the number of passages was a function not only of increased speed but also of reduced turnaround, which resulted from better facilities at both ends of the journey.[44]

The increased speed on the actual journey was largely a function of improvements in propulsion. The shift from collier brigs to steam colliers began in the 1840s; by the early 1870s "nearly all sea borne coal came in screw steamers."[45] This began a technological revolution which freed the coal carriers from the vagaries of wind and tide: previously, gluts and shortages in the London market were normal, depending on the direction and strength of the wind.

---

[44]Robert W. Johnson, *The Making of the Tyne: A Record of Fifty Years' Progress* (London, 1895), 225; *Shields Gazette and Daily Telegraph,* 13 January 1868; and George B. Hodgson, "The Genesis of the Screw Collier," *Nautical Magazine,* LXX, No. 3 (1901), 181.

[45]Smith, *Sea-Coal,* 334.

But most of the savings from this shift had been exhausted by the 1870s. Increasing speed and therefore a rise in the number of trips per year came from two directions. The design of steam engines improved as compound and triple-expansion engines were perfected. In addition, new boiler designs allowed higher steam pressures, which in turn produced greater speeds and reduced coal consumption.[46] Of equal importance were changes in loading and discharging cargoes. Larger cranes, capable of lifting bigger buckets, became more common. In the late 1890s the grab crane appeared, alleviating the necessity of the crew shovelling coal into the buckets. At the turn-of-the-century steam-powered and hydraulic cranes were being replaced by electric models.[47] In the 1900s William Doxford and Sons pioneered colliers with a conveyor system for discharging coal into lighters or railway trucks. These "self-discharging" colliers could unload 400 tons per hour.[48] The effect of these technological changes was to reduce the amount of labour required and to speed up the process. By the end of the nineteenth century the time for unloading had been drastically reduced, thus enabling a ship to make more voyages per annum.

Another innovation which had the double effect of speeding turn-around and reducing operating costs was the use of water rather than solid ballast. To load ballast cost money, since it had to be bought and labour paid to load it. Unloading also incurred a charge, since most ports levied ballast dues and some insisted on discharge only in certain designated areas. These operations also took time. Water ballast, which was free and readily available, was taken on board by opening cocks and disposed of by pumps. Moreover, it cost nothing to load or unload and could be done on the move, thus reducing costs and speeding up turnaround. The first collier, *QED,* in the Tyne-Thames trade to use water ballast was built as early as 1844, but for many years it remained a dubious technology. Various methods were tried, including "collapsible canvas bags," as on *John Bowes* in 1852, but these proved prone to puncture. By the 1870s double bottoms of cellular pattern were permanent, safe and did not intrude excessively into cargo space.[49] It soon became the normal method.

If it were important to speed unloading, it was also beneficial if loading could be made as rapid as possible. Getting coal on board had been quicker

---

[46]See Robin Craig, *Steam Tramps and Cargo Liners, 1850-1950* (Greenwich, 1980), 11-15, for more detail on improvements in marine engines.

[47]Smith, *Sea-Coal,* 337-338; and MacRae and Waine, *Steam Collier Fleets,* 52 and 64.

[48]Craig, *Steam Tramps,* 47.

[49]*Ibid.,* 5-8.

than discharging it, at least since keels had been replaced by spouts in the early nineteenth century. For instance, whereas it took about ten hours for *James Dixon* to unload about 1200 tons of coal in 1860, it took only about four hours to load.[50] Despite this, any reduction in loading time would allow a slightly quicker voyage and hence a few more trips per year. To this end, by the 1890s some staiths were equipped with hydraulic lifts to raise coal wagons sufficiently to allow loading even at high tide. Some had electric coal conveyors with anti-breakage devices installed to load more efficiently. In addition, colliers were constructed with larger hatches, which had the twin benefit of allowing faster loading and minimizing the amount of trimming needed.

Over time the number of staiths available increased; for example, when Dunston-on-Tyne opened in 1893 it had six staiths, thus minimizing waiting time for colliers.[51] Also, increasing the depth and size of docks ensured that even the larger colliers could be adequately served.

A combination of collier design and technical improvements in shore-based equipment thus made it possible to fill ever-larger colliers in much the same time, thus contributing to rapid turnaround. At the discharging end similar changes speeded that process, and hence colliers were able to pack in more trips per annum. At the same time, costs were contained by using larger ships with more fuel-efficient engines. The steam collier was able to cut operating costs and increase revenue, allowing it to continue to make positive returns despite low freights.

It might be argued that an obvious method of ensuring that the workforce maximized the number of voyages its collier performed was to pay them by the voyage. Hence, the more trips performed the greater the earnings. There is plenty of evidence that this tool was employed in the collier brigs. Simon Ville noted it for *Freedom* in the 1790s, and there are numerous contemporary sources to attest to its use in the 1850s and 1860s. Nor did it go out of favour on the sailing colliers after steam had captured the bulk of the trade, for the crew of the brigantine *Mary Annie* was paid by the voyage in both 1896 and 1906.[52] Still, I have found no evidence of steam colliers using this

---

[50]Johnson, *Making of the Tyne*, 225.

[51]MacRae and Waine, *Steam Collier Fleets*, 24.

[52]Simon P. Ville, "Wages, Prices and Profitability in the Shipping Industry during the Napoleonic Wars. A Case Study," *Journal of Transport History*, 3rd ser., II, No. 1 (1981), 42-51, reprinted in John Armstrong (ed.), *Coastal and Short Sea Shipping: Studies in Transport History* (Aldershot, 1996), 21-34; Hodgson, "Genesis," 174; H.Y. Moffat, *From Ship's-Boy to Skipper, with Variations* (London, 1911), 44; Walter Runcirnan, *Before the Mast – And After: The Autobiography of a Sailor and Shipowner* (London, 1924), 85; *North and South Shields Gazette*, 3 October 1860; and Great Britain, National Archives, Board of Trade 99/2469.

method. To the contrary, all evidence points in the opposite direction: steam collier crews appear to have been paid a weekly wage, certainly for the random sample of agreements I have examined. Why then was there this change? Why did steam colliers not continue to motivate mariners by paying piece rates? Two explanations seem possible: the technical and the financial. It might be that since steam colliers represented large sums of fixed capital, it was feared that piece rates might encourage the crew to maximize the number of trips by exceeding safe limits of boiler pressure and speed. Similarly, owners might have feared that crew would be tempted to cut corners and take risks which would increase the chance of accidents and hence damage the ship.

An alternative explanation is that customary wage rates existed on collier brigs. If these were applied to steam colliers making many more trips per annum, crews would have earned much enhanced wages. To avoid this, owners needed to break the link between the new technology and the old methods of payment. They did this by shifting to a new method: wages by the week. This had an added bonus for the owner: if things improved and the ship was able to make a larger number of trips per annum, there would be no extra crew costs to set against the increased revenue. The profit margin per trip would therefore be greater. In practice, this is what happened to succeeding generations of steam colliers. As shore-based facilities, ship speeds and designs altered to allow more rapid loading, unloading and transit, the number of trips (and revenue) that might be expected from a steam collier rose without any commensurate increase in labour costs. Given that freight rates were on a downward trend, increasing the number of trips was a necessity to maintain gross revenues. If wages had been determined by the number of voyages, there would have been increased costs, but with a fixed weekly wage bill this was impossible. Indeed, if the number of mariners could be reduced, costs would also be slashed. In addition, weekly wages meant an available labour force, obviating the need to recruit a new crew for each voyage with the possibility of delay to a very expensive piece of capital equipment.

Thus, the collier was able in the late nineteenth century to withstand significant reductions in freight rates, much more than the overall fall in prices. It thus contributed to a decline in the cost-of-living. The methods used to reduce operating costs included exploiting economies of scale by building larger ships with lower crew levels and costs per ton of cargo carried; improving the design of both ship and propulsion systems; and modernizing the means of loading and unloading. These changes allowed coasters to capture the bulk of the industrial coal traffic, especially for the riverside gasworks and power stations. After 1896 more coal reached the capital by sea than by rail.

What still needs to be investigated is the extent to which this situation was specific to the coal trade and how far it had general application. Many of the improvements in the cost structure of the steam collier were capable of being adopted by coasters in other trades and on various routes. Compound engines, higher boiler pressures, larger ships, more rapid methods of loading and unloading, smaller crew sizes and lower operating costs were available to all. There are one or two hints in the literature that suggest that coasters in general were engrossing a larger share of the traffic. For instance, between 1862 and 1880 the Aberdeen Steam Navigation Co. increased the tonnage of goods carried between Aberdeen and London from 26,500 to 56,500 tons, an increase of about 110 percent in eighteen years, or more than six percent per annum.[53] Further research is needed to ascertain if this was a flash in the pan or part of a wider resurgence of the coaster. What is clear is that coasters were crucial in providing heat, light and power to the capital at a lower cost per ton on a more reliable basis.

Appendix 1
Sources for Coastal Coal Freight Rates

Tyne to Thames

1870	*Newcastle Courant*
1870	*Newcastle Daily Chronicle*
1870-1879	*Iron and Coal Trades Review*
1877-1886	*Shipping and Mercantile Gazette*
1889-1911	Newcastle Central Library, L622.32 499273A, book of of press cuttings on the coal trade.
1891-1894	*Shipping Gazette and Lloyds List*
1897-1899	*Daily Freight Register*
1909-1913	*Coal Merchant and Shipper*

Cardiff to London

1871	*The Weekly Gazette* (Newport)
1874	*South Wales Coal, Iron and Freight Statistics*
1880	*Newport Weekly Mail*
1880,1883-1888, and 1890-1895	*Western Mail*
1881	*Shipping and Mercantile Gazette*
1896	*Swansea Gazette and Daily Shipping Register*
1897-1899	*Daily Freight Register*
1900-1913	J. Davies and C.P. Hailey (eds.), *South Wales Coal An- for 1916* (Cardiff, 1916), 104.

---

[53]Aberdeen University Library, Ms. 2479/12, Aberdeen Steam Navigation Co., Directors' Minute Book.

Chapter 10
Liverpool to Hull – By Sea?[1]

In 1884 Thomas Wilson and Son ran the steamship *Humber* between Hull and Liverpool carrying a mixed cargo of goods.[2] Wilsons has long been recognized as Hull's largest shipowner, involved in many overseas trades, although no full-scale history of its business has been published.[3] Passing reference has also been made to its occasional engagement in the coastal trade between Hull and Newcastle but not to its participation in the coastal trade between Hull and Liverpool.[4] Yet by the 1880s the railway network was at its peak, and the distance by rail between the two cities was about 120 miles, whereas the distance by sea was nearly 1000 miles.[5] On the face of it, this seems an absurd situation and a misallocation of resources. Why send the goods nearly eight times as far by a mode with a lower maximum speed, especially when the steamer took the northern route around the United Kingdom via the choppy and perilous passage to the north of Scotland, which included the Pentland Firth where "navigation is rendered difficult and dangerous by the rate of the tidal current – six

---

[1]This essay appeared originally in *Mariner's Mirror*, LXXXIII, No. 2 (1997), 140-168 (with Julie Stevenson).

[2]The authors wish to acknowledge the generous help of a number of other scholars, particularly David Starkey for sharing his research into the Wilson archive with us and drawing our attention to sources and ideas.

[3]On Wilson, see "Charles Henry Wilson," in David J. Jeremy (ed.), *Dictionary of Business Biography, Vol. V* (London, 1985), 846-849; *Dictionary of National Biography: Missing Persons Supplement* (Oxford, 1993); Joyce M. Bellamy, *The Trade and Shipping of Nineteenth-Century Hull* (York, 1971); James Taylor, *Ellermans: A Wealth of Shipping* (London, 1976), chap. 7; and B. Dyson, "The Wilson Line," *Business Archives*, LXIV (1992), 26-37.

[4]Bellamy, *Trade and Shipping*, 49; Arthur G. Credland and Michael Thompson, *The Wilson Line of Hull, 1831-1981: The Rise and Fall of an Empire* (Beverley, 1994), 12; Taylor, *Ellermans*, 252; and Gertrude M. Attwood, *The Wilsons of Tranby Croft* (Beverley, 1988), 150.

[5]Michael J. Freeman and Derek H. Aldcroft, *The Atlas of British Railway History* (London, 1985; New ed., London, 1988), 56 and 68-69; and Harold Fullard (ed.), *The Mercantile Marine Atlas* (London, 1959), 7-8.

to ten knots – and the existence of eddies and whirlpools"?[6] The length and peril of the journey could be abated if the ships went through the Caledonian Canal, which was opened in the early nineteenth century and properly navigable by the mid-1840s.[7] Such a route reduced the distance by over 150 miles, making the journey about 820 miles. However, the maximum length of the locks on the Caledonian was 160 feet;[8] the vast majority of the ships employed by Wilson in this trade were too big to traverse the canal and had to use the longer and less safe voyage around the north of Scotland.[9] Only the *Torpedo* at 151-feet-long was small enough to fit the Caledonian's locks.[10] This article outlines the dimensions of this problem and then seeks to provide an explanation for this apparently perverse use of transport modes.

I

First, let us stress that this was no flash in the pan or one-off speculation to ascertain if there was demand for the service. Wilsons ran steamers between Hull and Liverpool from March 1884 to the First World War, i.e., over thirty years, very much the long term.[11] In addition, this was a liner service with posted dates and times of departure and anticipated times of arrival.[12] It was not a tramp route, with vessels sailing when a full cargo was available, but regular and scheduled, implying that there was a substantial and steady flow of goods to be carried. Although the frequency of the service varied over time,

---

[6]John A. Hammerton (ed.), *The World Pictorial Gazetteer and Atlas* (London, 1932), 829.

[7]Anthony Burton, *The Canal Builders* (London, 1972; 4th ed., Stroud, 2005), 118-123; and H.J. Dyos and Derek H. Aldcroft, *British Transport: An Economic Survey from the Seventeenth Century to the Twentieth* (Leicester, 1969; reprint, Harmondsworth, 1974), 96.

[8]Alexander D. Cameron, *The Caledonian Canal* (Lavenham, 1983; reprint, Edinburgh, 2005), 149.

[9]Credland and Thompson, *Wilson Line,* 35, 38, 50, 51 and 55.

[10]*Ibid.,* 55.

[11]The records in Great Britain, National Archives (TNA/PRO), RAIL 318/1-12, Humber Conference, minutes, are not complete, and the service continued for years after the records end. March 1884 is confirmed as the date of the inauguration of the service by the *Hull News,* 17 March 1884, in which an advertisement appeared for the route; *Hull Trade and Transit,* 1912-1915, confirms that the service was still running.

[12]TNA/PRO, RAIL 318/9, 11 February 1886; and *Hull News,* 10 April 1884.

the ship performed the return journey at least once a month, in some periods once a fortnight, and sometimes the service was weekly.[13] This was year-round, including the winter months when the northern passage was at its most turbulent and testing.

Nor was it that the goods carried by the steamer were particularly obnoxious or threatening to human life and hence needed to be kept well away from human habitation. This ship was not carrying dynamite or poisonous chemicals. On the contrary, the cargoes it carried, such as wheat, meal, flour, barrels of sugar and bags of cotton seed, were eminently suitable for railway transport.[14] Although some chemical products were carried, they were the less dangerous kind, such as casks of soda and potash. In addition, the steamer was not only carrying bulky, low-value primary cargoes only but also semi-finished industrial products, such as the chemicals mentioned above and pigs of lead, as well as manufactured commodities such as iron nails, soap and machinery.[15] In addition, the steamer carried perishable foodstuffs, such as preserved meat, bacon, beef and fruit. This suggests that the service was quite rapid and able to compete with the railway on speed. This is confirmed by the time taken on the voyage, the steamboat usually completing the round trip in eight days.[16] Not merely could these commodities be carried by rail as well as by steamer, but on occasions they were conveyed by train either because the quantity of cargo exceeded the space available and goods were left over when the ship had been filled[17] or because strikes in the docks prevented the ship from functioning.[18] That the railways were keen to capture the trade is indicated by the persistent monitoring of the types, weights and quantities of goods carried on the steamer by the Humber Conference and its discussions on special freight rates for certain commodities in an endeavour to attract some of this traffic onto the rails.[19]

Thus, for over thirty years a regular liner service of steamships carried mixed cargoes comprising a wide variety of goods between Liverpool and Hull. The goods were perfectly capable of being carried by train, and indeed

---

[13]*Ibid.*, RAIL 318/9, 18 January, 10 June and 9 December 1886; and *Hull Trade and Transit*, 1913.

[14]TNA/PRO, RAIL 318/9, 19 January and 20 March 1889.

[15]*Ibid.*, 17 May and 5 June 1889.

[16]*Ibid.*, 20 March 1884.

[17]*Ibid.*, 11 February 1886.

[18]*Ibid.*, 23 May and 17 June 1893.

[19]*Ibid.*, 9 April 1885.

some were precisely the sort of cargoes the railways might be expected to carry according to what might be termed the "railway-centred view" of nineteenth-century transport history. Despite the distance by sea being eight times further than by rail, the coastal steamer was apparently able to compete and prosper on this route for a long period. It might be added, to support this rather odd pattern of goods traffic, that Wilsons was not alone in running ships on the long route around the north of Scotland. From at least 1870, Langlands ran a steamer between Liverpool and Dundee calling at Aberdeen,[20] and in 1874 or 1875 it continued the service on the east coast to Leith.[21] There is some disagreement over precise frequencies of sailing and the longevity of the service. Duckworth and Langmuir say it was bi-weekly and suggest the service was offered at least until 1890,[22] whereas Donaldson claims the ships operated a weekly service from 1874 until 1914.[23] The point is that such an indirect service could be run over a long time period at a reasonable frequency at a profit; there is also evidence that the steamer consistently carried greater tonnages than the more direct rail route.[24] The next section briefly outlines the limitations of the sources; the third examines the characteristics of this trade; and the fourth attempts to explain why this situation prevailed.

II

A word or two is required on the sources used. Essentially, the main source outlining the frequency of sailings and the type and quantity of goods carried by these coastal liners is the minutes of the Humber Conference. This was an association of railway companies formed to agree freight rates, pool receipts and determine train frequencies and other operating details for cross-country traffic which had to use the rails of more than one company between the Humber and the Mersey. The railway companies involved in the pool were the

---

[20]John Armstrong, "Railways and Coastal Shipping in Britain in the Later Nineteenth Century: Co-operation and Competition," in Chris Wrigley and John Shepherd (eds.), *On the Move: Essays in Labour and Transport History Presented to Philip Bagwell* (London, 1991), 94.

[21]Christian L.D. Duckworth and Graham E. Langmuir, *Clyde and Other Coastal Steamers* (Glasgow, 1939; reprint, Prescot, 1977), 58.

[22]*Ibid.*, 62.

[23]Gordon Donaldson, *Northwards by Sea* (Edinburgh, 1966; 2nd ed., Edinburgh, 1978), 24.

[24]Armstrong, "Railways and Coastal Shipping," 95-96.

North Eastern, the Lancashire and Yorkshire, the London and North Western and the Manchester, Sheffield and Lincolnshire (from 1898 the Great Central).[25] They expressed their purpose as "developing the traffic of the ports of Hull and Grimsby, and for a division of the receipts from all traffic passing between Liverpool, Manchester, Ashton, and Stalybridge, and Hull and Grimsby."[26] The Conference was in existence from at least 1855 to 1904, and as part of its information gathering it monitored the activities of Wilsons' steamers.[27]

However, there are weaknesses in the source, and these need to be appreciated. They are in no way novel but consist of the usual sorts of problems encountered by historians wishing to conduct long-term analysis. The series is not complete in its reporting of the quantities and types of goods carried. The coastal steamers are first mentioned in March 1884, but the details of the cargo carried are vague: "The cargo consisted principally of syrup, sugar, chemicals and soap" or "The cargo consisted principally of coal."[28] From November 1885 there is a detailed breakdown of both the types and quantity of goods carried, and this series continues until March 1902 when it simply fades away.[29] Thus, we have a run of sixteen complete years of fairly consistent information on the commodities and quantities carried. But the data are not always comprehensive: some small quantities are not recorded, and phrases such as "also sundry other goods"[30] or "a quantity of paints and colours"[31] are not unusual. Hopefully, the goods so dismissed were quantitatively insignificant and will not alter the general trends derived. Thus, this vagueness is annoying but no more. Excluding them from our calculations will slightly understate the actual quantity of goods carried, so our figures can be considered minima.

The main problem with the data is that the quantity of goods is not consistently recorded in common units, ideally tons or hundredweights. Rather, a whole plethora of units are used, such as bags, casks, bales, drums, cases, bundles and packages, to name but a few. There are in excess of seven-

---

[25]TNA/PRO, RAIL 318/1. This is confirmed by D. Brooke, "The Struggle between Hull and the North Eastern Railway, 1854-80," *Journal of Transport History*, New ser., I, No. 4 (1972), 223.

[26]TNA/PRO, RAIL 318/1.

[27]*Ibid.*, RAIL 318/1-12.

[28]*Ibid.*, RAIL 318/9, 20 March 1884.

[29]*Ibid.*, RAIL 318/12.

[30]*Ibid.*, RAIL 318/9, 18 January 1886.

[31]*Ibid.*, RAIL 318/9, 10 June 1886.

teen different units of volume employed. Some of these units have an accepted size, for example, a barrel is usually considered to hold thirty-two gallons, and a quarter of wheat is conventionally equivalent to about 480 1bs. By using a variety of sources, such as Zupko's dictionary and Harrison's *Freighter's Guide*,[32] these obsolete and arcane measures of volume were converted into a weight equivalent, though it has to be stressed that the calculated weights are approximations. Some of the units involved have no accepted consistent weight equivalent, for example, logs, bundles or packages. In these cases, it has been necessary to make a "guesstimate" of the average weight of such loads based on recorded examples or the closest equivalent that could be found. Therefore, there is a large element of approximation in the figures calculated below. They are not precise results but rather indicators and a leeway of plus or minus ten percent should be given. By being consistent in the application of weight equivalents, the trend over time should be accurate, and the main thrust of the results should be reliable. From 1897 the total of goods carried can be given greater credence since the source quotes them almost entirely in tons.

The other caveat is that the results arrived at are the net load, that is, the weight only of the actual commodity. In fact, the containers were not insubstantial, and the total burden of the cargo, the gross weight, included this tare weight. This therefore means that the steamers were carrying a slightly larger total tonnage than indicated in the following pages. This is another source of inaccuracy, but of minor proportions. Thus, although the main source has some weaknesses which rule out precise measurement, it is good enough to give a general idea of the types and quantities of commodities carried and the trends over time.

III

Because the conversion of a range of units for a number of commodities into a common weight equivalent is a slow and laborious process, it was decided initially to pick a sample of years, scattered throughout the data set, and to calculate the statistics for those years only. The five sample years chosen were 1886, 1889, 1893, 1897 and 1901. The total weights carried by the steamers in each direction for each month were calculated. This is the greatest level of disaggregation available in the source. The results are given in tables 1 and 2.

---

[32]Ronald E. Zupko, *A Dictionary of English Weights and Measures from Anglo-Saxon Times to the Nineteenth Century* (Madison, WI, 1968); George Harrison, *The Freighters' Guide* (Newcastle, 1834; 5th ed., Newcastle, 1848); J.R. McCulloch, *A Dictionary, Practical, Theoretical and Historical of Commerce and Commercial Navigation* (London, 1834; 2nd ed., London, 1839); and *Custom House Clerk's Guide* (2nd ed., Liverpool, 1864).

Table 1
Quantity of Goods Carried from Liverpool to Hull by Wilsons' Steamers
(Selected Years) (Tons)

	1886	1889	1893	1897	1901
December	320	470	492	591	598
January	855	199	183	578	235
February	277	369	364	650	470
March	854	526	384	420	290
April	928	531	ND*	180	410
May	470	538	0*	688	277
June	757	518	308	738	470
July	ND	369	218	590	455
August	ND	469	215	365	460
September	ND	235	587	700	565
October	ND	600	480	975	342
November	ND	949	508	676	691
Total	4461	5773	3769	7151	5263
Average per month	637	481	377	596	438

Note: * indicates that there was a dock strike in these months at Hull which prevented sailings.[33] ND = No data available.

Source: Great Britain, National Archives (TNA/PRO), RAIL 318/9-12.

These tables show that there are no clear trends or patterns in the data. At the annual level, there is no apparent trend over time in the Liverpool-to-Hull consignments, but there is steady growth in the cargoes going in the opposite direction. However, these annual figures are not strictly comparable since they are based on varying numbers of months as a result of data deficiencies or no sailing in a particular month. To allow for this, the annual totals have been divided by the number of months in which there were sailings to give an average amount carried per month. When these are calculated the Liverpool-to-Hull trade continues to be trendless, and sadly the return trade turns out to be the same. Thus, the best that can be said at the annual level is that average monthly consignments ranged from 377 to 634 tons on the Liverpool-to-Hull route and from 292 to 484 on the Hull-to-Liverpool leg. If the two series are summed to give the total amount of goods carried by Wilsons' steamers, there is a suggestion of a rising trend, but it is not particularly marked, and it is invalid because of the different number of months in the

[33]Bellamy, *Trade and Shipping*, 46; and Attwood, *Wilsons of Tranby Croft*, 151-152.

years. If this is standardized by the month, the trend disappears. Hence, all we can conclude is that there was some volatility in the trade but that on average between 770 and 1120 tons of goods were being carried each month between Hull and Liverpool in these years. These calculations are given additional credibility by a document in the Wilsons archive. This return was drawn up by the firm for the year 1908 giving, *inter alia*, the total tonnage of goods carried on each route.[34] For the Hull-to-Liverpool trade, 8606 tons went outward and 11,842 tons came back. This gives an average of 717 tons and 987 tons per month, respectively, which is in line with the figures calculated here.

Table 2
Quantity of Goods Carried from Hull to Liverpool by Wilsons' Steamers
(Selected Years) (Tons)

	1886	1889	1893	1897	1901
December	230	384	343	517	730
January	258	141	164	593	125
February	460	629	206	594	500
March	39	301	119	468	390
April	1424	147	ND*	170	613
May	563	275	0*	390	330
June	414	330	441	345	395
July	ND	469	249	474	460
August	ND	119	802	190	638
September	ND	305	667	402	360
October	ND	257	480	850	610
November	ND	146	654	501	570
Total	3388	3503	4125	5494	5721
Average per month	484	292	412	459	477

Note: See table 1.

Source: See table 1.

There is also no consistent direction for the "balance of trade." For three of the five years, the Liverpool-to-Hull route carried more goods, and the return route carried more in two years. Nor was there any consistency in the size of the annual difference in the quantity carried on the two legs, which varied between 356 tons and 2267 tons. This adds evidence to the previous observation that the trade was volatile and fluctuated significantly.

---

[34]University of Hull (UH), Brynmor Jones Library (BJL), DEW (2) 3/42. We thank David J. Starkey for this information.

The monthly figures point up the variability in the trade, with the Liverpool-to-Hull cargoes ranging between 180 and 975 tons and the Hull-to-Liverpool between 39 and 1424 tons. This suggests that the latter route was more volatile than the former. Nor is there any clear seasonal trend in the figures. About the only consistent monthly trend is that fewer goods were carried in January than in the preceding month and that there was an increase in February over January. In addition, January recorded the lowest quantity of goods carried in any month on four occasions out of a possible total of ten. This is not a particularly surprising finding given that the weather in January can be very adverse and that there may be a lull in economic activity as a reaction to the increased trade leading up to Christmas and New Year's.

Table 3
Principal Goods Carried, Liverpool to Hull
(Selected Years) (tons)

	1886	1889	1893	1897	1901
Chemicals	1201	680	316	765	520
Sugar, syrup, glucose	1445	1819	1200	910	392
Seeds	1085	1095	386	45	690
Meal and flour	391	271	142	20	55
Grain	60	538	270	190	50
Oil	2	69	409	175	310
Iron	-	65	-	1863	-
Soap	26	8	62	1585	1155
Tinplate	-	-	127	435	25
Rubber	-	-	-	-	665

Source: Derived from TNA/PRO, RAIL 318/9-12.

Tables 3 and 4 give a breakdown of the most important commodities being carried by the steamer on each of the two routes. For the Liverpool-to-Hull journey there appears to be some consistency. For instance, in all five of the sample years the groupings of sugar products, chemicals and seeds were among the top five commodities carried. In addition, some quantity of meal, flour and grain was carried in each year, as was oil. Over the five years the greatest amount carried was in the category of sugar products at 5800 tons, and chemicals and seeds followed with nearly 3500 tons each (see table 5). One noteworthy feature is the appearance of a new range of goods in the last two or three sample years. These were manufactured products, *viz.*, soap, iron, tinplate and rubber. The top five commodities comprised two-thirds of the total tonnage over the entire five years sampled. It is also interesting that of the top five commodities, four were semi-manufactures or final commodities, that is, they all had been partially processed, and two of them, soap and sugar, could be in their final form ready for the consumer, while iron and the chemicals were likely to be industrial inputs requiring further processing before consumption. The only raw material appearing in the top five was "seeds."

Table 4
Principal Goods Carried, Hull to Liverpool
(Selected Years) (tons)

	1886	1889	1893	1897	1901
Grain	1189	365	30	-	-
Timber	840	1373	1355	900	1030
Zinc	225	-	116	-	-
Oil	223	110	202	698	575
Beans and peas	150	-	5	-	-
Strawboards	-	241	1010	972	450
Empty sacks, bags and mats	106	930	341	-	30
Pig lead	-	-	460	200	-
Paints and colours	-	-	181	252	355
Meal and flour	20	69	20	1395	1620
Paper	1	30	87	620	800

Source: See table 3.

Explaining the composition of the top five commodities is not too difficult. Merseyside was famous for its links with the West Indies, and its leading sugar refiner, Henry Tate and Sons (established in 1869 though with earlier antecedents), was by the 1880s one of the largest in the country.[35] In 1884 there were nine sugar refineries in Liverpool, about one-third of the national total, and in 1900 six out of a country-wide total of sixteen.[36] The Humberside region, on the other hand, lacked sugar refiners, and hence the flow of sugar to Hull from Liverpool was an example of regional specialization; some of the commodity was consumed locally, while the rest was exported to Northern Europe. A similar set of explanations apply to both soap and chemicals. Merseyside boasted a number of important soap and bulk chemical works, such as

---

[35]Tom Jones, *Henry Tate, 1819-1899* (London, 1960), 23; Anthony Hugill, *Sugar and All That: A History of Tate and Lyle* (London, 1978), 36; and Philippe Chalmin, *The Making of a Sugar Giant: Tate and Lyle, 1859-1989* (Chur, Switzerland, 1990).

[36]Chalmin, *Making of a Sugar Giant,* 61-62; and Hugill, *Sugar and All That,* 30.

Crosfield, Lever Bros., Muspratts, Gambles and Deacon.[37] Hull did not specialize in these areas but rather was known for its flour mills and especially its seed crushing mills, which gave rise to oil and cattle cake and the associated industries of paint and colour manufacture. This explains the large quantities of seeds carried on this route. They were intended not for planting but for crushing to extract the oil from the fibrous, protein-rich residue.

Turning now to the commodities carried from Hull to Liverpool shown in table 4, we can see that it was not simply a mirror image. There was a similar degree of consistency in the main commodities carried, with timber and oil in the top five in each of the five sample years and strawboards in all but one. There was a similar intrusion towards the end of the period as for the Liverpool-to-Hull journey in that several manufactured or semi-manufactured products appeared in significant quantities, including paints and colours, paper, lead and meal and flour. Over the whole five sample years the largest quantity carried was of timber at 5500 tons, with meal and flour coming in second at 3100 tons and strawboards at nearly 2700 tons (see table 6). The top five commodities make up sixty-two percent of the aggregate of goods moved, just a little below that for the return leg. This was borne out by the number of different products carried on each trip, with rather more on the Liverpool-to-Hull route. The presence of large quantities of oil, meal and flour suggests that they were products in which Hull specialized; these were despatched to Liverpool for local consumption or for loading into ocean-going ships for export. However, Hull was not famous for its forests or for its strawboard manufacture. It seems much more likely that these commodities had been imported from the Baltic or Scandinavia, possibly on Wilsons' ships, and were being sent to Liverpool for local industrial use. Strawboards were used in a whole range of industries as packing and insulation for frangible products. The continuous advance in the amount of paper carried is also noticeable, and it was the only main commodity to show this characteristic. It too was unlikely to have been of local manufacture, most likely having been imported from the Baltic. Comparing the most important five commodities on each leg of the journey shows that there was no significant overlap (tables 5 and 6). None of the top five goods going to Hull was also brought back to Liverpool. The trade was essentially based on specialization and an exchange of surpluses for scarcities.

The next exercise is to group the cargoes carried by the coastal steamers into broad categories to ascertain what sort of goods were being shipped. Because of the guesswork involved in categorizing them by their intended use, i.e., whether they were destined for agricultural use, as cattle cake seems to

---

[37]Alfred E. Musson, *Enterprise in Soap and Chemicals: Joseph Crosfield and Sons Limited, 1815-1965* (Manchester, 1965); Charles Wilson, *The History of Unilever: A Study in Economic Growth and Social Change* (2 vols., London, 1954); and Rex Pope (ed.), *Atlas of British Social and Economic History since c. 1700* (London, 1989), 36-38.

have been, or for inputs into other industries, as was probable with soda, we chose simply to categorize them by their present status. Five groups were used: raw vegetable goods, such as wheat and timber; minerals; semi-manufactured foodstuffs, such as flour and syrup; semi-manufactured industrial goods, such as pigs of lead or strawboards; and finished products, such as nails and electric cable. The results of this exercise are given in tables 7-10.

Tables 5 and 6

Liverpool to Hull, Top Five Commodities Carried (Selected Years, tons)		Hull to Liverpool, Top Five Commodities Carried (Selected Years, tons)	
Sugar	5766	Timber	5498
Chemicals	3482	Meal and flour	3124
Seeds	3301	Strawboards	2673
Soap	2656	Oil	1808
Iron	1928	Grain	1584

Sources: Derived from tables 3 and 4.

Table 7
Hull to Liverpool, Analysis of Cargoes (tons)

	1886	1889	1893	1897	1901
Raw vegetable goods	2124	1741	1405	800	820
Minerals	80	0	0	0	200
Semi-manufactures: food	231	295	30	1522	1920
Semi-manufactures: industrial	690	458	1952	2030	1506
Finished products	151	80	293	1142	1205
Returned containers	112	929	341	0	30
Total	3388	3503	4021	5494	5681

Source: See table 3.

Table 8
Liverpool to Hull, Analysis of Cargoes (tons)

	1886	1889	1893	1897	1901
Raw vegetable goods	1210	2370	1066	901	2114
Minerals	2	209	409	435	325
Semi-manufactures: food	1923	2307	1710	1010	684
Semi-manufactures: industrial	1230	840	482	3093	680
Finished products	26	47	62	1609	1267
Returned containers	0	0	0	0	20
Total	4391	5773	3729	7048	5090

Source: See table 3.

There is a degree of arbitrariness in any form of categorization, and an element of judgement is necessary. However, these seem to be broad but meaningful categories, and the basis of allocation is reasonably objective. The totals in tables 9 and 10 do not always add to 100 because occasionally some goods were described as "sundries." The final category, "returned containers," requires some explanation. On the Hull-to-Liverpool route in some years, considerable quantities of empty bags, empty sacks and mats were listed. The use of the word "empty" suggests that these were sacks and bags which had once contained a commodity, probably grain, meal or flour, which were being returned. Hence, they have been categorized separately from the other finished products. It has also been assumed that the oil travelling from Hull to Liverpool was likely to have come either from Northern Europe via Wilsons' overseas steamers, in which case it was likely to be whale or fish oil, or to have resulted from seed crushing in Hull. In either case it was part-processed, while the oil from Liverpool to Hull was more likely to have come across the Atlantic and to have been mineral, rather than vegetable or animal derived, oil and hence in a different category.

Table 9
Hull to Liverpool, Analysis of Cargoes (percent)

	1886	1889	1893	1897	1901
Raw vegetable goods	62.7	49.8	34.1	14.6	14.3
Minerals	2.3	0	0	0	3.5
Semi-manufactures: food	6.8	8.4	0.7	27.7	33.6
Semi-manufactures: industrial	20.4	13.1	47.3	36.9	26.4
Finished products	4.5	2.3	7.1	20.8	21.1
Returned containers	3.3	26.4	8.3	0	0.5

Source: See table 3.

Looking at the figures in tables 7 and 9 depicting the situation on the voyage from Hull to Liverpool, a number of trends are apparent. Among the most marked was the decline in importance of raw vegetable goods, which were the largest single category by a huge margin in 1886 but which had dropped to a quarter of their importance by 1901. Nor was this only a relative decline; the actual amount carried in 1901 was less than half that moved in 1886. There seems to have been a fundamental shift in the nature of the trade, for two other trends are clear: a jump in the importance of finished products in the last two sample years and in the significance of semi-manufactured foodstuffs in the same two years. Both of these increases reflect a sharp rise in the amounts of such goods being carried and suggest the trade was moving away from raw materials to finished goods, part-processed industrial inputs and part-manufactured foodstuffs. Throughout the period, the trade in minerals was insignificant, and the quantity of returned bags and sacks was much greater in the earlier years when the trade in agricultural commodities such as grain was at its peak, suggesting a relationship between the two as indicated above.

Table 10
Liverpool to Hull, Analysis of Cargoes (percent)

	1886	1889	1893	1897	1901
Raw vegetable goods	27.1	41.0	28.3	12.6	40.2
Minerals	0	3.6	10.8	6.1	6.2
Semi-manufactures: food	43.1	40.0	45.4	14.1	13.0
Semi-manufactures: industrial	27.6	14.5	12.8	43.2	13.0
Finished products	0.6	0.8	1.7	22.5	24.1
Returned containers	0	0	0	0	0.4

Source: See table 3.

Tables 8 and 10, which give the details for the voyage from Liverpool to Hull, present a much less clear-cut picture. One consistent trend was the steady growth in importance of the finished goods category from less than one percent to nearly one-quarter of the total. This was matched by an increase in the amount of such goods being carried; the growth was essentially in the last two sample years, 1897 and 1901. The trade in minerals was also consistent and was a small proportion of the total in all five years. The pattern for semi-manufactured foodstuffs is also clear; it was the single largest group in the first three years, but its significance declined sharply to one-third of the previous level. There is no clear trend in either the vegetable raw materials category or in semi-manufactured goods, and the two categories were not strongly correlated. In each case there were sharp annual fluctuations in both the amounts carried and their proportion of the total trade. No clear trend is apparent.

What is clear in the figures for both trades is that a sharp dichotomy occurred between 1893 and 1897; before that date finished goods were insignificant on both routes, ranging between one-half of one percent and seven percent, but in the last two sample years their importance jumped to over twenty percent. At the same time, on the Hull-to-Liverpool route the importance of raw agricultural products declined sharply, and on the Liverpool-to-Hull leg the importance of foodstuffs declined equally markedly. This suggests a fundamental change in the nature of the trade with a much greater emphasis on manufactured commodities.

IV

To explain how the coaster could capture traffic on a route on which the railways seemed to have a natural advantage, one set of answers might be sought in deficiencies in the British railway system's goods-handling methods. These might be categorized as general inefficiencies in freight wagon operations, those peculiar to the particular companies involved in this trade and the problems associated with the specific route which freight wagons had to take between Hull and Liverpool.

Virtually all commentators agree that by the last quarter of the nineteenth century British railways were far from efficient in their handling of freight traffic.[38] The charges against them include that their wagons were of inadequate capacity. This was exacerbated by the fact that about half the freight wagon stock was privately owned by traders and hence beyond the control of the railway companies. These wagons were small and antiquated. The loading levels of wagons were low, and there was a lack of through trains; instead, trains were continually being put into marshalling yards where they were disassembled and reassembled, adding to the cost and the time taken to complete the journey. The train weights were too small, and there was a lack of continuous braking, leading to low speeds as trains frequently stopped at the summit before down grades to allow the brakeman to run along the train, manually applying the brake on each truck. And of course goods trains had to give way to all passenger services and hence were often diverted onto sidings, where they stood until the scheduled passenger train had passed. So bad was the congestion of small-sized, lightly-loaded, poorly-braked freight wagons that Sir George Paish calculated that "the cost of moving a ton of goods one mile" on the London and North Western Railway (LNWR) "rose by twenty-four percent between 1880 and 1900."[39] This was at a time when most prices were falling; by about eight percent in this twenty-year period, according to Feinstein's recent calculations.[40] Yet the LNWR ran the heaviest freight train loads bar only one company, the Lancashire and Yorkshire Railway.[41]

The poor performance of the British railway companies in handling freight was not confined to the period before 1914. Even after rationalization

---

[38]See, for example, Dyos and Aldcroft, *British Transport,* 151 and 17·4; and Roy Williams, *The Midland Railway: A New History* (Newton Abbot, 1988), 143-144.

[39]Dyos and Aldcroft, *British Transport,* 173.

[40]Charles Feinstein, "A New Look at the Cost of Living 1870-1914," in James Foreman-Peck (ed.), *New Perspectives on the Late Victorian Economy: Essays in Quantitative Economic History, 1860-1914* (Cambridge, 1991), 170-171.

[41]Dyos and Aldcroft, *British Transport,* 174.

and the formation of the four mainline companies in 1921, the same sorts of inefficiencies survived: "freight handling remained hopelessly antiquated and inefficient even by 1939,"[42] and as late as the 1960s the *Beeching Report* described the progress of freight trains as "slow and unpredictable."[43]

The net result of these failures to improve freight-handling techniques was that costs and therefore freight rates were high, speeds were low and arrival times were uncertain and haphazard. Consequently, the service provided by the railways gave rise to numerous complaints from customers.[44] This is not a particularly felicitous background to the rail traffic between Hull and Liverpool, but there were two specific problems on this route which aggravated the situation.

The first was the position of the railway companies in Liverpool. Kellett has argued that until 1865 there was much dissatisfaction with the Liverpool railways because they did not compete with each other and operated a comfortable duopoly with high charges and inadequate facilities.[45] Between 1865 and 1880 Edward Watkin of the Manchester, Sheffield and Lincolnshire Railway (MSL, later the Great Central) drove a new line into the city terminating at the Huskisson goods depot.[46] However, this new arrival provoked no bout of competition and hence reductions in prices and improvement in services, for the cost of land, engineering works, rail track, stations and terminals was so high that the MSL needed to charge high prices to earn a return on this outlay. As a result, it was "willing, as far as goods traffic was concerned to accept the L&NW's lead on charges and facilities."[47] Consequently, complaints continued to be made against the railway companies. These reached a head in March 1881 when a large, noisy and acrimonious public meeting was held to voice the accusations of excessive charges. Little seems to have resulted. Among the complainants were "the corn, timber, sugar, and cotton merchants' associations," who represented precisely the goods which were

---

[42]*Ibid.*, 311.

[43]*Ibid.*, 312.

[44]Peter J. Cain, "Traders versus Railways: The Genesis of the Railway and Canal Traffic Act of 1894," *Journal of Transport History*, New ser., II, No. 2 (1973), 65-84; and Cain, "The British Railway Rates Problem, 1894-1913," *Business History*, XX, No. 1 (1978), 87-99.

[45]John R. Kellett, *The Impact of Railways on Victorian Cities* (London, 1969), 188-192.

[46]*Ibid.*, 193.

[47]*Ibid.*, 194.

being carried in Wilsons' steamers.[48] Nor did the complaints about the poor service provided by the Liverpool railways end there. Part of the logic for building the Manchester Ship Canal was the inadequate and expensive service offered by the railways running between Liverpool and Manchester. Marshall Stevens, one of the promoters and first managers of the canal, fulminated against their "perfect camouflage of accommodation and services." The railway charged "the highest rate in the world" on raw cotton for a distance of thirty-five miles.[49]

Another problem was the railway service provided to Hull, which was not greatly appreciated by the local merchants and shipowners. As David Brooke says, in the period 1865-1880 the businessmen of Hull steadily increased the volume of their protests: "they condemned the NER's traffic policies and gave evidence against it at Parliamentary enquiries."[50] The NER had an effective monopoly of rail traffic into Hull and failed to provide adequate facilities for the trade. In the autumn of 1872 there was "an extraordinary series of blockages and delays to traffic...at the docks and on the lines out of Hull...the lines were choked with empty and full waggons...cargoes of all types were delayed."[51]

So annoyed was the mercantile community that it gave evidence against the NER at both the 1865 1867 Royal Commission on Railways and the Joint Select Committee on Railway Companies Amalgamation of 1872. When neither of these solved the problems, they supported the promotion and construction of alternate rail lines into the city in the hope of breaking the NER's monopoly or forcing it to provide better services.

Among the "very active centres of resistance to the railway" was the Steam Ship Owners' Association, and among the leaders of that Association was Charles Wilson.[52] Wilson translated his disquiet into action, for in 1872 he subscribed £20,000 towards the Hull, South and West Junction Railway, one of the alternative lines which were never constructed, and in 1879 he was one of the promoters of the Hull, Barnsley and West Riding Junction Railway (HB&WRJ), which was eventually built.[53] However, although the opening of the latter in 1885 provoked a brief rate war, the improvement did not last long.

---

[48] *Ibid.*, 195.

[49] D.A. Farnie, "Marshall Stevens," in Jeremy (ed.), *Dictionary,* V, 313-316.

[50] Brooke, "Struggle," 220.

[51] *Ibid.*, 226.

[52] *Ibid.*, 220.

[53] *Ibid.*, 227-230.

In a rather similar manner to what occurred in Liverpool, the HB&WRJ was soon "driven into an understanding with the NER on traffic matters" and was uniting with the NER to construct a new dock.[54]

All rail traffic entering or leaving Hull by rail before 1855 had to use the tracks of the NER; even when the Hull and Barnsley was opened, to go to Liverpool by a reasonably direct route meant running over NER tracks. Given Wilson's sustained dissatisfaction with that railway company's provision, he may well have been particularly keen to avoid using the railway route between the two cities to move goods.

A further complication with the railway route from Liverpool to Hull in the late nineteenth century was that the wagons had to transverse the lines of at least three different railway companies. The British railway network consisted mostly of lines radiating outwards from London to the major provincial cities. Thus, any cross-country journey entailed using a number of separate firms' tracks. For the journey from Hull to Liverpool, the NER provided the initial route to Selby, thence via the Midland Railway (MR) to Leeds, and from there via Manchester on the LNWR.[55] Alternatively, after the same commencement on the NER, the wagons could continue on the MR to Bradford and from there proceed to Liverpool via the Lancashire and Yorkshire Railway. However it was done, it involved a number of railway companies. The complication with this was that railway companies, quite naturally, gave their own trains and wagons precedence over those "foreign" vehicles using their tracks, and this further slowed the rate of progress and made the dates and times of delivery even less predictable.

This was particularly galling if some of the goods were intended for export and were required to arrive before the sailing date of the ship which was to carry them. We have seen in the previous section that this was the case. In addition, given that Wilsons' Hull-to-Liverpool steamships carried on average over 450 tons per month,[56] and that the average train load on English railways between 1880 and 1900 was a little over sixty-three tons,[57] it would have required an additional seven train loads, on an already overstretched system, further adding to delays and disappearance of wagons. Thus, on all three counts of price, speed and predictability, this route was not well served by the railway system.

---

[54]*Ibid.*, 233.

[55]Freeman and Aldcroft, *Atlas*, 56 and 68-69.

[56]See section III above.

[57]Peter J. Cain, "Private Enterprise or Public Utility? Output, Pricing and Investment on English and Welsh Railways, 1870-1914," *Journal of Transport History*, 3rd ser., I, No. 1 (1980), 16.

There were other reasons for Wilsons operating a coastal steamer rather than leaving the trade to the expensive and poor railway service. There was the question of control and the feeder role of such a shipping service. By operating its own coastal liners, Wilsons was in control of such traffic and therefore could affect and predict times of the arrival of the consignments, which was something it patently could not do if sent by railway. This became important if the goods had to arrive by a certain time, such as the sailing date of another ship. Thus, if Wilsons was using this route as a feeder to its more important transoceanic lines, such as from Hull to the Baltic and Northern Europe, and from Liverpool to North America, and these were liner trades with fixed sailing dates, time of delivery was important, and by using its own round-the-coast ships it could ensure that consignments reached the docks in good time. Wilsons was a very large-scale trader between Hull and Northern Europe, the Baltic, Scandinavia and adjacent areas.[58] In addition, the company ran liners from Liverpool across the Atlantic to North America.[59] Thus, it is conceivable that Wilsons brought into Liverpool some goods intended for re-export to Scandinavia or the Baltic and that the reverse also occurred: that some goods were brought into Hull from Northern Europe that were destined for re-export to North America. It was not economical for Wilsons to send its large deep water ships around the coast partly laden, but it was economical to tranship at Hull into a small ship and forward the goods to Liverpool for transfer into another deep-water liner. Another source of cost saving might have been that loading over the side, i.e., from ship to ship, avoided the charges for landing on quays or wharves which would have been incurred if the goods were carried by rail. Thus, it might be the case that these commodities were booked through from North America to Northern Europe or *vice versa* and that Wilsons chose to keep the transfer between the two British ports within its own domain rather than contracting it out to the railway network.

This theory is given some support by the nature of some of the commodities carried. For example, sawn boards or boxes of dried fish going from Hull to Liverpool could well be imports from the Baltic.[60] Similarly, bales of cotton and bags of meal or flour being shipped from Liverpool to Hull are likely to have originated in North America.[61] This is borne out in a few instances where it was explicitly stated that a particular commodity came from a

---

[58]Jeremy (ed.), *Dictionary*, V, 846-847.

[59]*Ibid.*

[60]TNA/PRO, RAIL 318/9, 18 January 1886.

[61]*Ibid.*, RAIL 318/10, 19 January 1889.

specific port, e.g., "500 boxes dried fish, ex Bergen"[62] or "eighty tons of sawn boards, ex Christiansand."[63] In addition, sometimes the ultimate destination of a good was stated, as for example "100 tons of soap, sugar and molasses for Drontheim, twenty tons of potash and seventy drums of soda for Riga."[64] Thus, it can be established that Wilsons' coastal steamers were transferring imports and exports in both directions. This was reinforced in some months by a subheading appearing in the entries "for export" or "from ship," indicating that the goods were intended for onward transmission by Wilsons' ships or had completed their first stage by them. Hence, there is no doubt that some of the goods carried by Wilsons' coastal steamers were exports being assembled or imports being distributed and that others were re-exports being carried on the middle section of their journey. In these latter cases it is conceivable that there was no free choice for those consignments that were booked through and Wilson chose to keep the activity in-house. If Wilson was economically rational, it would only perform this service if the costs of contracting it out were greater than its internal costs or there was some non-price disadvantage in using the external contractor. The latter could have been the slow speed and unpredictable time of arrival of goods consigned to the railways. This uncertainty was not a problem because the goods were likely to spoil or deteriorate but because if they were destined for export they were likely to be aimed for a particular ship with a specific sailing date. If the wagons were still in a marshalling yard when the ship left, waste and costs were likely to be incurred. Given the reliability and regularity of their coastal steamships, Wilsons felt it was safer to rely on them rather than on the railway system.

Although the coastal steamers carried some goods destined for export or re-export, not all of the cargo was destined for onward shipment. For some months the subheading "for Town" appears, making it clear that these consignments were going no further than Hull.[65] Since these were neither particularly perishable nor needed to arrive in time to be transhipped, it seems likely that these goods were attracted by comparable or lower freight rates charged by the coaster compared to the railway with no loss of speed. In general, coastal freight rates for long journeys were significantly lower than those charged by railways for carriage between the same two destinations.[66] Ideally,

---

[62]*Ibid.*, RAIL 318/9, 18 January 1886.

[63]*Ibid.*

[64]*Ibid.*, RAIL 318/9, 14 October 1886.

[65]See, for example, *ibid.*, RAIL 318/9, 11 February and 18 April 1886.

[66]John Armstrong, "Freight Pricing Policy in Coastal Liner Companies before the First World War," *Journal of Transport History*, 3rd series, X, No. 2 (1989), 180-

it would have been helpful to have discovered some examples of the freight rates charged by Wilsons for this coastal voyage and then compared them with the prevailing railway rates. However, an assiduous search of the firm's surviving business papers, the local press, the national *Daily Freight Register* and other potentially revealing sources shed no direct light on this question.[67] The only glimpse we have of freight rates was in 1908 when a chance survival in the Wilsons archive gives both the aggregate tonnage carried and the revenue earned.[68] From these figures we can calculate the average freight rates earned per ton. These work out at 8s 9½d from Hull to Liverpool and 9s 2½d on the return voyage. Fortunately, we know the standard railway rates for this year as well.[69] They ranged between 9s 7d for class B and 16s 3d for class C to 22s 6d for class 1 and 43s 9d for class 5. These rates were all greater than the average coastal rate. A crude arithmetic average of the lowest three class rates on the railway gives a price of 16s 1d. Thus, as far as published railway rates are concerned, the average coaster rate was lower than all of the railway class rates. However, since some of the goods that the coaster carried did not fall into the lowest categories of railway classification, the average on the railways for the goods that travelled on the coasters would have been well above sixteen shillings per ton. Of course, these were the published rates. All railway companies also had "exceptional" rates which were lower than the published rates but were kept secret from the public and arrived at by negotiation and bargaining. These we do not know, but to approach the coaster rate would have required a discount of at least forty-five percent on the published rates. It seems likely that the railways' exceptional rates were well above those of the coaster.

An alternative approach might be to recreate the cost structure of the coaster by breaking down the various expenses and then seeking information on each. The crew costs can be tackled from the crew agreements. In 1886 *Dynamo* made three return trips between Hull and Liverpool. For these voyages the ship's complement numbered thirteen and the total crew cost, excluding the captain, was £17 17s per week, with the crew having "to find their own provisions."[70] In 1890 *Torpedo* made eleven return trips between the Humber and the Mersey and carried a crew of thirteen whose total costs, ex-

---

197, reprinted in John Armstrong (ed.), *Coastal and Short Sea Shipping: Studies in Transport History* (Aldershot, 1996), 112-129.

[67]UH, BJL, DEW.

[68]*Ibid.*, DEW (2) 3/42.

[69]*Transit*, 15 August 1908.

[70]TNA/PRO, Board of Trade (BT) 99/1509.

cluding the captain, amounted to £19 8s per week.[71] In 1892 it was again involved in a number of voyages between the two port cities; although the crew size remained the same, at thirteen, the total weekly cost had grown slightly to £19 17s 4d.[72]

However, this is only one cost; there are many others which need to be taken into account, such as fuel, depreciation, maintenance and repairs and terminal costs, such as port dues, loading and unloading. Sadly, there is no direct evidence of most of these in Wilsons' business records other than that loading and unloading at Liverpool in 1914 cost one shilling per ton.[73] This observation is outside our time period and of little value in itself. We could try to extrapolate from other examples. For example, Craig calculated the new cost of steamships in the 1870s and early 1880s at about £16 per gross ton,[74] so we could multiply the known tonnage of Wilsons' ships by this figure to calculate a new cost and then depreciate this at five percent per annum (assuming a twenty-year life for each ship on average). Similarly, elsewhere it has been calculated that "disbursements at ports" made up fifty-five percent of total costs.[75] But this was for a small sailing vessel in the 1840s and there is no reason to believe this figure would apply to a steamer half a century later.

In any case, there is a more fundamental point. To use one or two pieces of indirect data may be reasonable, but the greater the number employed, the greater the distance from historical reality the results are likely to be. The historian's aim should be to find suitable evidence, not simply to take what is easily available. Otherwise, a circular argument may simply reinforce previous hypotheses rather than testing them. Hence, it seems impossible, given our present state of knowledge, to say anything precise about the relative cost of sea freight compared to railway rates on this route, but the indications are that sea freights were well below the railway rates.

Despite these difficulties, one snapshot exists of this service's profitability in 1908. A surviving return gives a partial breakdown of costs in 1908

---

[71]*Ibid.*, BT 99/1672.

[72]*Ibid.*, BT 99/1755.

[73]UH, BJL, DEW 6/2.

[74]Robin Craig, "William Gray and Company: A West Hartlepool Shipbuilding Enterprise, 1864-1913," in Philip L. Cottrell and Derek H. Aldcroft (eds.), *Shipping, Trade and Commerce: Essays in Memory of Ralph Davis* (Leicester, 1981), 185; reprinted in Craig, *British Tramp Shipping, 1750-1914* (St. John's, 2003), 345-376.

[75]Robin Craig, *et al.*, "Some Aspects of the Business of Devon Shipping in the Nineteenth Century," in Michael Duffy, *et al.* (eds.), *The New Maritime History of Devon* (2 vols., London, 1992-1994), II, 100.

and the aggregate revenue.[76] This shows that operating costs were £7446, but not how this breaks down. In addition, maintenance and insurance cost £1229, depreciation on the ships involved at four percent of capital cost an added £435, and various other incidental expenses added another £67. The total costs of £9177 when set against revenue of £9272 left a profit of only £95. This does not seem impressive. Three points need to be made. If we accept labour costs of about £20 per ship per week as calculated above, and, given a fortnightly service, two ships would be wholly occupied on the run, this gives total labour costs for the year of about £2080. This suggests labour costs made up about twenty-eight percent of operating costs, or twenty-three percent of total costs. The return of £95 for the year not merely seems small – it was. It represented a return of just over one percent on revenue. Given that the first cost of steamers for this route was about £9000, this gives a very poor return on capital employed of less than one percent. However, the period 1901-1911 has been characterized as an acute depression for shipping, and freight rates were lower than in the last quarter of the nineteenth century.[77] Not merely that, but 1908 in many of the series was among the worst years on record for freight rates. Thus, to make any profit may have been a reasonable achievement. This is borne out by the performance of the other trades in which Wilsons operated. Of the twenty-five routes, twelve made a loss, the largest being £20,248 on the New York and Boston run. Seen in these terms, the Hull-to-Liverpool route looks quite valuable. In addition, we know that in the previous year, 1907, the profit on voyages before insurance, depreciation and maintenance was about twenty percent higher. Given that these overhead costs were likely to be similar to the succeeding year, it could mean that net profits were as high as £430, equivalent to a return of about 4.5 percent. Thus, we might conclude that the route, although not particularly profitable in 1908, was at least not making a loss, like half the routes, and had the potential to be a small but steady earner.

V

What conclusions may be drawn from this study? First, it must be emphasized that the railway did not replace the coaster as a carrier of goods traffic in the nineteenth century. In the very last decades of that century, when the railway was at its peak in network mileage and perhaps efficiency, the coastal ship

---

[76]UH, BJL, DEW (2) 3/42.

[77]Frank Cyril James, *Cyclical Fluctuations in the Shipping and Shipbuilding Industries* (Philadelphia, 1927; reprint, New York, 1985), 24; Derek H. Aldcroft, "The Depression in British Shipping, 1901-11," *Journal of Transport History*, 1st ser., VII, No. 1 (1965), 14-23, reprinted in Aldcroft, *Studies in British Transport History* (Newton Abbot, 1974); and John Armstrong, "The Shipping Depression of 1901 to 1911: The Experience of Freight Rates in the British Coastal Coal Trade," *Maritime Wales*, No. 14 (1995), 89-112.

could compete with the train, despite the former traversing a route which was five times as long as the latter. Nor was the coaster simply carrying low-value, high-bulk commodities like coal, china clay or corn. Wilsons' steamers were moving semi-processed products and finished goods of high unit value, like chemicals, paint and tinplate. Nor was it the case that these were small consignments, too small to be of interest to the railway. The volume of goods transported varied between 9000 and 13,500 tons per annum, a significant amount made up of both locally produced goods and imports being distributed. In the course of these twenty years the nature of the goods carried changed drastically. By the 1890s manufactured and semi-manufactured goods were of growing importance.

To explain how the coaster was able to compete with the railway, two broad sets of factors can be invoked: the failings of the railway companies and the economic advantages of the coastal ship. The former included, on the Hull-to-Liverpool route, slow journey times, an inadequate stock of wagons and hence delays and uncertainty. Sadly, competition between railway companies did not guarantee an improvement in service. In part this was because the railways did not really compete but preferred a cosy collusion since the capital cost of construction was so great that high charges were needed to service the capital invested. In part, too, it was because of the need to run goods wagons across a number of independent railway companies' tracks, which was a guarantee of delay as "foreign" wagons were given low priority. All of this is borne out by the complaints of local business people. Among the leaders of those whose voices were raised in criticism in Hull was Wilson, who was able to demonstrate his dissatisfaction with the railway by running the coastal alternative.

Among the advantages enjoyed by Wilsons' steamers were regularity and reliability. Although covering a longer distance, once at sea they operated twenty-four hours a day continuously until they reached their destination, taking about four days for the port-to-port journey, which was usually quicker and certainly more predictable than the station-to-station journey by rail. In all probability the operating costs of the coaster, and hence the freight rates, were lower than those of the railway, but no direct evidence of this seems to have survived. Finally, the coasters enjoyed economies of size and scale. One of Wilsons' coasters could carry several hundred tons of cargo, whereas the average train load in 1900 was only about sixty-three tons. By these methods Wilsons' steamers were able to offer a competitive service into the twentieth century and demonstrate the continuing importance of coastal shipping as a means of long-distance, bulk freight carriage.

Chapter 11
Government Regulation in the British
Shipping Industry, 1830-1913:
The Role of the Coastal Sector[1]

Terry R. Gourvish has argued that the railway companies in Britain were "never the exemplars of Victorian private enterprise," as some now choose to characterize them.[2] Indeed, from the outset they were fairly tightly controlled in terms of maximum rates, service provisions, routes and other fundamental matters. He also explained why the railways needed to be involved with government from the outset: their capitalization was so large that no family or group of partners was likely to be able to raise such sums. Hence, the companies had to draw upon a large number of "professional" investors or *rentiers*. Most of these played no part in the management or direction of the business, and they therefore needed some protection for their capital from the possible profligacy of the executives and directors. To ensure that the maximum loss they were likely to incur was restricted to the capital they had invested, they needed limited-liability status for the firm. These two requirements, joint-stock form and limited liability, required the enterprise to obtain either a royal charter or a private act of Parliament before 1844 when the law on joint-stock companies became much more liberal. Even then it was not until the late 1850s that limited liability was granted as simply and cheaply as corporate form. Hence, railway companies needed to approach Parliament for this protective legislation, as well as for compulsory purchase powers to obtain the land they needed. From then on, the railways continued to be tightly constrained and regulated. Parliamentary trains; threats of nationalization; constraints on merger activity; rules regarding safety; constraints on pricing, especially from the 1870s; requirements for cheap, early workmen's trains; total government control during the First World War; and a complete restructuring by parliamentary edict just after the war were just some of the government's actions.

At first sight there appears to be no commonality between this experience and that of the British coastal shipping industry. The latter seems to be a

[1]This essay appeared originally in Lena Andersson-Skog and Olle Krantz (eds.), *Institutions in the Transport and Communications Industries: State and Private Actors in the Making of Institutional Patterns, 1850-1990* (Canton, MA, 1999), 153-171.

[2]Terry R. Gourvish, "The Regulation of Britain's Railways: Past, Present and Future," in *ibid.*, 117-132.

fine example of private enterprise untrammelled by government intervention – competitive, technologically dynamic, low cost and evolving new services and types of vessel to cater for emerging trades, commodities and needs. The government never threatened nationalization or price control, never insisted on special low-price services for workmen or any other groups, left coastal shipping largely decontrolled for much of the First World War and never contemplated interfering in the structure of the industry after the war.

Although the emphasis in this chapter is on the coastal shipping industry, when the government investigated marine matters and introduced legislation it did not restrict itself to this sector but rather looked at a specific problem as applied to all shipping. However, it was the coastal segment that frequently led to an inquiry and that felt the impact of government most forcibly. The emphasis on coastal shipping might also be justified as it was the only sector that could compete with the railways, and if government was intervening in railway issues, as a matter of equity it might have felt compelled to bring in countervailing legislation for the coastal trade. In practice this never seems to have been a consideration for nineteenth-century governments.

This chapter will explain why the coastal shipping industry was so much less interesting to government than the railways; demonstrate that nonetheless there was a degree of government inquiry and legislation; and argue that it was the coastal sector that led to much of the desire for intervention and that felt the impact of government most forcibly.

II

The apparent non-intervention of government can be largely explained by virtue of the quite different market characteristics of coastal shipping compared to railways. Unlike the railways, the coastal shipping business had moderate needs for capital. The smallest railway company needed to spend millions to purchase land, carry out civil engineering works, buy and lay ballast, sleepers and iron rails, pay for locomotives, wagons, carriages, goods vans and signalling and then complete stations, goods yards and bridges. By contrast, a coastal shipping company could start with only one ship that cost from a couple of thousand pounds[3] for a small wooden sailing vessel to a few tens of thousands later in the nineteenth century for a triple-expansion, steam-engined, steel-hulled liner.[4] Were we to compare the capitalization of the largest British

---

[3] A number of coastal schooners built in Barrow cost between £1100 and £2500 in the 1880s and 1890s; Tim Latham, *The Ashburner Schooners: The Story of the First Shipbuilders of Barrow-in-Furness* (Manchester, 1991), 49 ff.

[4] For example, a 600-gross register ton (grt) steam coaster cost £8000 in 1894; Charles V. Waine, *Steam Coasters and Short Sea Traders* (Wolverhampton, 1976; 3rd ed., Wolverhampton, 1994), 138. *Ban Righ* cost £28,000 in 1870 for the Aberdeen

railway companies with that of the largest coastal shipping firms, there would be no comparison. Whereas *all* of the British mainline railway companies appear in the list of the largest British businesses for 1896 or 1904/1905,[5] none of the British coastal shipping companies do. The capital value of firms like Burns, Laird, Sloan, Powell or the Aberdeen Steam Navigation Co. was in the region of £100,000,[6] compared to the figure of £190 million quoted by Gourvish for the Midland Railway in 1905 (or £137 million in Wardley).

Not merely were coastal shipping firms' requirements for capital small, but they did not need to buy large tracts of land and thus had no reason to acquire powers of compulsory purchase. In many cases coaster businesses owned no land at all.[7] They used port or harbour facilities provided by others for which they paid a fee. If the ship was a tramp – that is, running on no regular route but picking up cargoes where and when they were available – there was no sense in owning shore-based facilities, for the owners could not know where the ship would next berth. If the ships were liners, operating on a particular route to a fixed schedule, or were regular traders performing a shuttle service between two ports, as some colliers did between the North East and London, it might be an advantage to own a berth or quay so they could moor direct at a dedicated facility. In addition, where particular unloading equipment was required it could be installed rather than trusting to the harbour authorities to provide it. Where a rapid and predictable turnaround time was required – and this was likely for those ships running to a schedule such as liners or those representing large agglomerations of capital by virtue of their size, splendour or speed – it was economically advantageous to own a berth, but most firms did not. Thus, by virtue of their low demand for capital and land, coastal shipping companies did not need to seek parliamentary approval of corporate form or for compulsory purchase.

Railways were often perceived as natural monopolies despite Parliament's attempt to promote competing networks in the early years of railway

---

Steam Navigation Co., and a larger ship cost £40,000 in 1872; Clive H. Lee, "Some Aspects of the Coastal Shipping Trade: The Aberdeen Steam Navigation Company, 1835-80," *Journal of Transport History*, 2nd ser., III, No. 2 (1975), 104, reprinted in John Armstrong (ed.), *Coastal and Short Sea Shipping: Studies in Transport History* (Aldershot, 1996), 90-103.

[5]Peter Wardley, "The Anatomy of Big Business: Aspects of Corporate Development in the Twentieth Century," *Business History*, XXXIII, No. 2 (1991), 278. The top ten companies in capital value in 1904/1905 were all railway firms.

[6]The capital of the Aberdeen Steam Navigation Co. (ASN) in 1885 was £100,000; Lee, "Some Aspects," 107.

[7]This paragraph draws on John Armstrong, "Coastal Shipping: The Neglected Sector of Nineteenth-Century British Transport History," *International Journal of Maritime History*, VI, No. 1 (1994), 175-188.

construction, and in practice the railways' tendency to merge and amalgamate reinforced this perception. The coaster was never viewed in this light. Indeed, the case was quite the opposite. The coastal shipping industry was character-ized as highly competitive, a multitude of separate ships competing with each other to carry cargo. The sight of the large numbers of ships was reinforced by the apparently atomistic pattern of ownership, where individual families often owned a small sailing coaster and it was rare for them to own more than one or two such vessels.

It was also thought that competition was intense because the coaster was so flexible. If one trade or cargo needed extra transport and freight rates rose, ships could be deployed from less buoyant trades into those needing ca-pacity. The assumption was that coasters were not trade-specific but could be switched from one cargo or route into another where demand was high at short notice. Compare this to the railways where the tracks tied trade to specific routes and where territoriality effectively prevented much competition. The ocean seemed boundless and boundary-less, free to all and capable of sustain-ing an infinite number of routes.

Of course, the reality was a little different. From the early nineteenth century, if not before, there was a range of vessels built for specific trades or localities that were not easily redeployed to other areas or trades. The Humber keel,[8] for instance, was specially suited to the conditions in that and adjoining rivers and was not intended for deep-water trades. Similarly, the Mersey flat, although it did make coastal trips and even crossed the Irish Sea, was mainly suited to river, estuarial and coast-hugging journeys.[9] In the early Victorian period there were also some ships built specifically for particular trades, such as the strongly built ranterpikes which were used to carry heavy iron castings from the Clyde to the Mersey. Although these vessels could have been de-ployed into other trades, they had a competitive advantage in those for which they were intended, whereas in other trades they had no particular edge over other vessels.

During the course of the nineteenth century there was probably an in-crease in the proportion of ships that were built with a particular trade or cargo in mind. For instance, there was growth in the liner trade so that by 1900 or 1910 just about every major British port had a regular coastal liner service. These ships tended to be large, powerful and well appointed since they carried passengers, perishables and high-value items. They were not geared to a par-ticular commodity but catered for general cargo, excluding only the dangerous, dirty and offensive, and were built with rapid turnaround in mind as they

---

[8]See Frank G.G. Carr, *Sailing Barges* (London, 1931; reprint, Lavenham, 1989), 171-176.

[9]*Ibid.*, 214-221.

needed to load and discharge smartly in order to adhere to their schedules. Thus, they were inappropriate for goods requiring slow loading and unloading, or those commodities for which rapid delivery was of no particular benefit. In a similar vein, by the 1880s the largest single coastal trade, that in coal from the North East of England to London and other southern ports, was employing vessels that were dedicated to that trade and that, like liners, aimed at rapid loading and discharging in order to maximize the number of trips made in any given time period and so maximize revenue earned.[10] They used water ballast, were often of a huge size, for instance carrying 2000 or 2500 tons of coal, and some had hinged masts and funnels to allow them to go well upstream on the Thames, passing under the low bridges. Such vessels were geared to large-scale coal consumers, such as electric power stations, gas works or industries that used large amounts of steam like Tate and Lyle, the sugar refiners. Even to redeploy them within the coal trade was not quite as simple as it might at first seem, for matching their capacity to that of other users was not easy, and being large ships they had the concomitant requirement for deep-water, large-size berths. Compare these paragons of size, speed and scale to the small, slow, unpredictable sailing ship which still comprised about twelve percent of Britain's coastal fleet in 1900.[11] These schooners and ketches were often less than 100 registered tons and had cargo capacities of between 100 and 200 tons.[12] These craft could not be deployed in either the liner trades or to carry bulk coal. Similarly, liners and colliers would not be appropriate for the low-value, high-bulk, slow-loading cargoes, such as slates, bricks, manure, sand and shingle, which the sailing vessels were carrying. Thus, as ships were adopted to specialize in particular trades, their flexibility declined and there was less possibility of multiple deployments. However, even at the end of the century, compared to the railways coastal ships seemed and did have multiple trajectories available to them, while the railways remained enmeshed in their tracks.

In addition, size might limit the trades into which a ship could be deployed. One of the key requirements to operate a coastal tramp profitably was to match the cargo size to the ship. Ideally, to maximize revenue the vessel needed to minimize empty cargo space and thus preferred to have cargoes that nearly or actually filled its holds. Thus, it was wasteful to employ a large ship, such as a collier, in trades that required smaller consignments than those that

---

[10]See John Armstrong, "Late Nineteenth-Century Freight Rates Revisited: Some Evidence from the British Coastal Coal Trade," *International Journal of Maritime History*, VI, No. 2 (1994), 69 ff.

[11]Great Britain, Parliament, *Parliamentary Papers* (*BPP*), 1901, LXXV, Annual Statements of Navigation and Shipping, 521.

[12]Basil Greenhill, *The Merchant Schooners* (2 vols., London, 1951-1957; 4th rev. ed., London, 1988).

filled its cargo space. As average coaster size increased during the nineteenth century, the range of sizes widened, and hence the probability of a volume mismatch increased. This was but one restraint on the apparently unlimited flexibility of coastal ships.

The apparently intense competition that existed between coastal ships which negated the fear of monopoly and therefore a need for government to intervene was not quite as it might seem. For a start, from the earliest days of steam shipping, and perhaps even before, there were a number of confer-ences[13] or similar arrangements that restricted competition between coastal liner companies. Some regulated the frequency of sailing, others worked on a joint-purse, whereby receipts were divided between the parties in a pre-arranged ratio, and most fixed freight rates mutually in order to eliminate price competition. By these methods the more obvious aspects of competition were moderated. Of course, conferences waxed and waned as new entrants ap-peared, technology changed or patterns of trade altered, and their existence was kept as secret as possible. Thus, most of the public was ignorant of their existence, and there was no pressure on Parliament to act. Additionally, be-cause coasters charged lower rates than the railways, they appeared to be com-petitive and non-exploitative. In fact, since their cost structure was quite dif-ferent from the railways, the coasters could have been making better rates of return on capital employed at lower freight rates because of the huge disparity in relative capitalization. At the time this remained unknown, certainly to those outside the firm, and possibly to those in it. This hypothesis needs empirical investigation, if a sufficient body of records can be discovered to throw light on returns on capital employed.

Another restraint on competition was the tendency to merge into lar-ger units. This has been noted in the case of the railways as one of the causes of government concern.[14] There was a similar tendency to agglomeration in coastal shipping, but it was much less marked, much later and much less no-ticed. Hence, it never troubled Parliament. The evidence of merger activity was clearest in the coastal coal trade and among the liner companies. In the coal trade to London there had been mechanisms to prevent large quantities flooding the market and bringing down prices from at least the early modern period. Some of these operated at the point of supply to limit the number of colliers sailing; in addition, the London coal factors grouped together to regu-

---

[13]See John Armstrong, "Conferences in British Nineteenth-Century Coastal Shipping," *Mariner's Mirror,* LXXVII, No. 1 (1991), 55-65; and Armstrong, "Rail-ways and Coastal Shipping in Britain in the Later Nineteenth-Century: Cooperation and Competition," in Chris Wrigley and John Shepherd (eds.), *On the Move: Essays in Labour and Transport History Presented to Philip Bagwell* (London, 1991), 76-103.

[14]Gourvish, "Regulation of Britain's Railways."

late the arrival and unloading of colliers.[15] As a result of technological developments, the optimum-size of the screw collier increased dramatically in the last quarter of the nineteenth century and the early years of the twentieth, and the capital required in the trade rose sharply. The number of firms with access to such large quantities of finance was limited, and as a result this trade in the 1880s was more like an oligopoly than a competitive one.[16] Then there took place a number of mergers that in a decade or two reduced it to a virtual duopoly dominated by Cory and Wm. France, Fenwick and Co. which had absorbed a whole host of erstwhile famous names. Since the trend in freight rates on coastal coal was downward, there was no outcry against these mergers since they were more defensive than aggressive and did not obviously result in attempts to increase price.

In a similar vein, in the last decade or two of the nineteenth century there began a series of amalgamations in the coastal liner trade that was to result in a virtual monopoly on many routes by the late 1920s. Starting with the firms trading between Liverpool and London, the Powell, Hough and Bacon lines, and using Coast Lines as the absorbing entity, the Royal Mail Steam Packet group, which wholly owned Coast Lines, amalgamated sixteen firms by 1920 and an additional seven during the 1920s.[17] That this activity caused no outcry was partly a result of careful public relations, for the absorbed firms retained their operating names, colours, house flags and apparently their separate identities, partly a result of the coaster seeming less exploitative as its rates were lower than the railways, and partly because the coaster companies were never perceived as gigantic and overwhelming as were the railways.

Even before the coastal companies began amalgamating there were some social constraints on extreme competition. Because admiralty law permitted ships to be owned by up to sixty-four people, even before the general availability of joint-stock form it was sound economic logic to spread shareholding among a number of vessels rather than having a large number of shares in one ship. Adopting such a strategy enabled risks to be spread over a number of ships rather than having all one's eggs in a single basket. As a result, in many traditional shipowning areas individuals did not own ships but rather owned one or two sixty-fourths in a number of ships. This had the effect of muting price competition for the masters of such ships did not endeavour to undercut each other since the mix of owners might be very similar. Thus, to compete on price would drive down earnings and therefore profits for a similar

---

[15]Raymond Smith, *Sea-Coal for London: History of the Coal Factors in the London Market* (London, 1961).

[16]J.A. MacRae and Charles V. Waine, *The Steam Collier Fleets* (Wolverhampton, 1990), 49, 53 and 56; and Smith, *Sea-Coal for London*, 343.

[17]Peter Mathias and Alan W.H. Pearsall, *Shipping: A Survey of Historical Records* (Newton Abbot, 1971), 36-39.

set of owners and hence would be cutting off one's nose to spite one's face. The practice of registering single-ship companies, once limited-liability and joint-stock forms were legally and cheaply available, had a similar effect. Again, this was a device to spread and limit risks. By taking a small number of shares in each ship, local investors could ensure that if any one vessel were to meet disaster their loss would not be enormous.

Thus, while there were a number of features which constrained competition but were less obvious to the average merchant, the coaster was perceived as, and probably was, more competitive than the railway. In addition, the coastal ship in the later nineteenth century was much less important in carrying passengers because of its much lower top speed and the restricted range of routes on which it could operate. The coaster was used for holiday excursions, journeys across channels and to islands, but never as a large-scale people mover either for business or commuting. There was no pressure to introduce cheap fares, such as parliamentary trains in 1844 or workmen's tickets at early hours and low prices, for it was never a problem. As Ralph Turvey has shown, even in the early twentieth century on the Thames, which was a very sheltered and predictable stretch of water, steamboats as passenger vehicles were unreliable and lacking in comfort.[18] Thus, the coaster was more important for cargo than passengers and hence never was considered as a means of cheap workmen's travel and so was not subject to such legislation.

Despite the coaster being largely devoted to freight traffic, the coastal industry never suffered the imposition of any body equivalent to the Railway and Canal Commissioners (RCC). This was in spite of the dependence on the coaster that some trades demonstrated, for example the coal, china clay and iron ore trades. How can this lack of attention be explained? We are forced to resort to explanations already advanced. The establishment of the RCC was partly a function of the fall in prices after 1873 and the squeeze placed upon profits as foreign competition became much more effective, leading merchants and manufacturers to seek reductions in their costs, such as transport, and improvements in the service provided – hence, the spate of complaints about railways and their charges and "inadequate" frequencies and speeds. Although the same external conditions prevailed for merchants using coastal ships, the latter seemed less exploitative since over the last quarter of the nineteenth century coastal freight rates were falling in both money and real terms. For instance, the average rate on coal from the North East of England to London fell to less than fifty percent of its 1870-1874 level by 1905-1909 in money terms, and to just under fifty-five percent of its 1870-1874 level in real terms.[19] These

---

[18]Ralph Turvey, "The LCC's River Steamboat Service" (Unpublished paper presented at the Institute of Historical Research, London, February 1995.

[19]Armstrong, "Late Nineteenth-Century," 53-57 and 67.

were substantial reductions, and because there was no overt collusion, merchants and traders could not complain of conspiracies to raise or even maintain prices. As a result of the coaster's increasing productivity, based on improvements in the efficiency of the machinery, economies of scale, lower levels of crewing and speedier turnaround times, it was able to reduce its costs faster than average prices and hence head off any criticisms of exploitative pricing.

III

Given the drift of the chapter so far, it might be concluded that there is little substantive to say about the relationship between government and coastal ships because the former never took any interest in the latter. This would be incorrect for, as one of the leading British archivists on the records relating to maritime matters, Nicholas Cox, remarked, "[u]nlike most private industries, the Mercantile Marine has from an early date been closely regulated by the state."[20] To show this is not an aberrant judgment, another academic claimed that one Merchant Marine Act "regulated some of the minutest details in the operation of the shipowner's business."[21] This view is reinforced by the vast range of sources available for the study of maritime history in the United Kingdom that were originated by public servants at the behest of one of the many maritime acts. The amount of time and paper Parliament devoted to aspects of Britain's mercantile marine in the long nineteenth century (i.e., up until the First World War) was not much less than that devoted to the railway industry and significantly more than to any other individual industry. The available records speak for themselves: ship registrations, crew agreements and logs, shipwreck maps, harbours of refuge, certification of officers, crimping, grain ships, shifting cargoes, safety on steamboats and various censuses of seamen. Such a vast range of public records does not suggest an industry ignored by government. Quite the opposite, it indicates a high level of monitoring, enquiries and investigation at the least, and one might expect some regulation to ensue from these committees and commissions. One complication when talking about *coastal* shipping and regulation is that most of the legislation was aimed at the shipping industry as a whole and only rarely made specific reference to coastal or home trade shipping as a separate entity. However, coastal shipping inspired some of the more important debates and legislation.

The regulation of the shipping industry seemed to shift from a quasi-mercantilist philosophy to one of *laissez faire* in the first half of the nineteenth century. In 1800 the Navigation Acts were in force, and these had the effect of reserving the British coasting trade to British-owned ships manned by British

---

[20]Nicholas Cox, "Sources for Maritime History (II): The Records of the Registrar General of Shipping and Seamen," *Maritime History*, II, No. 2 (1972), 168.

[21]Jane H. Wilde, "The Creation of the Marine Department of the Board of Trade," *Journal of Transport History*, 1st ser., II, No. 4 (1956), 205.

crews. By 1854 the Navigation Acts, as applied to coastal shipping, had been repealed and the coastal trade was open to ships of any nation.[22] It might, therefore, be seen as a classic example of liberalization, rolling back the frontier of the state, moving away from mercantilism to free trade. Throughout the rest of the long nineteenth century, British trade remained open to the ships of other nations, but in practice the proportion captured by foreign ships was minuscule. It might be seen as an endorsement of the correctness of government policy in reducing its role in this industry in that it encouraged continued innovation in shipbuilding and ship-operating techniques.

Yet in the late 1840s and 1850s a number of acts were put on the statute book that created a Marine Department of the Board of Trade and also set up a Steamboat Department in that ministry, and throughout the nineteenth century there were a succession of committees, enquiries and commissions related to various aspects of maritime activity. These apparently contradictory trends can be reconciled by suggesting that although the early Victorian governments in principle and on fiscal grounds preferred minimal involvement and intervention, there were some issues that were of sufficient importance that they felt they had at least to investigate. Having done so, the evidence was such that they felt compelled to take action. In a number of cases the pressure then came from within the public service to amend or add to the existing legislation; in other cases the pressure came from inside Parliament but struck an answering response in the public at large. Paradoxically, the shipping interest, which was never as large inside Parliament as the railway interest, seemed to have more influence on legislation and not necessarily always to resist it. This chapter will now endeavour to isolate those variables that led the government to act and examine their actions in more detail.

Perhaps the most enduring aspect of maritime activity that inspired government enquiry and intervention was safety. This may be broken down into three sub-themes: safety of passengers travelling on the ships, safety of property, namely the ships and cargoes, and safety of the seamen manning the vessels. Undoubtedly the plight of passengers put in danger by collisions, shipwrecks or other incidents was the most evocative and seen as deserving the most sympathy. This partly explains the establishment of the Steamboat Department and the appointment of Captain H.M. Denham as its first inspector in 1846.[23] The early steamboats were largely confined to carrying passengers and

---

[22]Ronald Hope, *A New History of British Shipping* (London, 1990), 282; Adam W. Kirkaldy, *British Shipping: Its History, Organisation and Importance* (London, 1914; reprint, Newton Abbot, 1970), 27; and Sarah Palmer, *Politics, Shipping and the Repeal of the Navigation Laws* (Manchester, 1990).

[23]Hubert L. Smith, *The Board of Trade* (London, 1928), 104; and Roger W. Prouty, *The Transformation of the Board of Trade, 1830-1855: A Study of Administrative Reorganization in the Heyday of Laissez Faire* (London, 1957), 62.

high-value freight because the steamboat's advantage over the sailing ship was its speed and reliability. Its weakness was that both its capital and operating costs were higher because of the expense of machinery and the large consumption of coal by the inefficient early steam engines. Thus, the steamboat charged premium prices that could most easily be borne by passengers who valued speed and reliable times of arrival and by high-value goods such as parcels, post and perishables. Accidents involving steamboats in the 1820s, 1830s and 1840s put passengers at risk and caused death and injury, for example, the sinking of *Comet II* after a collision in 1825[24] and the loss of *Rothesay Castle* off Anglesey in 1831.[25] Both were accompanied by large loss of passengers' lives and caused letters to the press and parliamentary action – the Lord Advocate introduced a bill in 1826 to regulate steamboats in Scotland,[26] and a select committee was established in 1831 to look into "calamities by steam navigation."[27]

Eventually, after more disasters and more enquiries, the 1846 Steam Navigation Act embodied concern for the passenger by insisting that all passenger steamers be inspected, carry boats and fire-fighting equipment, have watertight compartments, carry a specified pattern of lights at night and obey a standard "rule of the road."[28] These simple measures, it was believed, would greatly reduce the chance of overcrowding and collisions and, if the latter did occur, minimize the loss of life. From the point of view of government intervention, it marked a significant intrusion into the costs of the industry and established the principal of inspection. Furthermore, by appointing a government official, it opened the possibility of further intervention engendered by "expert government," as Oliver MacDonagh has termed it.[29] This is where government officials in the course of their normal duties discover loopholes in the existing legislation, or the need for extensions of regulation, and initiate moves to bring in amending legislation. This is precisely what happened. As a result of Denham's professional activities, he noticed loopholes in the existing legislation and recommended additional statutes. These were introduced, for instance, in

---

[24]Brian D. Osborne, *The Ingenious Mr. Bell: A Life of Henry Bell (1767-1830), Pioneer of Steam Navigation* (Glendaruel, 1995), 204 ff.

[25]*BPP*, 1831, VIII, 1-201.

[26]Great Britain, Parliament, House of Commons, *Journal*, LXXXI (1826), 147.

[27]*BPP*, 1831, VIII, 1-201.

[28]*BPP*, 1846, V, 451-463; and 9 & 10 Vict. c. 100.

[29]Oliver MacDonagh, *A Pattern of Government Growth, 1800-1860: The Passenger Acts and Their Enforcement* (London, 1933; reprint, London, 1961).

1848 and 1851.[30] Thus, at a time when the hand of government via the Navigation Acts was being removed, it was being placed on one new form of technology.

It should also be stressed that although this legislation applied to all passenger steamboats, it had a much greater impact on the coastal, estuarial and cross-channel trades than on foreign routes. The bulk of steamboats before 1850 were restricted to these short-sea trades because their gargantuan consumption of coal necessitated that they ply close to ample sources of cheap fuel. To travel long, inter-continental distances would have required more coal than the early steamboat could carry or there would have been so little space left for cargo that it would have been quite uneconomic. In addition, virtually all of the pressure for reform originated from incidents involving ships on short-sea routes, partially again because that was where the bulk of steamboats were deployed, and partly because incidents on the British coast or in British rivers were more newsworthy and capable of being reported and hence arousing public and parliamentary outcry.

This concern for the lives of passengers is not surprising and is evidenced in the legislation on emigrant ships, so ably studied by Oliver MacDonagh.[31] It might be more surprising, initially, that an associated concern, not just in the debates over steamboat accidents, was the loss of property involved. In a number of the debates on the question of harbours of refuge in the 1850s, one of the arguments for government action was the huge value of the property that was being lost as a result of strandings and wrecks that could be avoided if there were an adequate number of carefully positioned harbours of refuge. For instance, Henry Liddell, MP, a representative of the North East coal trade, argued in favour of the need for "life harbours" on the east coast in 1858 on the grounds that "a million and a half of property" was being squandered each year.[32] Similarly, two years earlier Lloyd Davis, MP for Cardigan, arguing for the need for a harbour of refuge in Cardigan Bay, wondered *inter alia* that the government "placed in jeopardy so enormous an amount of the nation's wealth."[33] In a like vein, when discussion took place on the idea of a load line in the 1870s, the huge value of the losses of vessels and cargoes that occurred as a result of ships foundering or being wrecked was one of the ar-

---

[30]11 & 12 Vict. c. 81; and 14 & 15 Vict. c. 79.

[31]MacDonagh, *Pattern of Government Growth*.

[32]Baron F. Duckham, "Wrecks and Refuge Harbours, 1856-61: A Reform that Failed," *Transport History*, VI, No. 2 (1973), 155.

[33]*Ibid.*, 153.

guments deployed in favour of such a marking, especially when such ships were over-insured.[34]

While discussing the debate about harbours of refuge, it is worth making the point that, as with the legislation on steamboat safety, the drive for action arose mainly from accidents befalling coastal shipping, not those in overseas trade. Similarly, the greatest beneficial impact of any extra provision of harbours of refuge would also have been on the coastal trade, for the great majority of ships passing along the coasts were those in the coastal trade – especially on the east coast, the collier brigs – and thus it was those that suffered from gales and storms and became wrecked on a lee shore in an east wind. As an example, in January 1857 ninety-four ships, mostly colliers, were wrecked on the North East coast.[35] Similarly, if refuge harbours had been constructed, the coastal ships that hugged the coast would have been most affected, for foreign-bound ships tended to eschew such dangerous practices and head for deep water well clear of threatening rocks and promontories as soon as possible after clearing harbour. So here again legislation apparently intended for the whole merchant marine was in fact mainly inspired by and effective for the coastal trade.

The third feature that inspired parliamentary enquiry was the loss of life among seamen. I cannot agree with Ronald Hope who wrote that "Parliament cared little or nothing for the welfare of seafarers."[36] He may well be right that it was not a disinterested philanthropy, but concern was expressed through the committees, commissions and debates of the nineteenth century. Agitation for reform, such as to extend harbours of refuge, was partly motivated by the huge loss of seamen in the coal trade; collier brigs carried few passengers but many crew. Lloyd Davis, urging such action in 1856, claimed the government had "placed in jeopardy the lives of so many of the bravest of Her Majesty's subjects."[37] Liddell in 1858 bemoaned "a thousand lives a year...being squandered on our coasts," and Bentinck in the same debate claimed the government was "trifling with the lives of thousands."[38] These numbers were too large for passengers and referred to the great number of seamen who drowned every year. These examples might be multiplied, but it would be tedious. The point is that Parliamentarians *did* care about "the welfare of seafarers." Why they did so is more important.

---

[34]Geoffrey Alderman, "Joseph Chamberlain's Attempted Reform of the British Mercantile Marine," *Journal of Transport History*, 2nd ser., I, No. 3 (1972), 175.

[35]Duckham, "Wrecks and Refuge Harbours," 152.

[36]Hope, *New History of British Shipping*, 256.

[37]Duckham, "Wrecks and Refuge Harbours," 153.

[38]*Ibid.*, 155.

In part, and for some Parliamentarians, there may well have been a genuine concern for human life and the welfare of the seamen. In addition, however, there were more practical reasons. For many decades the merchant service was perceived as being of strategic importance because of its role in providing a reserve army of seamen for the Royal Navy in time of war, when the demand for warships and naval crews rose sharply. Thus, part of the concern for the loss of mariners in wrecks, founderings and collisions was the waste of a valuable potential resource in periods of warfare. This is borne out by some of the other activities of government. One of the enduring themes was the establishment of a register of merchant seamen. In 1835 the Merchant Shipping Act "provided for the registration of seamen, to create a means of manning the navy in time of war,"[39] and gave birth to the voluminous Agreements and Crew Lists which have been such a valuable source for maritime history. These were then collected and retained throughout the rest of the pre-First World War period, although the formal register was abandoned in 1856.[40] Concern about the register of seamen was partly ameliorated by the belief that a snapshot was obtained every decade when the regular census was taken as it included British seamen, partly by the more regular employment conditions of seamen on the steam-driven liners that were increasing their share of the overall trade, which made it easier to trace seamen. But that there was still concern is shown by the special censuses of seamen that were taken between 1896 and 1911, showing their age, grade and nationality, and that were inspired by a concern that the British merchant marine was being taken over by foreign seamen who would be unavailable for naval service if there were a national emergency.[41] Arising out of similar concerns was the committee appointed by the Board of Trade in 1894 to enquire into the manning of British merchant ships.[42] There was an abiding concern with the number and nationality of the crews of British ships because of the need for naval recruits.

In many ways the coastal trade was crucial in this respect, for it was the coal trade from the North East that was claimed as early as 1615 to be "if not the only, yet the special nursery, and school of seamen."[43] This sentiment

---

[39]Cox, "Sources," 174.

[40]*Ibid.*, 175 ff.

[41]*BPP*, 1897, LXXVIII; 1902, XCII; 1908, XCVI; and 1912/1913, LXXVI.

[42]*Ibid.*, 1896, XL.

[43]Ralph Davis, *The Rise of the English Shipping Industry in the Seventeenth and Eighteenth Centuries* (London, 1962; reprint, Newton Abbot, 1972), 114.

was repeated down the years.[44] The belief was that because the voyages were of relatively short duration and did not take them far from home, "landsmen" were more willing to sign on as members of such a crew than for the long overseas ocean voyages. On the other side, shipowners were often reluctant to recruit greenhorns for long voyages where they were not easy to dispose of, and in some cases their contracts with the shippers required them to have an experienced crew. Thus, the coastal trade was crucial in training seamen, and this role was important to Parliament as it created potential naval recruits, hence their concern to monitor and register seafarer numbers. This took more formal shape in the creation of the Royal Naval Reserve in 1859,[45] to which many coastal seamen belonged, seeing it as an additional source of income.[46]

Another area in which the government intervened in merchant shipping was that of labour relations. In the eighteenth century, for instance, it instituted a compulsory superannuation scheme via the "seamen's sixpences" fund,[47] although it was not much better administered than some more modern versions and of course imposed criminal law upon contracts between employee and employer.[48] In the nineteenth century further intervention took place. For example, the crew agreements were in part intended to reduce the grounds for crew indiscipline by providing written evidence of the wage rates at which they agreed to serve and their acknowledgment of this via their signature.[49] Similarly, acts established minimum dietary scales, for example the 1867 Act,[50] though these were more significant in deep-water trades because the crews of most coasters found their own rations. In the 1880s, when land-based industrial relations were being taken out of the realm of criminal law and put into civil law, a parallel move occurred at sea with desertion becoming subject to

---

[44]For example, in 1671 by Colonel Birch and again in the early nineteenth century; see Hervey Benham, *Once upon a Tide* (London, 1955; 2nd rev. ed., London, 1986), 69.

[45]Cox, "Sources," 183.

[46]Rita T. Pope (ed.), *Down to the Sea in Ships: The Memoirs of James Henry Treloar Cliff* (Redruth, 1983), 39; and W.J. Slade and Basil Greenhill, *West Country Coasting Ketches* (London, 1974), 22.

[47]Jon Press, "The Collapse of a Contributory Pension Scheme: The Merchant Seamen's Fund, 1747-1851," *Journal of Transport History*, 2nd ser., V, No. 2 (1979), 91-104.

[48]Alderman, "Joseph Chamberlain," 171.

[49]Cox, "Sources," 176.

[50]Geoffrey Alderman, "Samuel Plimsoll and the Shipping Interest," *Maritime History*, I, No. 1 (1971), 74.

civil rather than criminal action in the 1880 Act.[51] The government also acted to ban advance notes in 1881,[52] though again these were not normally used in the coastal trade and only applied to deep-water ships. The certification of ship's officers, which might be seen as a sort of professionalization of seafaring by establishing entry exams and hence guaranteeing technical competence and thus another intrusion into working conditions,[53] was not applied to voyages in the home trades, which included all coastal voyages.[54] In general, the government seems to have been less concerned with labour relations on the coasters than on the deep-water ships.

As Gourvish has noted, one of the modern models of government regulation embodies the idea of "capture" whereby the body or individuals acting as regulator are "captured" by the industry. In these cases the relationship between regulator and regulated becomes too cosy and is likely to be less effective. This notion is opposed to the concept of "expert government," as advanced by Oliver MacDonagh, in which once administrators are appointed they begin to acquire expert knowledge of the weaknesses of current regulation and other problems within the industry, and hence the nature of, and justification for, further legislation to correct these abuses.[55] There were a number of "experts" appointed to regulatory roles, such as Captain Denham of the Steamboat Department of the Board of Trade, who was appointed by the 1846 Act,[56] and Thomas Farrer, secretary of the Maritime Department of the Board of Trade and Thomas Gray, his deputy, and in 1865 his successor.[57] The evidence on Denham suggests that he conformed to MacDonagh's idea, for within two years of his appointment he had brought about the introduction of a further act to extend the coverage of the legislation and tighten up some clauses.[58] To

---

[51]Alderman, "Joseph Chamberlain," 171.

[52]David M. Williams, "Advance Notes and the Recruitment of Maritime Labour in Britain in the Nineteenth Century," in Lewis R. Fischer (ed.), *The Market for Seamen in the Age of Sail* (St. John's, 1994), 81-100.

[53]Clifford Jeans, "The First Statutory Qualifications for Seafarers," *Transport History*, VI, No. 3 (1973), 248-267.

[54]Greenhill, *Merchant Schooners*, 118 ff.

[55]MacDonagh, *Pattern of Government Growth*.

[56]9 & 10 Vict. c. 100.

[57]Wilde, "Creation," 204; Alderman, "Joseph Chamberlain," 172 ff.; and Alderman, "Samuel Plimsoll," 78 ff.

[58]11 & 12 Vict. c. 81.

further demonstrate his impartiality, three years after this, in 1851, a further act was introduced raising the penalties for noncompliance and making the appointment of local inspectors independent of local interests.[59] In this regard, if a degree of capture had occurred it was soon reversed.

The Marine Department of the Board of Trade is a less clear-cut case. Certainly Plimsoll felt it was an obstacle to achieving a compulsory load line and accused it of not wishing "to be disturbed" over reform,[60] implying that it had been captured by the shipping interest. Some allowance has to be made for Plimsoll's lack of scruples in raising a storm of protest and his tendency to hyperbole. In addition, although both Farrer and Gray were reluctant to act, this arose from a combination of technical and philosophical reasons. They were reluctant to see regulations introduced that might stultify design at a period of rapid technological change and hence prevent long-term innovations. Both were also dedicated adherents of private enterprise, opposed in principle to government intervention *per se*, especially where they thought that "healthy competition...self interest and emulation" would lead to improvements.[61] In their view, publicity and exposure were enough to bring about reform.

IV

This has been a rather rapid passage through some complex waters. A number of important topics have been ignored, such as the role of the Post Office in pioneering steam navigation on a number of cross-channel routes and hence acting as a model for private operators. Given that these experiments took place from 1821 to the 1840s, it might be argued that the Post Office demonstrated the speed and reliability of steamboats and then withdrew once private firms could match its speed and security.[62] Thus, the state played an important developmental role in this new technology on coastal routes. Another positive role played by the government was establishing a Meteorological Department of the Board of Trade in 1854 which introduced a system of storm warnings and later daily forecasts.[63] This was of most value to coastal ships because the area covered was limited in extent and was essentially the British coasts.

---

[59]14 & 15 Vict. c. 79.

[60]Alderman, "Samuel Plimsoll," 77 ff.

[61]*Ibid.*, 78.

[62]Philip S. Bagwell, "The Post Office Steam Packets, 1821-36, and the Development of Shipping on the Irish Sea," *Maritime History*, I, No. 1 (1971), 4-28; and Brian Austen, "Dover Post Office Packet Services, 1633-1837," *Transport History*, V, No. 1 (1972), 29-53.

[63]Wilde, "Creation," 196.

The role of the "shipping interest" is also worth exploring, partly to compare it with the actions of the railway interest, and partly because it appears to have had significant influence upon the course of legislation. Chamberlain in 1884 described the shipping interest in Parliament as "well organized and with great parliamentary influence,"[64] a judgment echoed nearly thirty-five years later by Walter Long when he was First Lord of the Admiralty: "There was no vested interest in the House of Commons so powerful as the shipping interest."[65] So effective was this lobby that on a number of occasions it prevented legislation from being introduced. For example, to raise money for harbours of refuge by charging tolls on ships passing the harbour was not unusual in the eighteenth century. In the mid-nineteenth century, when it was proposed to extend this practice, the shipping interest prevented it and was able to have existing tolls removed, so putting the whole of the burden of upkeep on the ports themselves. Similarly, the shipowners were able successfully to oppose the compulsory certification of masters and mates between 1836 and 1851.[66] When voluntary examinations were introduced in 1845, their administration was effectively "captured" by local shipowners.[67] The shipping interest also caused Plimsoll's compulsory load lines to be watered down to voluntary markings in 1876, and it also played a role in the withdrawal of Chamberlain's Bill of 1883 to establish a shipping council.[68] Thus, it had teeth and was prepared to use them. It would be interesting to explain why it was so much more effective than the railway interest.

To conclude, this chapter has endeavoured to show that in a classic period of *laissez faire* the state was prepared to act positively to try to save life and property at sea and to pioneer new technology. When it opposed legislation, it was often because no one best form of technology had yet emerged, and it was reluctant to cause design to stultify when improvements might otherwise emerge. Much of the pressure for enquiries and ultimately legislation came as a result of tragedies involving ships in the coastal trade, whether wrecks, collisions or explosions, and many of the reforms and intended reforms affected coastal ships much more than those in the deep-water trades.

---

[64]Alderman, "Joseph Chamberlain," 177.

[65]Great Britain, *War Cabinet Minutes*, DXXXIV, 19 February 1919.

[66]Jeans, "First Statutory Qualifications," 253 ff.

[67]Wilde, "Creation," 194-204.

[68]Alderman, "Joseph Chamberlain," 169; and Alderman, "Samuel Plimsoll," 73.

Chapter 12
An Estimate of the Importance of the British Coastal Liner Trade in the Early Twentieth Century[1]

Introduction

Of late there has been a revival of interest in the role of coastal shipping in the British economy and a preliminary attempt at cross-European comparisons.[2] Nonetheless, there are still large gaps in our knowledge; indeed, it would be fair to say that it is a largely blank slate on which a few words have been written faintly. One thing which seems to be established is that the idea of a single coastal trade is misleading, for there were a range of coastal shipping services catering for a number of different types of customers.[3] In other words, the coaster segmented the market by offering different sorts of services varying in speed, price, regularity, reliability and frequency. At the top end in terms of reliability and speed were the coastal liners. These were large, modern ships running to a regular timetable on a specified route, taking mixed cargoes, some of them small consignments, and charging premium prices.[4] The ships were not merely modern but also well appointed, since they usually carried passengers as well as cargo. For those shippers with regular consignments of bulky goods in large enough volume to fill an entire vessel, there were the regular traders, ships dedicated to a limited range of routes or commodities. These craft frequently returned in ballast and relied on speed and frequency to

[1]This essay appeared originally in the *International Journal of Maritime History*, X, No. 2 (1998), 41-63 (with John Cutler and Gordon Mustoe).

[2]See John Armstrong (ed.), *Coastal and Short Sea Shipping: Studies in Transport History* (Aldershot, 1996), introduction; and Armstrong and Andreas Kunz (eds.), *Coastal Shipping and the European Economy, 1750-1980* (Mainz, 2002).

[3]The next few sentences are based on John Armstrong, "Coastal Shipping: The Neglected Sector of Nineteenth-Century British Transport History," *International Journal of Maritime History*, VI, No. 1 (1994), 182-185.

[4]C. Ernest Fayle, *A Short History of the World's Shipping Industry* (London, 1933; reprint, Abingdon, 2006), 253-254, defined liners as providing "a fixed service, at regular intervals, between named ports...as common carriers of any goods or passengers."

deliver large cargoes efficiently. The best examples of this segment of the market were the screw colliers which plied between the coal fields of the North East and the consuming regions of the South East.[5] For large consignments which could fill a ship but which required movement less frequently and regularly, there was the steam tramp. They were much less predictable than regular traders, as cargoes had to wait until a ship of suitable size became available in the vicinity of the despatching port. Even less reliable was the sailing coaster, since not only was there no guarantee of a vessel being available when the cargo needed to be moved but also even when safely stowed the ship might be delayed by contrary, insufficient or dangerous winds. The schooners and ketches, which still comprised over ten percent of entries and clearances in the coastal trade at the turn of the century, charged the lowest freight rates and carried the cargoes which were least perishable, of lowest value or which needed slow and careful handling, such as bricks, slates and earthenware.[6]

Although we are aware that these different sectors existed, we know nothing about their relative importance within the coastal shipping industry. Indeed, we still have only one estimate of the total volume of work performed in the long nineteenth century by the entire coastal shipping fleet, a judgment that was made some years ago and involved several fairly heroic assumptions about average hauls and carrying capacities.[7] Indeed, there is still relatively little known about the coastal liner firms. No full-length scholarly book on them has ever appeared, although there are a number of articles of varying quality and rigour on some of the more important lines.[8] Given that the coastal

---

[5]J.A. MacRae and Charles V. Waine, *The Steam Collier Fleets* (Wolverhampton, 1990), 47-66; and John Armstrong, "Late Nineteenth-Century Freight Rates Revisited: Some Evidence from the British Coastal Coal Trade," *International Journal of Maritime History*, VI, No. 2 (1994), 45-81.

[6]Great Britain, Parliament, *Parliamentary Papers (BPP)*, "Trade and Navigation Accounts," various years.

[7]John Armstrong, "The Role of Coastal Shipping in UK Transport: An Estimate of Comparative Traffic Movements in 1910," *Journal of Transport History*, 3rd ser., VIII, No. 2 (1987), 164-178, reprinted in Armstrong (ed.), *Coastal and Short Sea Shipping*, 148-162.

[8]Clive H. Lee, "Some Aspects of the Coastal Shipping Trade: The Aberdeen Steam Navigation Company, 1835-80," *Journal of Transport History*, 2nd ser., III, No. 2 (1975), 94-107, reprinted in Armstrong (ed.), *Coastal and Short Sea Shipping*, 90-103; A.M. Northway, "The Tyne Steam Shipping Company: A Late Nineteenth-Century Shipping Line," *Maritime History*, II, No. 1 (1972), 69-88; Ian Bowman, "The Carron Line," *Transport History*, X, Nos. 2 and 3 (1979), 143-170 and 195-213; and Gordon Mustoe, *Fisher Renwick: A Transport Saga, 1874-1972* (East Nynehead, 1997).

liners offered the most up-market service and were most able to compete with the railways in speed and reliability, there are a range of areas where more information is needed. This article tries to address some of them by estimating the total volume of work performed by the coastal liners just before the First World War; by attempting to determine which routes had the greatest frequency and capacity; and by establishing a hierarchy of routes.

Methodology

The methodology used in this analysis is not original. It derives from the methods used by John Chartres and Gerard Turnbull to calculate the growth in the horse-drawn road carrying trade in the seventeenth and eighteenth centuries. They used local directories to ascertain which carriers serviced which routes and how often they called at various inns to get a measure of frequency.[9] After ascertaining whether the service was pack horse, cart or wagon, and by making assumptions about the carrying capacity of such vehicles at different dates, they were able for a number of routes into London to estimate the total annual capacity of horse-drawn vehicles. By doing this over time, they were able to calculate the annual growth rates for this transport service. Although their findings have been revised sharply downward by later research, and although on reflection the growth rate of eleven percent per annum over eighty-odd years seems rather improbable, the methodology was innovative and sound.[10]

This study in effect substitutes coasters for wagons and ports for inns. By a process of multiplication it is then possible to calculate the total coastal shipping capacity on each route and hence to construct a hierarchy. By calculating the distance by sea between each pair of ports and then multiplying this figure by the capacity, total ton-mileage on each route can be calculated, and by summation we can estimate the total volume of work performed in tons and ton-miles by the coastal liner shipping industry.

This methodology was operationalized as follows. First, the names of all firms operating in the coastal liner trade in 1914 were ascertained from the *Directory of Shipowners, Shipbuilders and Marine Engineers (DSSME).*[11] This

---

[9]See, for example, John A. Chartres, "Road-Carrying in England in the Seventeenth Century: Myth and Reality," *Economic History Review,* 2nd ser., XXX, No. 1 (1977), 73-94; and Gerard L. Turnbull, "Provincial Road Carrying in England in the Eighteenth Century," *Journal of Transport History,* New ser., IV, No. 1 (1977), 17-39.

[10]Dorian Gerhold, "The Growth of the London Carrying Trade, 1681-1838," *Economic History Review,* 2nd ser., XLI, No. 3 (1988), 392-410; and John A. Chartres and Gerard L. Turnbull, "Road Transport," in Derek H. Aldcroft and Michael J. Freeman (eds.), *Transport in the Industrial Revolution* (Manchester, 1983), 85.

was then checked against the secondary literature on the coastal liner trade to discover omissions or incorrect designations. The aim was to identify all the coastal liner firms active just before the Great War and then to determine which routes they served, the frequency of the service and the capacity of the ships used. For example, according to the *DSSME*, the Rea Shipping Co. was allegedly in the liner trade. Yet the routes it served, mostly South Wales to ports such as Southampton, Devonport and Portland, seemed suspicious. Other sources confirmed that Rea ran colliers to provide bunkering facilities for ocean-going liners and the navy; hence, its ships were "regular traders" rather than liners.[12] This was given some support by a stamped warning on the crew agreement of the *Queensgarth* stating that "the crew of vessels loaded with coal are warned that taking naked lights into, or striking matches in, the hold or places below deck is attended by very great danger."[13] In other words, the vessel carried coal. The *DSSME* also gave the routes served by each coastal operator and, in some cases, the frequency of service and the names of the ships. Thus, we had a list of all relevant companies, the routes each served and the names of most of the vessels employed by each coastal line.

This information was compared to a variety of other sources to fill in the numerous gaps as well as to verify the data. The main source was the *Shipping and Mercantile Gazette (SMG)*, which by the early twentieth century was published six days per week. It reported the names of the coastal ships and sometimes their masters and ports of call. It was then possible to see how often a particular ship entered or cleared from each port and hence to determine the frequency of service. To reduce the burden of checking an entire year's entrances and clearances, one month (July 1914, the last month before war disrupted services) was taken as representative and the frequency in this sample was then multiplied by twelve to arrive at an annual total. Using only one month might be seen as a weakness in the methodology because there may have been seasonal fluctuations in the frequency of service. Indeed, it might be thought that the ships would have sailed more frequently in the summer because of the adverse weather conditions in the winter. While in some cases this might have been true, two points need to be made. First, by 1914, thanks to triple-expansion engines and steel hulls, coastal liners were mostly large, powerful ships which were little deterred by inclement weather. Rather than weather, it was the nature of the traffic that determined the level of operations. Second, some liner firms increased frequency in the winter rather than in the

---

[11] *Directory of Shipowners, Shipbuilders and Marine Engineers* (London, 1914).

[12] MacRae and Waine, *Steam Collier,* 149.

[13] Great Britain, National Archives (TNA/PRO), Board of Trade (BT) 99/3057.

summer. For instance, in the 1870s the Aberdeen Steam Navigation Company (ASN) ran two steamers per week in the summer and three in the winter.[14] As the chairman explained to a general meeting, this was because "during the winter months the traffic is greatest."[15] This probably resulted from the large traffic for London in cattle, meat and oats which were available between November and February. For example, in 1896 the ASN carried over 100,000 sacks of oats from Aberdeen to London, of which 52,000 were handled during the winter. By contrast, in the four summer months only 22,000 sacks were shipped.[16] Thus, although there may have been seasonal variations in service, there was no clear bias toward one season, and the fluctuations might well cancel each other out. Alternatively, it may be that the results here understate the significance of the coastal trade.

Unfortunately, some ships did not appear in the *SMG*, particularly those serving smaller ports or on the Dublin-Belfast route. For the latter, the Dublin bills of entry were used to confirm the names and frequency of ships in this trade. For the trade to the Islands and Highlands, served by David MacBrayne, the Clyde bills of entry were used to fill in the gaps.[17] In addition, some entries were checked against local newspapers. For instance, the *Aberdeen Free Press* ran advertisements every day in July 1914 for the Aberdeen Steam Navigation Co. which stated that it had ships sailing every Wednesday and Saturday evening for London, returning from the capital every Wednesday and Saturday. This leads one to expect that there should be about eight sailings in each direction for this company in the data base, which in fact is what has been recorded.

Given that by definition a liner trade requires regularity of service, we interpreted this in practical terms to mean that a service needed a minimum frequency of once a fortnight to qualify. If a route did not have two return voyages in the sample month, it was considered to be something other than a liner trade, even if it was served by a vessel normally employed as a liner. In such cases, the vessel was not included in our calculations. In other cases, it was clear that a voyage was missing. For example, a ship arrived at Plymouth and was then recorded as leaving London. The vessel must have sailed from Plymouth to London, even though it was not picked up in the records. To

---

[14]*Aberdeen Directory,* 1877-1880.

[15]Aberdeen University Library, Ms. 2479/12, Aberdeen Steam Navigation Company, Minute Book of General Meetings, 28 February 1876.

[16]Guildhall Library, Ms. 1667/1, "Return of Grain Brought into the Port of London," 1896.

[17]*Dublin Bill of Entry and Shipping List,* July 1914; and *Clyde Bill of Entry,* July 1914.

minimize over-counting, the Plymouth-London passage was assumed to have been made in ballast and was excluded from our calculations.

Another complication was that most liners carried passengers as well as cargo. This might be seen as a problem, since the space taken by passengers would diminish the area available for cargo, but in practice space was not transferable between the two uses, and passenger space could not in the short run be converted to cargo capacity. For most routes this was not a major problem because the number of passengers carried was small, both absolutely and as a proportion of total available space. There is some support for this in the crew agreements, which were required to declare the number of passengers if more than twelve were on board. The vast majority of masters did not complete this section, suggesting that at most only a few passengers were carried. On the other hand, it could be that the master simply neglected to fill out all sections of the form, and not much can be read into this omission. The balance of evidence from the secondary literature is that on most routes few passengers were carried, and that those that were detracted little from the space available for cargo. It is more worrisome for those routes where no alternative means of moving people was available, such as the trans-Irish Sea service or voyages from the British "mainland" to the Isle of Man or the Isle of Wight.

The *DSSME* reported both gross registered (grt) and net registered tonnage (nrt) for the ships. But neither is synonymous with carrying capacity, which is more accurately measured by deadweight tonnage (dwt). Grt, for instance, included engine and bunker space, neither of which was available for cargo. To more closely approximate cargo capacity, space which could not be used for goods was deducted from grt to give nrt. Although nrt was closer to actual carrying capacity, it too has problems, since many of the deductions were done by formula and holds were not measured.[18] Fortunately, Maywald has calculated the factors required to convert nrt to dwt which, he believed, depended upon hull material and date of build.[19] As the *DSSME* gives both nrt and grt, Maywald's factors were used to convert nrt to dwt, except in those few cases where the *DSSME* gave dwt as well.[20]

In the end, then, we have a list of each ship employed in the coastal liner trade, with the route(s) it worked and its carrying capacity. It is only a matter of simple arithmetic to obtain the total liner capacity available on each

---

[18]Yrjö Kaukiainen, "Tons and Tonnages: Ship Measurement and Shipping Statistics, c. 1870-1980," *International Journal of Maritime History*, VII, No. 1 (1995), 29-56.

[19]K. Maywald, "The Construction Costs and Value of the British Merchant Fleet, 1850-1938," *Scottish Journal of Political Economy*, III, No. 1 (1956), 47-49.

[20]The nrt of each ship was multiplied by the following to arrive at dwt. For ships built in the 1880s, 2.1; the 1890s, 2.2; the 1900s, 2.3; and the 1910s, 2.4.

route. The best measure of work performed by a transport mode is the ton-mile, which takes into account the distance of the voyage as well as the volume of the cargo. The distance between pairs of ports was derived from an appropriate marine atlas; this allowed the ton-mileage for each route to be calculated.[21]

We also needed to ensure that there was no double-counting where a ship sailed from port A to port B with intermediate calls at ports C and D. The mileage would have been exaggerated if the calculation were made as A to C, plus A to D, plus A to B. To avoid this, such a route was recorded as A to C, plus C to D, plus D to B. In this way, the distance recorded for such a "stopping service" was very similar to that from A to B direct. Only direct voyages from A to B are so recorded in the data base. A journey from A to B via C and D will not show as a trip from A to B, although some part of the cargo could well have made this voyage. This means that the importance of some long-distance routes, such as Liverpool-London, are understated because they will be shown instead as a succession of shorter journeys.

Aggregate Output

The results of this study suggest that in 1914 the total capacity of the British coastal liner trade was 20,082,600 tons and that the total work performed was 5,447,644,956 ton-miles. The average haul was 271 miles. These figures, however, are in nautical miles; to make them comparable to other forms of transport they need to be multiplied by 1.15 to convert them into statute or land miles. This then gives a ton-mileage of 6,264,791,600 and an average haul of 312 miles.

These results can be compared to two other estimates of the work performed by coasters. A decade ago John Armstrong suggested that the whole coastal trade of the United Kingdom in 1910 amounted to 81.5 million tons, with a total ton mileage of 20.4 billion.[22] The average haul was 252 statute miles. This was further broken down into the coal trade and all other commodities. Since liners almost never carried coal, the liner trade would look more like the "other" category, where the figures were 57.3 million tons, 13.2 billion ton-miles and an average haul of 231 miles. What this suggests is that liners performed a little less than one-third of the total work of the British coastal fleet in the last years before the First World War and was about equal to the total ton-mileage in the coal trade. All other trades, namely steam tramps and sailing vessels with cargoes other than coal, accounted for the final third of activity.

---

[21]Harold Fullard (ed.), *The Mercantile Marine Atlas* (London, 1959).

[22]Armstrong, "Role," 176.

At 312 miles, the average haul in the liner trade was greater than either the coastal colliers (296 miles) or the tramp ships with non-coal cargoes (231 miles). This might be explained by the relative values of the cargoes carried. Liners carried the most valuable cargoes, such as manufactured or semi-processed goods, while the tramps and sailing vessels specialized in bulky low-value cargoes like coal, china clay, iron ore and grain.[23] The higher-valued goods could bear longer journeys and higher freight rates since transport costs were a smaller proportion of their total value. Yet the differences between the average distances are not so great as to cause any concern over the methodology. On the contrary, this study reinforces the findings of the earlier work.

The other examination of work performed by coasters was carried out by Percy Ford and John Bound. Published in 1951 and based on 1948 statistics, their findings can only be used with extreme caution to verify the figures in this study because of the great temporal difference. Moreover, the radical economic changes wrought by the two world wars and the intervening depressions obviously had an especially serious impact on the coastal trade, as did the rise of road transport. Nonetheless, the Ford and Bound study does offer the possibility of measuring change over time. They found that about 8.8 billion ton-miles were worked in 1948 by the coastal fleet, of which 7.6 billion were in coal voyages.[24] This is a much reduced total compared to the 1910 figure, although the coastal coal trade in 1948 looked very similar to that in 1910. What seem to have declined precipitously are the liner and general tramp trades. To some extent this is not surprising. In 1948 Ireland was no longer part of the United Kingdom; hence, commerce between Eire and the UK had been reclassified as foreign trade, and voyages between two Irish ports were no longer included. The staple industries, which provided much of the bulk cargoes for coasters, had declined, and road haulage was taking away the more valuable commodities which had been carried previously by the coastal liners. Thus, Ford and Bound's study seems intuitively compatible with the estimate arrived at in this article, albeit with a sharp long-run decline.

The average hauls calculated by Ford and Bound – 368 statute miles for coal traffic and 359 overall – are higher than found in the two pre-First World War studies, but not by enough to cast any serious doubts. It might well be that the motor lorry was most able to compete on the shorter runs, taking these away from coasters at an early date. As a result, coasters retreated into the very long-haul market, a niche in which neither the railways nor the lorries were keen to compete.

The data base also provides some indication of other dimensions of the British coastal liner trade just before World War I. There were 138 differ-

---

[23]Armstrong, "Coastal Shipping," 181.

[24]*Ibid.*, 22.

ent routes, plied by fifty separate companies and serving seventy-seven ports. This reinforces the view advanced elsewhere that the British coastal trade involved a large number of ports.[25] This is brought out clearly in the map of routes and ports (see figure 1). Even where there was no direct liner service between two ports, the journey could be made indirectly. For instance, while there was no direct line between Bristol and Hull, there was a service offered by George Bazeley between Bristol and London, and another between London and Hull operated by the General Steam Navigation Co. Thus, by using London as a transhipment point, goods could be sent between these ports. Indeed, most combinations of ports were possible. Even when a town was not on the liner network, small consignments could be forwarded by irregular traders from the closest port on the network. Most coastal and riverine communities could be served by the coaster expeditiously and reasonably regularly.

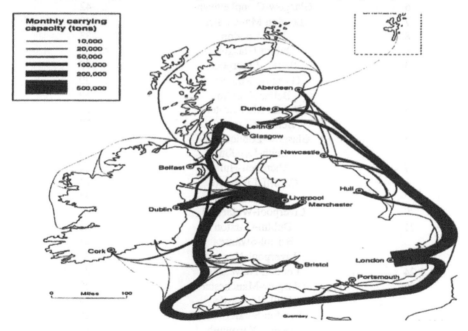

Figure 1:   The British Coastal Liner Trade in July 1914

Source: See text.

Relative Importance of Routes and Ports

The calculation of aggregate tonnage and ton-miles performed by British coastal liners also enables us to determine the relative importance of various

---

[25]Armstrong (ed.), *Coastal and Short Sea Shipping*, xvii.

routes. There are three different measures: frequency of service (how often a ship left each of a pair of ports per month); capacity (dwt of the ships times the number of calls by each at that port); and the extent of the liner trade in each port (the aggregate capacity of all routes ending or beginning at a port).

Table 1
Top Coastal Liner Routes by Frequency of Service, July 1914

Rank	Route	Frequency
1	Dublin-Holyhead	133
2	Aberdeen-Leith	87
3	Glasgow-Liverpool	57
4	Liverpool-London	45
5	Belfast-Liverpool	43
6	Glasgow-Campbeltown	42
7	London-Manchester	40
8	Hull-London	33
9	Dublin-Glasgow	32
10	Bristol-Liverpool	30
	Glasgow-Rothesay	30
	Belfast-Dublin	30
13	Dublin-Manchester	29
14	Belfast-Glasgow	27
15	Glasgow-Londonderry	26
	Leith-London	26
17	Aberdeen-Hull	24
	Cardiff-Liverpool	24
	Cork-Liverpool	24
20	Liverpool-Waterford	24
21	Dublin-Garston	22
	Bristol-Swansea	21
	Liverpool-Newry	21
24	London-Newcastle	21
25	Glasgow-Manchester	20
	Hull-Newcastle	19
	London-Sunderland	19
28	London-Yarmouth	19
29	Cardiff-Swansea	18
	Bo'ness-London	17
	Douglas-Liverpool	17
	London-Goole	17
	London-Middlesborough	17

Source: See text.

These measures enable us to rank the routes and to determine the most important ports in the coastal liner trade. Table 1 shows the top thirty-three routes based on frequency of service. The frequency figures are of one-way voyages in the course of a single calendar month. Thus, each figure needs to be divided by two to give the number of return voyages. The most frequent service was across the Irish Sea between Holyhead and Dublin, where on average there were two sailings in each direction every day. Next in popularity was between Aberdeen and Leith, which averaged more than one sailing from each port per day. In third place came Glasgow-Liverpool with approximately one sailing per day each way. An analysis of trade by broad region suggests that the highest frequency of service was across the Irish Sea, which accounted for ten of the top thirty-three routes, with an average of thirty-eight sailings per route per month. The next most frequently served was the east coast, where ten routes averaged twenty-seven sailings per month, followed by the west coast with eight routes averaging thirty sailings per month. What is remarkable is the frequency of service for the long route from Liverpool and Manchester to London, which was in excess of 600 nautical miles and averaged forty-two sailings per month. It might have been thought that the railways had captured all this traffic, since the mileage between London and either of these North West cities was just over 200 land miles, much shorter than by sea. In practice, however, this was not the case.

Table 2
Top Coastal Liner Routes by Carrying Capacity (Tons), July 1914

Rank	Route	Monthly Carrying Capacity
1	Dublin-Liverpool	80,785
2	London-Manchester	70,064
3	Liverpool-London	65,554
4	Aberdeen-Leith	63,267
5	Dublin-Holyhead	56,437
6	Dundee-London	56,310
7	Belfast-Liverpool	48,693
8	Dublin-Glasgow	40,982
9	Leith-London	40,784
10	Hull-Newcastle	38,231
11	Cork-Liverpool	35,208
12	London-Newcastle	34,639
13	Glasgow-Liverpool	33,379
14	Hull-London	29,689
15	Grangemouth-London	28,176
16	Bristol-Liverpool	27,487
17	Glasgow-Londonderry	26,678
18	Dublin-Manchester	25,705
19	London-Goole	25,175
20	Aberdeen-London	23,677

Source: See text.

When carrying capacity is taken into account, as well as service frequency, the results are a little different (see table 2). The route with the largest capacity was Dublin-Liverpool and the number two and three positions were filled by long-distance trade from London to Manchester and Liverpool. The leading route is not too surprising given that there was no alternative at this date to ship goods between Ireland and Great Britain. The second and third places are more unexpected and point up the ability of the coaster to continue to provide a valued service, even on routes where it had a natural disadvantage, such as the long sea journey from London to the North West.

Table 3
Top Twenty Ports Served by Coastal Liners by
Frequency of Service (Sailings and Arrivals), July 1914

Rank	Port	No. per Month
1	Liverpool	520
2	London	468
3	Glasgow	386
4	Dublin	379
5	Aberdeen	196
6	Leith	150
7	Bristol	138
8	Holyhead	133
9	Manchester	125
10	Belfast	117
11	Hull	114
12	Cardiff	82
13	Dundee	79
14	Newcastle	79
15	Swansea	77
16	Cork	55
17	Waterford	45
18	Campbeltown	42
19	Plymouth	39
20	Llanelly	36

Source: See text.

If the top twenty routes are aggregated into broad groupings, the rank order is rather different from table 1. The east coast was home to nine services with a capacity of 340,000 tons per month, compared to seven routes across the Irish Sea with an aggregate capacity of 314,000 tons. The west coast had only two of the top twenty routes, with a monthly capacity of 61,000 tons. This was totally eclipsed by the long-distance routes from London to the North West of England, with a monthly capacity of 136,000 tons. In this case, the amount of traffic generated between major cities, such as Manchester, Liverpool and London, was much greater than between west coast ports such as Glasgow, Liverpool and Bristol. The pull of London as an entrepôt, assembler of exports and distributor of imports may have been crucial in this regard.

Table 3 gives the twenty most important ports involved in coastal trade by frequency of service. It is dominated by the major ports, such as Liverpool, London, Glasgow and Dublin, which were nodes for overseas as well as coastal trade, or were terminals for ferry lines running across the Irish Sea. Perhaps the least expected is Aberdeen, in fifth place, which seems higher than its importance ought to justify. The appearance in ninth place of Manchester is explained by the completion of the ship canal, which gave direct access to larger vessels, and the operations of the Fisher Renwick Line in providing frequent large steamers for London.[26]

When total carrying capacity is calculated, a slightly different ranking emerges (see table 4). This list reflects rather more accurately the order of importance of the urban centres in terms of population, commerce and manufacturing. It is no surprise to see London at the top of the rankings, or Liverpool, the second largest port in the UK, Dublin, the capital of Ireland and hence the centre for much cross Irish Sea traffic, and Glasgow with its huge shipbuilding and heavy engineering base following closely behind.[27] Again, the fifth place showing of Aberdeen is surprising given that it ranked only twenty-second in terms of population in Britain.[28] We believe that Aberdeen's high ranking was partly a result of its role as the staging post from the mainland to the Orkneys and Shetland, since direct sailings to these places comprised eleven percent of Aberdeen's total sailings and twelve percent of its tonnage. Yet removing them would make no difference to Aberdeen's position in the frequency table and would push it down only one place in tonnage. The largest single route for Aberdeen was to Leith, which contributed forty-four percent of sailings and forty percent of tonnage. On this route the lion's share of services was provided by the North of Scotland and Orkney and Shetland Steam Navigation Co., which contributed nearly two-thirds of total tonnage. This was a feeder service to the route to the islands: each ship started its voyage at Leith and called at Aberdeen en route to Stromness, Kirkwall or Lerwick.[29] Given

---

[26]D.A. Farnie, *The Manchester Ship Canal and the Rise of the Port of Manchester* (Manchester, 1980); and Mustoe, *Fisher Renwick*, chap. 2.

[27]W. Hamish Fraser and Irene Maver (eds.), *Glasgow, Volume II: 1830-1912* (Manchester, 1996).

[28]B.R. Mitchell with Phyllis Deane (comps.), *An Abstract of British Historical Statistics* (Cambridge, 1962; reprint, Cambridge, 1976), 25-27.

[29]Gordon Donaldson, *Northwards by Sea* (Edinburgh, 1966; 2nd ed., Edinburgh, 1978), 28.

that all visitors to these islands had to go by sea at this time, that there was some tourist trade in the summer and that all of the North of Scotland's ships were equipped to carry passengers as well as cargo, it may be that the tonnage and ton-mileage for this company – and hence for Aberdeen – are slightly inflated. The carrying capacity of the ships may be less than the calculated dwt because more space was devoted to passengers than in other ships. Hull's rise in the rankings from eleventh by frequency to eighth by carrying capacity is understandable given its population size, extensive docks and foreign trade, and large manufacturing sector. Since in general these rankings correlate well with a port's rank in British trade and industry, we can conclude that the coastal liner trade reflected the economic levels of individual ports and was an indispensable part of that activity.

Table 4
Top Twenty Ports Served by Coastal Liner Companies by Tonnage of Carrying Capacity per Month

	Port	Tonnage per Month
1	London	577,719
2	Liverpool	442,474
3	Dublin	265,473
4	Glasgow	239,708
5	Aberdeen	161,224
6	Leith	145,820
7	Manchester	138,476
8	Hull	111,568
9	Bristol	110,682
10	Dundee	108,329
11	Newcastle	100,681
12	Belfast	96,187
13	Cork	65,574
14	Cardiff	56,932
15	Holyhead	56,437
16	Plymouth	50,376
17	Swansea	46,491
18	Waterford	39,832
19	Southampton	38,120
20	Grangemouth	33,412

Source: See text.

Coastal Liner Companies

Another benefit of this data base is the ability to examine the relative importance of individual firms in the coastal liner trade. As already mentioned, there were fifty different lines running regular services in 1914. Although this suggests a fairly competitive industry, when compared to the railway sector it looks *more* concentrated, for in 1914 there were over 100 separate railway companies in the UK, twice the number of coastal liner firms.[30] Mere numbers, however, are not a good guide to levels of concentration. We also need to know the proportion of the total market served by these companies. This information is provided in tables 5 and 6. The former lists the top twenty companies in the coastal liner trade measured by aggregate ton-mileage. The firms that dominated operated long-distance routes, such as Liverpool-London, Manchester-London, Glasgow-eastern Scotland, Glasgow-Ireland and Glasgow-south coast. It was this long-haul traffic which created large ton-mileages.

Table 5
Top Fifteen Coastal Companies in 1914 by
Ton-Mileage Worked

Rank	Company	Ton Mileage Per Year
1	Powell, Bacon and Hough Lines Ltd.	824,221,488
2	Fisher Renwick Manchester-London Steamers Ltd.	554,906,880
3	M. Langlands and Sons	520,725,528
4	Clyde Shipping Co. Ltd.	367,364,052
5	Dundee, Perth and London Shipping Co. Ltd.	320,121,504
6	Carron Line	261,362,484
7	Tyne-Tees Shipping Co. Ltd.	244,040,472
8	City of Cork Steam Packet Co. Ltd.	202,367,964
9	London and Edinburgh Shipping Co. Ltd.	185,271,168
10	George Bazeley and Sons Ltd.	178,916,568
11	G. and J. Burns Ltd.	155,163,132
12	Aberdeen Steam Navigation Co.	128,419,968
13	Thomas Wilson Sons and Co. Ltd.	124,737,840
14	London Welsh Steamship Co. Ltd.	119,664,552
15	General Steam Navigation Co. Ltd.	111,272,856

Source: See text.

It would be ideal, albeit impossible, to have revenue figures for each firm. On the other hand, ton-mileage is not a bad surrogate. Apart from some tapering of charges over the long haul, freight rates were roughly proportional to distance and certainly to tonnage carried. The company performing the most work was the Powell, Bacon and Hough Line, which operated out of Liverpool and traded to many west and south coast ports en route to London. As its name implies, it was a merger of three separate companies and was soon to change

---

[30]Gerald W. Crompton, "'Efficient and Economical Working?' The Performance of the Railway Companies, 1923-33," *Business History*, XXVII, No. 2 (1985), 222.

its name to Coast Lines, in which form it would absorb many other coastal liner companies in the interwar years.[31] It was easily the premier company at this time, performing nearly fifty percent more work than the next largest company, Fisher Renwick.[32]

Table 6
Concentration Ratios for Coastal Liner Firms in 1914
(Percent of Total)

Top	Frequency	Carrying Capacity	Ton Mileage
1	7.5	10	15
2	14	17	25
3	20	23	35
4	25	28	42
5	30	32	48
15	60	68	79

Source: See text.

The degree of concentration in an industry can be measured by concentration ratios. These try to ascertain the proportion of the total market served by the top firms in an industry. Table 6 lays these figures out by all three measures which we have used – frequency of service, carrying capacity and ton-miles – for the leading firms. Two observations stand out. First, the degree of concentration was greater in carrying capacity than in frequency and was greatest of all in ton-miles. Second, there was a significant degree of concentration in this industry. If all fifty firms had been roughly equal in size, each would have taken about two percent of the total market. In practice the single largest company accounted for more than seven percent of all sailings, ten percent of capacity and fifteen percent of ton-miles. Similarly, the top five firms contributed one-third of total capacity and nearly one-half of all work, and the top fifteen provided nearly two-thirds of capacity and four-fifths of total ton-miles. As might be expected, this was a very uneven industry: ten percent of firms accounted for nearly half of all route mileage and seventy percent of the firms performed only twenty percent of total work.

These levels of concentration can be compared to those in the railway industry, the chief rival to the coastal liner in terms of regular and reliable goods transport. Terry Gourvish has argued that as early as 1870 the four largest railway companies – LNWR, Midland, GWR and NER

---

[31]E.R. Reader, "World's Largest Coaster Fleet," *Sea Breezes,* VIII (1949), 88-105.

[32]Mustoe, *Fisher Renwick.*

– earned forty-four percent of all revenue, while the largest fifteen firms earned over eighty percent of total revenue.[33] He believes that these levels of concentration changed hardly at all down to 1913. The railways have long been acknowledged as being rather concentrated, and the merger manias were well documented and caused much public concern. By the standards of the railway industry, the coastal liner business was less concentrated, but not by much. Whereas the top four railway firms took forty-four percent of revenue, the top four coastal liners provided forty-two percent of all ton-miles, an insignificant difference. Similarly, while the top fifteen railway firms took just over eighty percent of total revenue, the top fifteen coastal liner businesses provided just less than eighty percent of all ton-miles. Again, the difference here is well within the margin of error and is not significant. The difference in degrees of concentration in the two industries was negligible, although railways have long been considered "concentrated" while coastal liners have been assumed to be much more competitive. Of course, we should not lose sight of the enormous differences in the two industries in their capitalization, revenue and employment. The railways were Brobdingnagians compared to the coastal Lilliputians, but within their own industries both were surprisingly similar in degree of concentration.

Crew Agreements

Ideally, an historian prefers to check any source by comparing it with another which throws light from a different angle, but it is rare to find such a surfeit of documentation. The frequency data used here were derived from the *SMG,* and it is impossible to know how reliable and comprehensive it is as a source. One way in which these figures might be checked is by looking at the crew agreements for those ships identified as operating in the coastal liner trade.[34] These documents, when completed each six months by ships in the coastal trade, have a section on the last page in which the master can record the journeys made by the ship in the half year. In theory, it should be easy to compare the data presented here with that recorded in the *SMG.* In practice, however, there are some problems in using the crew agreements. For a start, they are in three main locations – the National Archives, National Maritime Museum and Memorial University of Newfoundland – and identifying individual ships is not easy. Yet there is a more fundamental drawback: in many cases the master did not specify the number of voyages. For instance, the crew list of *Blackrock,*

---

[33]T.R. Gourvish, *Railways and the British Economy, 1830-1914* (London, 1980), 10.

[34]Keith Matthews, "Crew Lists, Agreements, and Official Logs of the British Empire, 1863-1913," *Business History,* XVI, No. 1 (1974), 78-81.

owned by the Tedcastle Line, simply states "Trading regularly between Liverpool and Dublin, Dublin and Liverpool for a period of six months" and gives no indication of the number of voyages.[35] Others are even less informative, sometimes omitting the route(s) being served. For example, the crew agreement of *Lady Gwendolen,* owned by the British and Irish Steam Packet Co., simply recorded "in the coasting trade of the UK."[36]

In some cases, there is an apparent difference between the two sources. This was the case for *Wirral Coast,* owned by the Powell, Bacon and Hough Line. The *SMG* states that this ship made four return voyages in July between Bristol and Manchester, while the crew agreement suggests a more complicated pattern in which the vessel sailed from Swansea to Manchester, Manchester to Bristol, Bristol to Cardiff and Cardiff to Swansea.[37] What has happened here is that the *SMG* did not pick up the intermediate calls made en route from Bristol to Manchester. In practice, however, the effect on ton-mileage is minimal: the data base will under-record distance only slightly. In this respect, the aggregate figures reported above may understate the actual case, but the difference would be small.

Some of the crew agreements are more informative and do give a breakdown of the routes and the dates of departure and arrival. For instance, *Kelvinside,* owned by the General Steam Navigation Co., lists six voyages from Lowestoft to London and five from London back to Lowestoft in July.[38] Where such detail is given it is possible to compare the routes and number of voyages made according to the crew agreement with those in the data base derived from the *SMG.* In this case, the crew agreement records more voyages than the *SMG* picked up – eleven compared to eight. Where direct comparisons were possible, and this was only for a small sample of ships, the *SMG* appears to under-record the number of voyages compared to the crew agreements. The difference between the two sources varied considerably, in some cases being in close accord, as with *Virago,* where the crew agreement recorded one more voyage between Liverpool and Glasgow than did the *SMG,*[39] and in others being in significant disagreement, as with *Kelvinside* where the *SMG* recorded four return trips between London and Lowestoft and the crew

---

[35]TNA/PRO, BT 99/3001.

[36]*Ibid.*, BT 99/3070.

[37]*Ibid.*, BT 99/3049.

[38]*Ibid.*, BT 99/3002.

[39]*Ibid.*, BT 99/3045.

agreement showed five return trips and one extra on the southbound leg.[40] In general, it looks as though the *SMG* under-recorded between ten and twenty percent of the voyages recoverable from the crew agreements. It is thus possible that the aggregate figures above underestimate the amount of work performed by coastal liners by between ten and twenty percent. As a result, the figures should be treated as minima rather than maxima.

Conclusion

This paper has estimated the amount of work performed by the British coastal liner trade just before the First World War. The results propose a total goods capacity of about twenty million tons and an output figure of approximately 6.25 billion ton-miles, with an average haul of just over 300 miles. These figures are compatible with those calculated a decade ago by Armstrong for the entire coastal fleet and suggest that the total coastal trade divided into three roughly equal sectors: liners, coal and the remainder. The relative importance of ports and routes has been calculated from the data base. The ports with the largest coastal liner capacity were generally those one would expect: cities with large populations active in overseas trade, such as London, Liverpool, Dublin and Glasgow. The routes that had the greatest capacity usually linked these cities, so that Dublin-Liverpool, London-Manchester and Liverpool-London were the most served routes, with other major ports also having relatively intensive traffic. The most important coastal liner companies in ton-mile terms were those that operated long-distance routes, such as the Powell, Bacon and Hough Line between Liverpool and London and Fisher Renwick, which linked Manchester and London.

What is clearly demonstrated by this research is that the British coastal liner trade just before the First World War was extensive. All major UK ports were served, and ships owned by fifty separate firms operated on 138 different routes. While no investigation has been made in this article of the cargoes, this is a topic that would repay future research. It appears that the liners competed for higher-value goods, not just low-value bulky items, by offering scheduled departure and arrival times and by making rapid passages at economical rates. They continued to play a crucial role in the British economy down to the Great War.

---

[40]*Ibid.*, BT 99/3002.

Chapter 13
The Role of Coastal Shipping in UK Transport: An Estimate of Comparative Traffic Movements in 1910[1]

I

The aim of this article is to estimate the amount of work performed by coastal shipping in the period just before World War I and to compare its contribution with that of the railways and canals. Until recently, with honourable exceptions,[2] the coastal trade in the railway era was either ignored[3] or merited but scant coverage.[4] The impression was given that the railways, like juggernaut's chariot, swept everything else away, rendering obsolete the coach, wagon, canal barge and coaster. By 1910 the railway network was virtually complete compared with, say, 1875, when many branch lines had yet to be constructed, or 1850, when only the basic network had been built, and it might be assumed that little scope was left for the carriage of freight in coastal ships. Thus, the year 1910 should provide the least favourable case for the coaster and the most favourable for the railway in terms of total tonnage of goods carried. Passenger traffic will not be considered here; this is not to imply that coastal shipping

---

[1]This essay appeared originally in *Journal of Transport History*, 3rd ser., VIII, No. 2 (1987), 164-178, and was reprinted in John Armstrong (ed.), *Coastal and Short Sea Shipping: Studies in Transport History* (Aldershot, 1996), 148-162.

[2]Philip S. Bagwell, *The Transport Revolution from 1770* (London, 1974; New ed., London, 1988), chapter 3, pioneered the study of coastal shipping in the nineteenth century.

[3]Savage, writing in 1959 and 1966, made no mention of coastal traffic in the nineteenth century; Christopher I. Savage, *An Economic History of Transport* (London, 1959; 2nd rev. ed., London, 1966).

[4]Dyos and Aldcroft devoted two pages out of six chapters to it, though acknowledging that it did play a significant role in freight traffic; H.J. Dyos and Derek H. Aldcroft, *British Transport: An Economic Survey from the Seventeenth Century to the Twentieth* (Leicester, 1969; reprint, Harmondsworth, 1974), 208-210. Barker's major revision of Savage's book also devoted two pages to coastal shipping in the early nineteenth century; T.C. Barker and Christopher I. Savage, *An Economic History of Transport in Britain* (London, 1974), 70-72.

did not carry people, but it seems indisputable that where coaster and railway competed, passengers preferred rail to sea travel.[5] The main strength of the coastal passenger ship in 1910 was in services the railway could not provide in ferrying passengers to places like the Isle of Wight, Isle of Man, Channel Isles and the Scottish islands. This article will confine itself to considering the use made of the various types of transport in providing cargo-carrying services.

One measure of freight traffic which might be used is tonnage carried. At first sight this seems the obvious criterion; it is commonly used for foreign trade or modern lorry traffic, and the capacity of railway wagons or canal barges seems crucial. However, a more informative and accurate measure of total cargo movement performed is ton-mileage, as this takes into account not merely the tonnage of goods moved but also the distance consignments were carried. The value of ton-mileage has long been recognized by railway managers and statisticians. Historians, too, have emphasized its importance, notably Hawke, who used it as the measure of total railway output,[6] and Cain, who said that "[t]on miles...are the only reasonable measure of railway output."[7] Different forms of transport may be better suited for different lengths of journey. For example, the horse and cart was acceptable for short-distance town transport, while patently neither the railways nor the coasters were suitable.[8] Thus, an investigation of the average haul may shed light on the different market segments at which various forms of transport aimed, as well as giving a more comprehensive measure of the total carrying activity performed by each. I have therefore tried to reach some conclusions on three quantitative characteristics of the three forms of transport: average haul, tonnage carried and total ton-mileage performed. This article is confined to the canals, railways and coastal shipping in the belief that long-distance road traffic by horse had largely died out by 1910 and the motor lorry had not yet proved itself.

II

The source used for canal traffic is the *Report of the Royal Commission on Canals and Inland Navigation of the United Kingdom* (chaired by Baron Shut-

---

[5]Dyos and Aldcroft, *British Transport*, 209.

[6]G.R. Hawke, *Railways and Economic Growth in England and Wales, 1840-1870* (Oxford, 1970), 63-69 and 91-92.

[7]Peter J. Cain, "Private Enterprise or Public Utility? Output, Pricing and Investment on English and Welsh Railways, 1870-1914," *Journal of Transport History*, 3rd ser., I, No. 1 (1980), 9.

[8]Bagwell, *Transport Revolution*, 138-144; and John R. Kellett, *Railways and Victorian Cities* (London, 1979), 349.

tleworth). Its *Fourth Report,* issued in 1910, contains much material on the goods traffic of the canals in the period 1905-1906.[9] It calculated the average haul for fourteen canals, representing about sixteen percent of total UK mileage and thirty percent of total tonnage, as 17.5 miles, with a range for individual canals of between 3.5 and twenty-three miles.[10] It emphasized that most canal traffic was short-haul[11] and said of the Trent and Mersey Canal that "the average distance over which traffic passes is stated to be about twenty miles"[12] and of the Birmingham canal system that "most of this traffic is carried for very short distances on the waterways."[13] The Royal Commission concluded that "the great bulk of the trade borne by inland waterways is of a local character and that long-distance traffic – any traffic, say, involving a journey of more than fifty miles – is relatively small in its proportion."[14] The Commission examined the reasons for the canals' failure to retain a larger share of long-haul traffic. It laid some blame on the railway companies, which owned about a third of canal mileage and seemed unwilling to maintain the system, let alone improve it, but this issue need not detain us here.[15] The important point is that the average haul by canal seems to have been less than twenty miles.

The Royal Commission's working definition of long-distance traffic as that over fifty miles begs the question of why it chose this figure. If its implicit comparison was with the railways, then it would be interesting to know its source, for no such statistics were systematically collected prior to World War I.[16] Munby, using the railway returns in the *Parliamentary Papers*, gives average hauls for railway traffic in the 1920s broken down into coal and coke, minerals, merchandise and livestock. His figures averaged 54.4 miles for all goods traffic in 1920-1925 and 52.8 miles for the period 1922-1925, if 1920

---

[9]Great Britain, Parliament, *Parliamentary Papers (BPP)*, 1910, XII, 1-757.

[10]*Ibid.*, para. 280.

[11]*Ibid.*, para. 264.

[12]*Ibid.*, para. 288.

[13]*Ibid.*, para. 290.

[14]*Ibid.*, para. 291.

[15]See Bagwell, *Transport Revolution*, 158-159 and 166-167, for a discussion of this.

[16]Cain, "Private Enterprise," 9.

and 1921 are excluded as being abnormal years of boom and slump.[17] Within this average figure the range was from forty-four miles for coal and coke to eighty-four miles for "merchandise and livestock." This postwar figure was probably not radically different from the prewar average, for there had been no significant change in railway transport technology or the pattern of freight services. The motor lorry had not yet begun to capture long-distance goods traffic and indeed was not even supreme in short-distance delivery services.[18] It is likely that postwar average haul figures were higher than those of prewar, for the logic of the government's control of the railways during World War I was to reorganize the traffic of the numerous separate companies to establish something closer to a unified national system, thereby improving efficiency by reducing the movement of empty wagons and increasing "through running."[19] This increased efficiency was further encouraged after the war when the railways were handed back to private hands. The government grouping of the hundred-plus companies into four major regional networks[20] made through running of wagons easier and more frequent. The average haul figures for 1922-1925 are therefore likely to overstate the case for the railways in 1908-1912. Cain has suggested that the gain was of the order of forty percent.[21] If this is correct, it implies an average haul in 1910 of about thirty-eight miles.

Although this adjustment to Munby's post-World War I figures to provide a pre-1914 average haul is rather rough and ready, the result is supported by figures calculated for the later nineteenth century. In his pioneering quantitative work on English railways, Hawke suggested that the average haul for goods traffic in 1865-1867 was thirty-three miles.[22] Dorsey believed the average haul in 1884 was about thirty-five miles and that the English railway company with the longest haul, the London and South Western, had an average of only fifty-nine miles for goods traffic,[23] although it was not predominantly a "goods" line. The economist George Paish provided the figures for a calcula-

---

[17]D.L. Munby, *Inland Transport Statistics: Great Britain, 1900-1970, Vol. 1: Railways, Public Road Transport, London's Transport* (Oxford, 1978), table A 16.

[18]Bagwell, *Transport Revolution*, 225.

[19]Dyos and Aldcroft, *British Transport*, 283-284; and Bagwell, *Transport Revolution*, 236-238.

[20]Bagwell, *Transport Revolution*, 247-249.

[21]Cain, "Private Enterprise," 10, 12 and 13.

[22]Hawke, *Railways and Economic Growth*, 61.

[23]Edward B. Dorsey, *English and American Railroads Compared* (New York, 1887; reprint, Whitefish, MT, 2007), 82.

tion, undertaken by Cain, of the average haul in 1900-1901 for freight on five of the most important English railway lines – the London and North Western, Midland, Great Western, North Eastern and Lancashire and Yorkshire. Hauls ranged between twenty-one and thirty-eight miles,[24] and the average was 30.4 miles.[25] Given that these lines were among the country's largest freight carriers, their average haul was probably not too unrepresentative of that of the other major companies, though it may have been lower than the average haul on the railway system as a whole since through traffic would have caused the figure to be higher. Paish's figures reinforce the idea of average freight hauls being significantly less than fifty miles. Munby's figures for the 1920s, therefore, were most probably higher than the average haul on the railway system in 1908-1912, which lay more in the range of thirty-eight to forty miles. This confirms part of Dyos and Aldcroft's characterization that "[t]he pattern of traffic on British railways [was] *short hauls* and small consignments" (my emphasis).[26]

This figure appears to conflict with Cain's assumed prewar average haul of twenty-eight miles for England and Wales in 1911, derived from Paish.[27] However, both he and Munby explain that railway traffic statistics were collected in two different ways.[28] The tonnage recorded was either a summation of "carryings," that is, all goods carried on each separate railway, although, as some travelled on more than one railway as through traffic, there was an element of double counting; or was the "traffic originating" basis in which freight was counted only once when it came on to the railway system. For any given ton-mileage the former method gives a lower average haul than the latter. Before 1913 the figures were normally collected on a "carryings" basis and after 1920 on "tonnage originating." Fortunately, for at least one year (1913), both methods were employed, so the two series can be compared. Cain chose to calculate his pre-1913 average haul on the "carrying" formula,

---

[24]George S. Paish, *The British Railway Position* (London, 1902; reprint, Whitefish, MT, 2008), 28, 172, 175-176, 191, 196-197, 220 and 281.

[25]Cain, "Private Enterprise," 12-13.

[26]Dyos and Aldcroft, *British Transport*, 174.

[27]Cain, "Private Enterprise," 16, table 5.

[28]*Ibid.*, 10 and 12; and Munby, *Inland Transport Statistics*, 188. However, they disagree on which term applies to which method of collecting the statistics and which was the norm in the pre- and post-World War I periods. Cain states that "[b]efore 1913 [tonnage] figures were collected on a tonnage-originating basis" (10); this appears to be an error. Munby states correctly that "[u]p to 1913, the figures represent 'carryings;' from 1913 onwards traffic originating." (188). *The Coal Merchant and Shipper*, 1 February 1913.

whereas this article uses "traffic originating." Cain's result, therefore, represents the average haul of goods within a railway company, mine on the railway system as a whole. Inevitably, the latter will be greater than the former, as it takes through traffic into account and seeks to eliminate double counting. This is a fairer basis for comparison with coastal shipping, where ownership of the route mileage was not a factor.

Table 1
Coal Coming into London by Coaster, 1910

	Tons of Coal Despatched	Nautical Miles	Ton-Miles (Nautical)
Grimsby	68,334	214	14,623,476
Goole	1,390,604	235	326,791,940
Hull	1,038,499	226	234,700,774
Middlesbrough	617	279	172,143
Hartlepool	424,777	276	117,238,452
Seaham	893,305	297	265,311,585
Sunderland	1,473,813	301	443,617,713
South Shields	2,209,719	308	680,593,452
North Shields	10,580	308	3,258,640
Newcastle	1,489,249	314	467,624,186
Total	8,999,497		2,553,932,361

Note: Average: 238.79 = 326.35 land miles.

Sources: Great Britain, Parliament, *Parliamentary Papers* (*BPP*), 1911, CI, 754-756; and Harold Fullard (ed.), *The Mercantile Marine Atlas* (London, 1959).

For coastal shipping no contemporary estimate of average haul was made. However, an attempt can be made to calculate it from a variety of sources. *The Coal Merchant and Shipper* gives the amount of coal coming by coastal ship into London in 1910 as 8,982,046 tons,[29] and this is broadly supported by the figure in the annual return of the Commissioners of Mines and Quarries in the *Parliamentary Papers*.[30] The *Parliamentary Papers* also give the amounts of coal carried coastwise from each port of the UK.[31] If all coal received by London came from a coal-producing area, i.e., it did not come from ports south of the city but from ports to the north on the east coast, it is possi-

---

[29] *The Coal Merchant and Shipper*, 1 February 1913.

[30] *BPP*, 1911, CI, 756, gives a figure of 8,999,497 tons, a difference of less than 0.2 percent.

[31] *Ibid.*, 754-755; these figures are confirmed by the "Coal Tables" supplement to the London Bills of Entry published annually from 1902 to 1913.

ble to calculate a *minimum* average haul for London-bound coal. It is assumed that all coal exported coastwise from the closest port to London did in fact go there, and each east coast port is taken in sequence until a total tonnage figure is reached equal to that received by London, as in table 1.

Table 2
Coal and Coke Moved by Coastal Shipping, 1910

		Bills of Entry (i)	*Parliamentary Papers* (ii)	Variation of (i) from (ii) %
Bristol	(imports)	1502	367,561	-99.6
Dublin	(imports)	1,362,950	1,218,512	+11.9
Grangemouth	(imports)	250	-	-
Grangemouth	(exports)	221,102	223,287	-1.0
Greenock	(imports)	37,477	113,795	-67.1
Greenock	(exports)	23,229	12,650	+83.6
Liverpool	(exports)	1,796,244	1,913,242	-6.1
Port Glasgow	(imports)	1160	n.a.[a]	-
Troon	(imports)	780	-	-
Troon	(exports)	226,006	232,178	-2.7

Note: [a] The *Parliamentary Papers* do not give separate figures for Port Glasgow.

Source: British Library (BL), Customs Bills of Entry, 1910: and *BPP*, 1911, CI, 754-756.

Table 3
Coal and Coke Average Hauls, 1910

		Tonnage	Ton-Mileage	Average Haul (Nautical Miles)
Bristol	(imports)	1502	43,560	29
Dublin	(imports)	1,371,410	221,730,270	162
Grangemouth	(imports)	250	38,750	155
Grangemouth	(exports)	226,867	96,766,945	426
Greenock	(imports)	37,477	13,257,230	354
Greenock	(exports)	23,545	2,484,802	107
Liverpool	(exports)	1,798,731	352,911,867	196
Port Glasgow	(imports)	1160	430,360	371
Troon	(imports)	780	118,560	152
Troon	(exports)	226,559	25,558,280	113
Total		3,688,281	713,340,624	

Notes: Average 193.41 nautical miles = 222.42 land miles.

Source: BL, Customs Bills of Entry, 1910.

It may be objected that it is unrealistic to assume that all the coal from each of these ports was going to London; furthermore, coal may have been coming into London from ports in South Wales or from the Scottish coalfields. This is quite true. However, the impact would be to *increase* the average haul, since the Scottish and Welsh ports are all farther away than the farthest port in the calculation (Newcastle is 314 nautical miles compared with Swansea at 483 or Leith at 404). This calculation therefore errs on the side of caution, since it is not intended to overstate the case for the coaster. The figure calculated by this exercise is 283.8 nautical miles or 326.4 land miles. This figure is more than seven times the average haul for coal on the railways in 1922-1925. The London coal trade accounted for about a third of all UK coastal coal shipments in 1910 and was thus a large proportion of total coal traffic on the coastal system. However, as it was only one commodity and one route, it would be unwise to use this figure as a measure of average haul for all commodities on all coastal routes.

Another method of calculating average haul is to use the bills of entry. Many of these are now preserved in the British Library Newspaper Collection at Colindale. Although they were not compiled for every port and were partial in their coverage of years and commodities, a number exist for a range of ports in 1910.[32] There are sufficient data on the goods carried and their despatching port or destination to calculate the average haul for a particular commodity from or to a specific port. This has been done for those bills of entry which provide sufficient information. It may be argued that the range of ports covered is very limited. This is true, but there is no systematic bias in the sample and no reason to think that it is not a random sample of ports and types of trade. Certainly it cannot be said that the ports covered are all of one type, ranging as they do from a major port, such as Liverpool or Dublin, to relatively small ports like Troon. Thus, although the selection is not systematic it may well be random in terms of the trades covered. The same argument applies to the commodities included in the bills of entry. The bills certainly did not record all commodities involved in the coastal trade, even for those ports which compiled regular returns. The information collected on the coastal trade was not dictated by the Customs House in London but left to local discretion and so varies from port to port. For instance, the Liverpool bills recorded only imports of grain and meal. No other imported commodities were recorded even though it is known from other sources that considerable shipments of, for example, china clay were sent into Liverpool for transhipment to the potter-

---

[32]Edward Carson, "Sources for Maritime History (I): Customs Bills of Entry," *Maritime History*, I, No. 2 (1971), 176-189. These records originally were housed in the Customs House Library in London.

ies.[33] Similarly, the figures for Workington and Harrington record only ship-ments of various types of iron products and make no pretence of recording any other coastal shipments either received or despatched.[34] In some cases it is possible to carry out a crude check of the figures in the bills by comparing the statistics of coal received and despatched coastwise with the data in the *Annual Report of the Commissioners of Mines and Quarries*.[35] The figures are given in table 2. From this it is obvious that the bills of entry are a very imperfect source. For some ports – for example, Grangemouth, Troon and Liverpool – there is a great agreement between the two sources; for others, such as Bristol and Greenock, there is a marked discrepancy. Thus, while the bills of entry are far from comprehensive, there is no evidence to indicate any particular bias which would lead either to over- or under-estimating the average haul, and the sample size is sufficiently large in terms of ship movements recorded to give confidence in their use.

From the sample bills of entry average hauls were calculated for coal and coke, and all commodities other than coal and coke. These are shown in tables 3 and 4. The average haul for coal and coke works out at 222.4 miles. This figure is much higher than for either railway or canal average hauls and closer to the minimum haul calculated for the London coastal coal trade. If the figure for the provincial coal traffic is combined with that already calculated for coal carried into London, an overall minimum average haul can be calcu-lated for coal carried by coaster. This works out at 296.1 miles because the sheer size of the London trade dominates the sample and pushes the overall average closer to that for London. If these figures seem high, they can be sup-ported by qualitative evidence. For example, Slade's reminiscences mention some of the routes he worked just before World War I in a small sailing coaster. His trips with cargoes included Waterford to Cardiff, Teignmouth to Cork, Appledore to Liverpool, Gloucester to Ramsgate, London to Gloucester, Glasgow to Topsham and Teignmouth to Runcorn,[36] all trips of over 200 miles and in some cases more than 400. The data calculated from the bills of entry and the *Parliamentary Papers* indicate an average haul for coal and coke of about 296 miles and for all other commodities of about 230 miles. These seem reasonable figures. The conclusions are that the average haul by coaster was

---

[33]W.J. Slade, *Out of Appledore: The Autobiography of a Coasting Shipmaster and Shipowner in the Last Days of the Wooden Coasting Ships* (London, 1959), 12, 13, 23 and 29; and Lionel T.C. Rolt, *The Potters' Field: A History of the South Devon Ball Clay Industry* (Newton Abbot, 1974), 129.

[34]TNA/PRO, Customs 83/7, "Workington Iron Returns."

[35]*BPP*, 1911, CI, 754-756.

[36]Slade, *Out of Appledore*, 5, 7, 12, 14, 28 and 29.

between five and seven times that of the railways, depending on the commodity carried, and between six and sixteen times that of the canals. This throws considerable doubt on Kellett's statement that "[c]oastal shipping and canals were competitive...for industrial goods on very short distances"[37] and warns against lumping together canal barges and coastal ships just because both were waterborne.

Table 4
Average Haul for All Commodities Other than Coal and Coke, 1910

		Tons	Ton-Mileage	Average Haul (Nautical Miles)
Bristol	(imports)	78,877	19,887,067	252
Dublin	(imports)	186,034	49,421,847	266
Dublin	(exports)	218,142	40,045,012	184
Grangemouth	(imports)	503,405	92,837,196	184
Grangemouth	(exports)	12,312	3,877,954	315
Greenock	(imports)	23,083	7,296,864	316
Greenock	(exports)	942	159,739	170
Harrington	(exports)	102,410	12,513,915	122
Liverpool	(imports)	13,039	3,074,994	236
Liverpool	(exports)	406,364	79,833,436	240
Port Glasgow	(imports)	33,107	6,618,615	200
Troon	(imports)	27,760	4,507,340	162
Troon	(exports)	24,391	7,823,803	321
Workington	(exports)	158,974	30,766,397	193
Total		1,788,840	358,664,179	

Note: Average 200.5 nautical miles = 230.6 land miles.

Source: See table 3.

III

The total tonnage of goods carried on the UK canal system was recorded, as already noted, by the Royal Commission on Canals of 1906-1909. For the fourteen canals studied in detail, it was just over twelve million[38] and for the total system about 39.5 million tons.[39] This provides a benchmark for canal freight just before World War I.

---

[37]Kellett, *Railways*, 351.

[38]BPP, 1910, XII, para. 280.

[39]*Ibid.*, para. 279.

Munby provided figures for the total tonnage of goods carried on the railway system for the 1900s, as well as for the 1920s, based on the railway returns.[40] The average for all merchandise carried in 1908-1912 emerges as 500.5 million tons per annum. The corresponding figure for 1922-1925 was just over 346 million. This surprising decrease in freight traffic was not real, of course. As Munby explained, and we have already observed, the figures up to 1913 were recorded on a different basis from those collected after 1913, and the earlier figures contain a measure "of duplication between companies."[41] Fortunately, for 1913 the statistics were collected in both ways and thus "a measure of the degree of duplication in the earlier returns"[42] is revealed. The two figures for 1913 indicate that the "carryings" statistics need to be deflated by about thirty-four percent to obtain "tonnage originating" figures comparable to those collected in the 1920s, which eliminates the duplication.[43] If this factor is used, the average annual tonnage carried by the railways between 1908 and 1912 becomes nearly 330 million tons. There is again no conflict here with Cain's figure of total tonnage carried on English and Welsh railways in 1911 of 448 million.[44] Cain's figure is based on the "carryings" method which, as explained previously, aggregated the tonnage of each separate railway company even though it meant double counting of goods carried on more than one line. My estimate eliminates this duplication and gives net aggregate tonnage for the railway system as a whole. The figure is over eight times the tonnage carried by the canal system at roughly the same time. It reinforces the views of the Royal Commission that total traffic on the canals remained stationary between 1880 and 1905[45] and that railways had captured the lion's share of goods traffic from the canals in the late nineteenth and early twentieth century.[46]

No figures were collected contemporaneously of total tonnage carried by all coastal shipping. It has already been noted that the bills of entry are far from comprehensive in their coverage, and Parliament showed no interest in receiving such statistics. The *Parliamentary Papers* do, however, contain an-

---

[40]Munby, *Inland Transport Statistics*, tables A10 and A11.

[41]*Ibid.*, 188.

[42]*Ibid.*

[43]The figure by the "old" method was 557,063,000 tons in 1913; the "new" figure was 367,125,000 (367,125,000 ÷ 557,063,000 = 65.9 percent).

[44]Cain, "Private Enterprise," 12.

[45]*BPP*, 1910, XII para. 276.

[46]Bagwell, *Transport Revolution*, 157.

nual returns of the registered tonnage of ships entering each designated customs port in the coastal trade, broken down by arrivals and departures and by vessels in ballast and those carrying cargo.[47] The total tonnage of ships arriving with cargoes in 1910 was nearly thirty-two million, which was very similar to that for ships departing.[48] The figures are for the registered tonnage of the ships, not the cargoes they contain. If, however, it was possible to arrive at some relationship between a ship's registered tonnage and the amount of freight it carried, it would be possible to use the figures for national coastal entrances to calculate the amount of cargo carried by coasters. There are two related measures which need to be distinguished: the maximum potential amount of cargo which could be carried relative to registered tonnage and the actual amount normally carried. The former is often given in advertisements of ships for sale. For example, the Dublin and Liverpool Steamship Company advertised its three steam screw schooners for sale: "*Dublin,* 154 tons register, carries 200 tons cargo. *Liverpool,* 148 tons register, carries 200 tons cargo. *Waterwitch,* 212 tons register, carries 300 tons cargo."[49] For an earlier period Farr used the Bristol shipping registers combined with newspaper claims to exemplify this: "*Wellington,* built 1811 at Whitehaven; 225 register tons has delivered 340 tons of coal. *Mazy,* built 1834 at Bristol, 259 register tons; 375 tons capacity."[50] Another example, drawn from the copper ore trade of South West England, was for *Mary Simmons,* a schooner built in 1860, with a registered tonnage of 110 but a capacity of 165 tons.[51]

Although these claims need to be treated with some caution as they may have been inflated to overstate their freight-earning capacity to potential purchasers, there is a fairly direct relationship between registered tonnage and maximum cargo-carrying capacity. This relationship changed over time. Improvements in ship design, e.g., the use of new materials such as iron and later steel and the opening up of cargo holds by replacing pillars with cantilevered brackets around 1900,[52] increased cargo capacity relative to registered tonnage,

---

[47]These are in the *Annual Accounts of Navigation and Shipping.*

[48]*BPP,* 1911, LXXIX, 280-281.

[49]Dublin Bills of Entry, 27 March 1850.

[50]Grahame E. Farr (ed.), *Records of Bristol Ships, 1800-1838: Vessels over 150 Tons* (Bristol, 1950), 8.

[51]Peter H. Stanier, "The Copper Ore Trade of South West England in the Nineteenth Century," *Journal of Transport History,* New ser., V, No. 1 (1979), 27, reprinted in Armstrong (ed.), *Coastal and Short Sea Shipping,* 130-147.

[52]J.C. Robertson and H.H. Hagan, *A Century of Coaster Design and Operation* (Glasgow, 1953), 213.

while the introduction of steam propulsion initially worked in the opposite direction by requiring, for engines, boilers and bunkers, space which could otherwise have been devoted to freight.

Table 5
Carrying Coefficients, 1910

		Cargo (Tons)	Registered Tonnage	Carrying Coefficient
(a) *Coal*				
Dublin	(imports)	1,384,520	487,160	2.8
Grangemouth	(exports)	222,652	99,788	2.2
Greenock	(imports)	37,477	12,700	2.9
Greenock	(exports)	23,479	12,140	1.9
Port Glasgow	(imports)	1160	435	2.7
Troon	(exports)	226,181	67,396	3.4
Total		1,895,469	679,619	Mean c.c. = 2.789
(b) *Coke*				
Dublin	(imports)	8460	2915	2.9
Grangemouth	(exports)	5795	4673	1.2
Greenock	(exports)	316	137	2.3
Troon	(exports)	553	234	2.4
Total		15,124	7959	Mean c.c. = 1.9
(c) *All goods, excluding coal and coke*				
Dublin	(imports)	187,494	53,065	3.5
Grangemouth	(imports)	503,415	213,718	2.4
Grangemouth	(exports)	12,462	8191	1.5
Greenock	(imports)	23,083	18,232	1.3
Greenock	(exports)	1258	4493	0.3
Port Glasgow	(imports)	36,482	12,994	2.8
Troon	(imports)	27,760	11,220	2.5
Troon	(exports)	24,391	9123	2.7
Total		816,345	331,036	Mean c.c. = 2.466

Source: See table 3.

The second relationship mentioned above, the ratio of cargo actually carried to the registered tonnage of a ship, is slightly different since it takes into account the variations in cargo capacity resulting from different types of commodity and the effect of part cargoes. It is a measure of the actual cargo carried, not of the maximum that could be carried in favourable conditions. As such it is a more accurate figure to use in any calculations based on registered tonnage. However, the figures quoted in the newspaper advertisements are not

acceptable to calculate this. To arrive at an estimate of the actual carrying coefficient, the bills of entry can be used. Some of them give not only the name
of the ship, its destination or despatching port, type and quantity of cargo, but
also its registered tonnage. Thus, for these ports it is possible to calculate carrying coefficients for each commodity and for the port as a whole. If they are
aggregated, an overall carrying coefficient by commodity can be calculated.
This has been done for the small number of ports which give the requisite information in 1910 (see table 5). The same sorts of criticisms may be levelled
against this method as for the average haul, *viz.*, that it uses a small selection
of ports. The riposte would also be similar, namely that the number of ship
movements recorded is large, as is the total range of trades covered, and that
there is no reason to believe that the sample of ship movements is not random
and no reason to believe that there is any bias either to under- or over-state the
carrying coefficients. The carrying coefficients calculated for the three trades
for coastal traffic for 1910 are 2.8 for coal, 1.9 for coke and 2.46 for all goods
other than coal and coke.

   These figures are significantly higher than those quoted previously for
the earlier nineteenth century, which were in the range 1.25 to 1.5. This is to
be expected. Improvements in ship design increased the carrying coefficient,
and as the average size of the coasting ship grew larger it was likely to benefit
from increased cargo space relative to registered tonnage. These coefficients
are not inconsistent with those calculated by Maywald, who believed that steel-
hulled ships at the peak of their development – that is, built after the 1890s –
had carrying coefficients of about 2.7.[53] They are also supported by Crammond's estimates of carrying capacity for late nineteenth-century steamships of
about 2.0 for ships built before 1888, rising to 2.5 for ships built after 1898.[54]
It thus seems reasonable, lacking any better measure, to use these carrying
coefficients to try to calculate the total tonnage of goods carried by coastal
ships in 1910.

   From the *Parliamentary Papers* the total amounts of coal and coke
carried coastwise in 1910 are known.[55] They are 24,113,396 and 113,722 tons,
respectively. By using the respective carrying coefficients, the total tonnage of
shipping required to move these goods can be calculated, *viz.*, 8,611,927 and
59,854 registered tons, respectively. If these figures are summed and the result
subtracted from the known entrances of coastal ships in all ports of the UK

---

[53]K. Maywald, "The Construction Costs and Value of the British Merchant
Fleet, 1850-1938," *Scottish Journal of Political Economy*, III, No. 1 (1956), 47 and 49.

[54]Edgar Crammond, *The British Shipping Industry* (London, 1917), 13.

[55]*BPP*, 1911, CI, 754-756.

given in the *Shipping and Navigation Account*,[56] (31,965,088), the residue is the registered tonnage of coastal shipping carrying all cargoes other than coal and coke. The figure is 23,293,302 register tons. If it is multiplied by the carrying coefficient for all goods other than coal and coke calculated above (2.46), an estimate of the total tonnage of goods other than coal and coke carried by coaster in 1910 is deduced, namely 57,301,522 tons. Adding these three figures gives an estimate of the total volume of cargo carried by coaster in 1910: just over eighty-one million tons. This figure lies between the Royal Commission's estimate of canal traffic and Munby's for the railways. The cargo carried by coaster is approximately twice that moved by canal and about a quarter that transported by the railways. Certainly in 1910 UK coastal shipping was playing an important part in domestic freight transport and had in no sense been knocked out by the railway system.[57]

IV

As explained previously, total tonnage carried is only one measure of the transport service provided. A superior measure is the ton-mileage worked, which takes into account the variations in average haul on different transport modes. The total ton-mileage worked by the canals in the period about 1905 is given in part by the Royal Commission on Canals. For the fourteen canals which the commissioners investigated intensively, they concluded that the total ton-mileage was about 210 million.[58] Although they reported the total tonnage carried on the whole system as 39.5 million tons,[59] the commissioners did not go on to calculate the aggregate ton-mileage for the total network. If, however, their figure for the average haul on the fourteen canals is taken as representative of the haul on the whole network – and, as explained above, that does not seem unreasonable – the figure for aggregate ton-mileage can be calculated at a little over 691 million ton-miles.

For the railways, Munby gives statistics of total ton-miles worked by the system only for the period 1920 onwards.[60] However, figures for total tonnage carried by the railways have already been calculated based on Munby, and his estimates of average haul for coal and other goods made in the 1920s have been amended to take account of prewar conditions. By combining these

---

[56]*Ibid.*, LXXIX, 280-281.

[57]As Cain has pointed out in "Private Enterprise," 13-15 and 18.

[58]*BPP*, 1910, XII, para. 280.

[59]*Ibid.*, para. 279.

[60]Munby, *Inland Transport Statistics*, table A12.

two sets of data an estimate of the total ton-mileage worked by the railways before World War I can be computed. Taking the Munby figures for 1908-1912, deflated by the factor calculated previously, gives a total average tonnage of 260 million tons of minerals and nearly sixty-nine million tons of other goods. Multiplying these by the maximum haul previously postulated (forty miles) gives an aggregate ton-mileage of 13.2 billion. This figure is fairly close to Cain's estimate of total ton-mileage worked by English and Welsh railways in 1911 of 12.5 billion.[61] Using his methodology, the aggregate figure for Great Britain, including Scotland, works out at 14.5 billion ton-miles. Neither Munby's nor Cain's adjusted figures include Ireland, whereas both the coastal and canal statistics are for the UK as a whole. Total freight tonnage carried on Irish railways in 1910 was 6.5 million.[62] If the average haul on Irish railways was about the same as on the British, this gives 260 million ton-miles of goods traffic. It seems reasonably safe to put the aggregate ton-mileage worked by UK railways just before World War I in the range of thirteen to fifteen billion ton-miles. This is about twenty times the figure computed for canal ton-mileage and gives some idea of the dominance of the railway over the canal as a provider of goods transport.

Table 6
Summary of UK Coastal Trade, 1910, with Railway and Canal Comparisons

	Average Haul	Tonnage Carried	% Total	Ton-Mileage	% Total
Coastal:					
Coal and coke	296.1	24,227,118		7,173,649,940	
All other commodities	230.7	57,301,522		13,219,461,425	
Total	251.7	81,528,640	18	20,393,111,365	59
Railways	40.0	336,332,800	73	13,453,312,000	39
Canals	17.5	39,500,000	9	691,250,000	2
Total		457,361,440	100	34,537,673,365	100

Source: See table 3.

For coastal traffic the same method was used to calculate total ton-mileage worked. The figures for tonnage carried by coaster were multiplied by the respective average haul. For coal and coke the tonnage figures were taken

[61]Cain, "Private Enterprise," 16.

[62]B.R. Mitchell with Phyllis Deane (comps.), *An Abstract of British Historical Statistics* (Cambridge, 1962; reprint, Cambridge, 1976), 229.

from the *Parliamentary Papers*[63] and multiplied by the minimum average haul calculated previously; the average haul calculated in section II for "all other commodities" was similarly used to multiply the figure derived in section III for total tonnage carried by coastal shipping other than coal and coke. The figures are shown in table 6. The total ton-mileage worked out by this method came to just over seven billion for coal and coke and over thirteen billion for all other commodities, making a total of over twenty billion ton-miles.

Thus, the total amount of work performed by coastal shipping in 1910 exceeded that performed by the railways at the same time and was nearly thirty times that of the canals. The estimates shown in table 6 indicate a distribution as follows: coastal shipping, fifty-nine percent; railways, thirty-nine percent; and canals, only two percent. Admittedly the figures for coastal traffic are only rough estimates and contain a significant degree of approximation. However, the figure for average haul of coal coming into London is a minimum. Much of the coal came from more distant ports, and the actual average haul for coal received by London was likely to have been several miles greater, which would in turn push average haul up a few more miles and aggregate ton-mileage concomitantly. The average haul figures used for the railways may overstate the case, for reasons explained previously, and thus wherever possible the minimum case has been made for the coaster while the maximum has been made for the railway. Allowing for the crudeness of the method of calculation and accepting that the figure calculated may be out by ten percent does not significantly alter the overall conclusion. In terms of ton-mileage, coastal shipping in 1910 provided at least as great a freight transport service for the United Kingdom as did the railway system. It did so, however, in a different manner: by concentrating on bulk cargoes which needed to move long distances but which by virtue of their non-perishability and low intrinsic value did not need the speed of journey or perhaps higher cost of railway transport. Certainly the canals had been relegated to an inferior role, perhaps because of railway ownership of a large part of the system; equally certain is that the coasters provided a vastly greater amount of transport than the canals. The two should *not* be perceived together as transport modes virtually obliterated by the advent of the railway.[64] The coaster, at least until World War I, carved out a very important niche in the transport market by concentrating on long-haul bulk cargoes.

---

[63]*BPP*, 1911, CI, 754-756.

[64]Kellett's characterization that "long-distance [freight] transport virtually *was* railway transport" (his emphasis) is clearly wide of the mark. Kellett, *Railways*, 351.

Chapter 14
Climax and Climacteric:
The British Coastal Trade, 1870-1930[1]

The period 1870-1930 saw the continued rise and precipitous decline of the British coastal trade. Coastal shipping had been a crucial component in industrialization, transporting low-value, high-bulk goods such as coal, corn, bricks and slates which were essential to the growth of towns and industry. In the early nineteenth century its volume increased steadily as the economy expanded, and specialization led to an increased movement of goods from low-cost areas to centres of consumption. The coastal trade continued to expand until the First World War and in many ways was at its zenith just before that conflict. However, the Great War brought an abrupt end to this long period of expansion. In the 1920s, although there was growth from the very low wartime levels, the prewar volume of activity was never achieved, and the interwar period saw the British coastal trade stagnate. This chapter will outline and explain the trends in each of these three sub-periods. In so doing the role and economics of the coastal ship will become apparent.

Apogee, 1870-1914

The period 1870-1914 saw the British coastal trade at an apogee. This high point can be demonstrated in a variety of ways. The number and tonnage of ships with cargoes entering the harbours of the UK in the coasting trade rose continuously between 1870 and 1913. A dense network of coastal lines had developed which linked virtually all major ports with regular, fast and scheduled services. The coastal collier was moving an increasing quantity of coal from the producing regions, especially the North East of England and South Wales, to the consuming areas, particularly London and the South East. In addition, the screw collier took on the competition offered by the railways and from the 1890s regained the lion's share of the coal trade to London, carrying more than the railway companies each year save two from then until the First World War. In the last years of the nineteenth century and the early years of the twentieth, the coaster was performing about as much work as the railway

[1]This essay appeared originally in David J. Starkey and Alan G. Jamieson (eds.). *Exploiting the Sea: Aspects of Britain's Maritime Economy since 1870* (Exeter, 1998), 37-58.

system as a whole if measured in ton-miles. Finally, the coaster was extensively used for passenger traffic, both as a means of business and pleasure travel, and increasingly in the last decade or so of the nineteenth century as part of the nascent mass leisure industry.

There is much evidence to support these points. Firstly, the volume of coastal shipping activity grew steadily in the period 1870-1913. Contrary to the view implicit in some transport history textbooks, the nineteenth century did not see the inexorable rise of the railway juggernaut at the expense of all other forms of transport. Hoffman pointed out many years ago that "coastal shipping expanded in a surprisingly steady manner right up to...1914."[2] Reference to the figures for entries and clearances of ships in the coastal trade at British ports confirms this. In 1870, 142,000 ships entered UK ports carrying cargoes in the coastal trade. They comprised some 18.4 million registered tons. By 1892, halfway through our sub-period, this had increased to 206,000 ship entrances of a total of 29.1 million registered tons, and by 1913 the two figures were 169,000 ship entrances totalling 34.8 million tons.[3] The crude tonnage figures suggest a near doubling of activity in about forty years to give a simple per annum growth figure of two percent. If we allow for compound growth using an end-point ratio, this comes down a little but is still more than 1.5 percent per annum, suggesting that there was a steady if unspectacular growth in coastal shipping activity between 1870 and 1914.[4]

The second piece of evidence to attest to the vigour of the coastal trade was the dense network of coastal liner companies which operated regular, frequent and fast scheduled services between all the major cities and towns of the kingdom. No systematic study has been made of these businesses, but

---

[2]Walther G. Hoffman, *British Industry, 1770-1950* (Oxford, 1955), 47.

[3]Great Britain, Parliament, *Parliamentary Papers* (*BPP*), Trade and Navigation Accounts, various years.

[4]It should be stressed that these figures are for the registered tonnage of the ships entering rather than the cargoes they carried. However, they exclude ships in ballast, and the likelihood of ships consistently running lightly loaded in the long run is very low. Hence, they offer a reasonable guide to activity. Admittedly, there were one or two changes to the basis on which the figures were collected, so the statistics are not strictly comparable. Overall, however, these changes may have balanced each other out, as of the two major alterations one was to include vessels previously excluded, while the other was to exclude some which had until then been counted. The fall in numbers of ships entering in the coastal trade between 1892 and 1913 was essentially a function of the decline of the sailing ship fleet and the fact that the average steam coaster was much larger than the average sailing coaster. See Philip S. Bagwell and John Armstrong, "Coastal Shipping," in Michael J. Freeman and Derek H. Aldcroft (eds.), *Transport in Victorian Britain* (Manchester, 1988), 171-217.

the number of companies and services was very large.[5] Along the east coast several firms, including the Aberdeen Steam Navigation Co. (ASN), Dundee, Perth and London Shipping Co., Carron Co., Tyne Tees Steam Navigation and the General Steam Navigation Co., offered liner services from many towns.[6] It was similar on the west coast where the Bacon, Hough and Powell Lines competed in England, while firms like Burns, Laird and Sloan ran services between Scotland and England.[7] These companies operated fast, large, well-appointed ships which carried both passengers and general cargo. They avoided dangerous and anti-social goods, such as dynamite, coal or raw hides, and required no minimum quantity. They were fast by comparison to railway freight traffic, reliable and ran to a timetable. A glance at the advertisements in any local newspaper for a port town will indicate the range of routes served and the frequency of sailings.

Coal was the largest commodity carried by coastal ship in this period, and the quantity being shipped increased steadily. In 1870 about 10.9 million tons of coal were carried coastwise, a figure that had risen to fifteen million by 1892 and 20.5 million by 1913.[8] Once again, the amount of work performed by the coastal ship roughly doubled in just over four decades. This was bettered by the performance of the screw colliers operating to London. In 1870 about three million tons of coal was carried to London by ship, by 1892 that

---

[5]J.R. Couwper, "British Coasting Trade and its National Importance," *Journal of the Institute of Transport,* XV (1934), 223-232, suggests that in 1931 there were 308 services "not including services to the western islands of Scotland." The pre-1914 number would have been much greater.

[6]Clive H. Lee, "Some Aspects of the Coastal Shipping Trade: The Aberdeen Steam Navigation Company, 1835-80," *Journal of Transport History,* 2nd ser., III, No. 2 (1975), 94-107, reprinted in John Armstrong (ed.), *Coastal and Short Sea Shipping: Studies in Transport History* (Aldershot, 1996), 90-103; Gordon Jackson, "Operational Problems of the Transfer to Steam: Dundee, Perth and London Shipping Co., c. 1820-1845," in T.C. Smout (ed.), *Scotland and the Sea* (Edinburgh, 1992), 154-181; Ian Bowman, "The Carron Line," *Transport History,* X, Nos. 2 and 3 (1979),143-170 and 195-213; and A.M. Northway, "The Tyne Steam Shipping Co.: A Late Nineteenth-Century Shipping Line," *Maritime History,* II, No. 1 (1972), 69-88.

[7]John Armstrong, "Freight Pricing Policy in Coastal Liner Companies before the First World War," *Journal of Transport History,* 3rd ser., X, No. 2 (1989), 180-197, reprinted in Armstrong (ed.), *Coastal and Short Sea Shipping,* 112-129; and Graham E. Langmuir and Graeme H. Somner, *William Sloan and Co. Ltd., Glasgow, 1825-1968* (Kendal, 1987).

[8]*BPP,* Annual Report on Mines and Quarries, part III.

had roughly doubled to 5.8 million, and by 1913 it was nine million tons.[9] The coaster also improved its competitiveness. Whereas in the 1870s and 1880s the coaster brought about thirty-eight percent of all coal coming in to London, from the 1890s it provided the majority of coal to the capital, and from 1898 to 1913 this averaged over fifty-two percent.[10] The increased competitiveness of the coasters sprang from a market in which they had an economic advantage, namely the industrial segment of the coal market: those firms located near the sea or navigable rivers which needed large quantities of coal delivered regularly, such as gasworks, electric power stations, stores of bunker coal and large industrial users. The coastal colliers responded to this segment of the market by providing large ships which benefited from economies in crew and capital costs as their size increased.[11] The evidence from the coal trade confirms not merely that the coastal ship was increasing its level of activity in this area but also that it was competing successfully with the railway system and recapturing market share from it.

There is further evidence to suggest that the coaster was playing a vital role in internal transport in an analysis of the amount of work performed in 1910 by coastal ships in comparison to railways and canals.[12] This suggests that the railway system and coastal shipping performed about the same quantity of work if measured in ton-miles, whereas the canal network did not approach the other two modes. However, the way in which the railways and coasters achieved their output was quite different. The coaster had a very large average haul, around 250 miles, whereas the railway's was only about fifty miles. The coaster carried a smaller tonnage than the railways, but each ton was carried a much longer distance. This is explicable because the coaster benefited from the economies of a long haul, as much of its costs were incurred as terminal charges, i.e., payments made at the ports at each end of its voyage, whereas once at sea its running costs were meagre. Compare this to the railway where the running costs were more proportional to the distance covered, as it owned

---

[9]B.R. Mitchell with Phyllis Deane (comps.), *Abstract of British Historical Statistics* (Cambridge, 1962; reprint, Cambridge, 1976), 113; *BPP*, Annual Report on Mines and Quarries, part III; and London Bills of Entry, Coal Table Supplements, various years.

[10]*The Coal Merchant and Shipper*, XXVI (1913), 111.

[11]John Armstrong, "Late Nineteenth-Century Freight Rates Revisited: Some Evidence from the British Coastal Coal Trade," *International Journal of Maritime History*, VI, No. 2 (1994), 45-81.

[12]John Armstrong, "The Role of Coastal Shipping in UK Transport: An Estimate of comparative Traffic Movements in 1910," *Journal of Transport History*, 3rd ser., VIII, No. 2 (1987), 164-178, reprinted in Armstrong (ed.), *Coastal and Short Sea Shipping*, 148-162.

the land on which it ran, had to prepare and maintain a permanent right of way, provide signal box operatives, police and level crossing attendants. The coaster did not incur these costs. Hence, the coaster was particularly suited to conveying large quantities over long distances, such as coal from Newcastle to London and china clay from Cornwall to Runcorn.

The final dynamic aspect of the coaster's performance was its role as a passenger carrier in this period. In providing a regular, scheduled service between the main port cities of the country, the coastal liner companies offered passenger accommodation, often for several classes of traveller. The ships were large, fast and well appointed with grand saloons attended by stewards and stewardesses in pale imitation of the crack trans-oceanic liners. Such passenger traffic could be substantial: for example, the ASN took £7236 in 1894, reportedly a poor year.[13] This was at a time when first-class fares were thirty shillings and second-class fifteen shillings, implying that several thousand passengers were carried.[14] Part of the appeal of the coastal steamer lay in its speed and reliability. The ASN claimed to do the 500-mile journey in thirty-six hours in 1875.[15] Although slower than the best passenger trains, the coaster offered the advantages of a brief sea cruise with attendant bracing air and ozone. That this was a lure is borne out by the seasonal surge in passenger traffic and the complaints of the company that poor summers led to a fall in those paying fares.[16]

In addition to this part-business, part-recreational travel, in the last decade of the nineteenth century there was a great growth in the pure leisure role of the coastal steamer. The holiday trips of the Clyde river and coastal steamboats – trips "doon the watter" – are well known,[17] but in addition the Bristol Channel, Solent, north Welsh coast and the Scottish islands were suitable locations for day cruises or longer trips. A combination of bank holidays, slightly larger real incomes, the secularization of Sundays and the appreciation

---

[13]Aberdeen University Library (AUL), Ms. 2479/17, Aberdeen Steam Navigation Co. (ASN), Directors' Report, 1894.

[14]*Ibid.*, Ms. 2479/29, ASN, handbill.

[15]*Ibid.*

[16]"The cold and wet weather which prevailed during August and September, always the best passenger months, told adversely on this branch of the company's business." *Ibid.*, Ms. 2479/17, ASN, Directors' Report, 1896.

[17]Andrew McQueen, *Clyde River Steamers, 1872-1922* (Stevenage, 1990); Alan J.S. Paterson, *The Golden Years of the Clyde Steamers, 1889-1914* (Newton Abbot, 1969); and James Williamson, *The Clyde Passenger Steamer: Its Rise and Progress during the Nineteenth Century, from the Comet of 1812 to the King Edward of 1901* (Glasgow, 1904; reprint, Whitefish, MT, 1987).

of the commercial possibilities of the desire of people to escape from the towns in search of diversion and scenery led to a boom in recreational coastal sea travel in the late Victorian and Edwardian eras.

If it is accepted that the coaster was alive, well and expanding its activities in the late nineteenth and early twentieth centuries, we need to explain the economic basis of its role and how it was able to compete against the apparently all-devouring railways. The first necessity is to cease talking about coastal shipping as though it was a homogenous entity. One of the secrets of the coaster's success was the provision of a range of services to suit different types of trade. It segmented the market and then provided appropriate services to each segment. At the top end of this range of services was the liner trade.[18] This offered speed, regularity and reliability, with no minimum quantity required. In return for this premium quality, the liner charged the highest freight rates of any coaster. These were still significantly less than those charged by the railways for long-distance routes, and the service offered by the coaster was not inferior. For manufactured goods, such as Bovril, bird seed, soap, cigars or pianos, whose value-to-bulk ratio was high and whose starting point and ultimate destination were close to the port, river or canal, this was the most appropriate method of transport. A crucial determinant was the final location, for land carriage by horse and cart was very expensive, and shippers endeavoured to minimize this element. Thus, if either the despatching point or the final destination was closer to a railway goods yard than to a wharf or quay, the railway stood a good chance of gaining the traffic.

The second segment served by the coaster was those firms that needed regular deliveries of a bulky product sufficient to allow some firms to specialize in this trade and devote particular coasters to it exclusively. The clearest example of this was the coal trade to London with screw colliers built for that trade alone, not seeking a return cargo but going in water ballast and maximizing the number of coal-carrying journeys they could make.[19] By the 1890s and 1900s fifty or sixty return trips each year between the Thames and the Tyne was not unusual. Some of these vessels specialized to the extent that they had hinged funnels and masts to allow them to navigate under London's bridges and penetrate upriver.[20] Other trades with a similar need for regular bulk deliveries were china clay from Cornwall to Runcorn for transhipment into canal barges for despatch to the Potteries, iron ore from Cumberland to South Wales for smelting, followed by a cargo of Welsh coal to a port on the east coast of Ireland where pit timber could be picked up for the iron ore mines. These

---

[18]See Armstrong, "Freight Pricing Policy."

[19]See Armstrong, "Late Nineteenth-Century Freight Rates."

[20]J.A. MacRae and Charles V. Waine, *The Steam Collier Fleets* (Wolverhampton, 1990), 54.

"regular traders" offered a reliable service for firms which needed a steady flow of raw materials or fuels. They kept prices low by speedy turnaround and by suiting the ship to the trade.

Below this segment was that which needed bulky cargo moved less frequently and where reliability and regularity was less important. For this the coastal steamer provided a tramp service. It could be reasonably fast once at sea as it travelled steadily at nine knots or so. It required a reasonably full load to be profitable, which meant matching the ship to the cargo, and hence the goods might have to wait until a suitable vessel became available. Its freight rates were lower than those charged by the coastal liner. Thus, non-perishable, bulk commodities whose despatch did not have to be instantaneous, such as iron and steel products, bags of grain, barrels of oil and hogsheads of sugar, were despatched in steam tramps.

Finally, there was the sailing ship. Even in 1870 it made up less than half the total tonnage of entries in the coastwise trade and was declining. By 1900 it comprised only one-eighth.[21] It was unreliable, depending on wind strength and direction and affected by currents and tides. It could make a fast passage if the weather was with it, but it could also spend days skulking in port. Its main economic advantage lay in its provision of a very cheap form of bulk transport – a floating warehouse, cheaper than other forms of coaster and all modes of land transport. Schooners of 100 registered tons could carry about 160 tons of goods with a crew of three. Improvements in rig design and the use of winches and self-furling gear increased labour productivity and reduced crew size. Brigs and brigantines had largely been superseded by schooners and ketches because the shift from square to fore-and-aft rigging not only reduced labour requirements but also made the ship easier to handle when wind conditions were adverse.[22] Thus, for bulky goods with low unit values, which did not deteriorate and which were not needed at a particular date, the sailing coaster offered the cheapest form of transport. Thus, sand, stone, ballast, manure and grains were normal cargoes. In addition, those goods which required careful and therefore lengthy loading usually went by sailing ship, for the demurrage costs were much lower than on a steamer because the latter's original capital cost was so much greater. In the early twentieth century goods like slates (which cracked easily), bricks (prone to chipping) and tiles and fireclay products (frangible) tended to be transported in sailing coasters.[23]

Thus, one of the explanations of the success of the coaster was its ability to provide a range of services at a range of prices, so catering for a

[21]Christopher I. Savage, *Inland Transport* (London, 1957).

[22]Basil Greenhill, *The Merchant Schooners* (2 vols., London, 1951-1957; 4th rev. ed., London, 1988).

[23]Great Britain, National Archives (TNA/PRO), RAIL 226/110.

number of types of commodity and trade. A second reason was technological change. The coaster was not a technologically static mode of transport, and while the developments of the first three-quarters of the nineteenth century may be better known – the application of steam to propulsion, the use of iron for hulls, the paddle being superseded by the propeller and the use of water ballast – there were also a number of changes in the last quarter of the century. The use of steel for hulls, boiler pressures exceeding 100 lbs. per square inch, the water-tube boiler, triple-expansion steam engines and eventually the turbine, first fitted commercially in 1901 in *King Edward*,[24] were all important innovations which offered strength, economy and speed. Moreover, all were pioneered on coastal craft. The diffusion of these innovations took time and had a lasting effect upon improving the efficiency of the coastal fleet. In addition, coastal ships were becoming larger in most trades, and this brought economy in crew numbers and costs[25] because crew size did not rise proportionately to ship size. We have already mentioned that there were improvements in the sailing ship, partially spurred by the need to keep costs down in order to undercut the rates charged by steam tramps. Also, the shift from sail to steam in itself resulted in productivity gains, for the steam sailor was nearly four times as productive as the crew member of a sailing ship.[26] Hence, shifting tonnage from sail to steam reduced labour requirements and increased labour productivity. Thus, in a multitude of ways the coaster increased its efficiency and productivity in this period.

The third point to stress when explaining the coaster's ability to expand is the economics of operations. The coaster's main competitor, the railway, had a totally different cost structure. The coaster normally owned no wharf or quay, needed no land or fixed capital other than the ship itself and had to pay the wages only of the crew, which was relatively small in number when compared to the tonnage the ship could carry. Once the coaster was at sea its costs were minimal – the crew's wages and fuel for steamers. For the sailing ship the latter was avoided, and although there might be wear and tear on the sails and ropes, repairing them and maintaining the ship was a normal part of the sailor's duties. Some sailing ships also passed on to the crew members the uncertainty of wind and weather by paying them piece rates by the trip rather than regular weekly wages.[27] This method acted as an incentive to mari-

---

[24]J. Graeme Bruce, "The Contribution of Cross-Channel and Coastal Vessels to Developments in Marine Practice," *Journal of Transport History*, 1st ser., IV, No. 2 (1958), 65-80, reprinted in Armstrong (ed.), *Coastal and Short Sea Shipping*, 57-72.

[25]Armstrong, "Late Nineteenth-Century Freight Rates."

[26]*BPP*, Trade and Navigation Accounts, various years.

[27]TNA/PRO, Board of Trade (BT) 99/2569.

ners to speed up their voyage and not fritter away time in port since they earned the same whether the voyage took three days or three weeks. Most of the operating expenses of a coaster were incurred as terminal costs, that is, at the beginning and end of a voyage when entering or leaving port.[28] Then port or harbour dues became payable, there might be pilotage charges and fees for towage were incurred by sailing ships, though many coastal skippers took it as a matter of pride – as well as of pocket – to avoid these by expert handling and navigation. As well, costs for unloading and loading might be incurred, though for most sailing ships and some steam tramps this was the responsibility of the crew members, and possibly trimming fees if it was a bulk cargo capable of shifting. In addition, there were often other incidental expenses consequent upon obtaining a cargo, such as the agent's fee, telegrams back to the home port and miscellaneous running expenses. Hence, a high proportion of costs other than fuel and labour were incurred at the ports, and these charges were levied irrespective of the length of the preceding or subsequent voyage. Their cost per ton-mile declined as the journey length increased, and coasters therefore had cost advantages on long hauls. The depreciation of coasters was also relatively low. Providing they did not meet an untimely end via a collision, stranding or freak storm, coastal ships could and did last for twenty to twenty-five years and even longer.[29] Thus, their rate of depreciation was four or five percent per annum. In addition, some of the ravages of time and weather could be kept at bay by the regular maintenance which was part of the normal duties of the crew. As a result, servicing the capital was not too onerous.

Compare this to the cost structure of the railways. The railways needed to buy expensive land, not just for the tracks but also for stations, goods yards, sidings and marshalling yards, some of it in prime locations such as city centres where costs were at their highest. The stories of landowners extorting huge sums for small strips of territory essential to a railway route are apocryphal but contain elements of truth. In addition to this outlay on fixed capital, the costs for engineering works, such as viaducts, bridges, tunnels and cuttings were high. Then there was further capital cost in terms of rails, sleepers, points, signal boxes and stations, plus, of course, the rolling stock, including locomotives, wagons, carriages and goods vans. As a result, the capital cost of a short railway ran into millions of pounds before a bag of cotton had been carried. Compare this to £20,000 for a large coastal liner in the 1870s. Worse was to follow. Once the railway had laid down all this capital, it needed an army of workers to operate it. Not just the drivers, firemen and guards on the trains, but those who operated signals and level crossing gates, as well as

---

[28]See, for example, W.G. Rickman, *The Shipowners' Register of Port Charges* (2 vols., London, 1913; 3rd ed., London, 1929).

[29]Myrvin Ellis-Williams, "The Sloop *Jenny*, 1787-1919," *Maritime Wales*, No. 10 (1986), 138-141, gives an even longer-lived example.

police to guard the lines, station staff and workshop staff to fettle, repair and build the complex rolling stock. In 1898 in the UK over half a million people were employed on the railways,[30] whereas in 1896 there were slightly less than 30,000 mariners employed in the British coasting trade, and even if those engaged in the "Home Trade" are included, the figure is still less than 37,000.[31] If it is borne in mind that it has been estimated that in 1910 the railways and coasters performed about the same amount of work in moving goods when measured in ton-miles,[32] the difference in the labour force involved was enormous. It might be objected that the railways also carried many more passengers than the coasters, and hence the railway labour force was also catering for this type of traffic. This is a fair comment. However, passenger traffic generated less than half of all railway revenue in 1900, while freight traffic contributed fifty-four percent (£51.7 million out of a total of £95.1 million).[33] If we assign fifty-four percent of the railway labour force to the goods traffic, this gives 288,400 employees, somewhere between eight and nine times as many as the coastal fleet. The rates of pay of the average mariner, the able seaman or fireman, and the railway employee were not greatly different. The skipper and first engineer of a steam coaster were paid more than the drivers of locomotives, but only double, not eight or nine times, and therefore the total wage bill of the railway system assigned to freight carriage must have been many times greater than that of the coastal fleet. Hence, these operating costs, as well as the charges to service the huge capital outlays, were much larger on the railway than on the coaster; these are disparities which explain why railway freight charges were so much higher than those on the coaster.

This section has attempted to explain some of the more important aspects of coaster operations and economics, which in turn help us to understand why its charges were lower than those of the railway and why the coaster was able to carry an increasing amount of freight. We now need to turn to a much gloomier period, that of the First World War.

Climacteric, 1914-1918

The First World War was a climacteric for British coastal shipping. Not only did it have short-term deleterious effects but it initiated a forty-year depression in the industry, inasmuch as the registered tonnage of entries and clearances

---

[30]D.L. Munby, *Inland Transport Statistics: Great Britain, 1900-1970, Vol. 1: Railways, Public Road Transport, London's Transport* (Oxford, 1978), 46-47.

[31]*BPP*, 1897, LXXVIII, 407.

[32]Armstrong, "Role of Coastal Shipping."

[33]Munby, *Inland Transport Statistics*, 22.

with cargoes in the coastal trade achieved in 1913 were not surpassed until 1952.[34]

Firstly, let us look at the short-term effects during the period of conflict. The war caused the volume of coastal shipping activity to fall to less than half its prewar volume. Whereas in 1913 approximately thirty-five million registered tons of coastal shipping with cargoes entered British ports, in 1918 the equivalent figure was less than seventeen million.[35] The contraction is not too surprising, for enemy action was directed at all of Britain's maritime trades and accounted for about forty percent of its total fleet. Coastal shipping was no exception, with numerous coasters falling victim to mines, shell fire, torpedoes and bombs. Heroism and tragedy co-existed in the coastal as much as in the deep-water trades. In addition to destruction, the coastal fleet was subject to the depredations of the government. It requisitioned coasters to carry troops, munitions and coal for the war effort and in some cases armed them and used them as Q ships, minesweepers and escorts, further reducing the tonnage available for commercial activities. To cite one instance, Supple estimates that "the Admiralty commandeered eighty percent of the colliers normally used to ship coal between the north-east and the south."[36] In addition, the government encouraged the redirection of some larger coasters into the foreign trade to help the hard-pressed deep-water fleet.[37]

For those ships that continued to work in the coastal trade there were a number of additional difficulties. Firstly, there was the problem of waiting while convoys were formed and escorts provided, which slowed down turn-around times and reduced the amount of work performed in a given period. The latter was compounded by the deterioration in port facilities as more ships endeavoured to use the same ports whose equipment suffered from a lack· of repair, replacement and renewal.[38] Secondly, the costs of operation increased drastically. Insurance premiums rose in view of the additional war risks, and wages climbed, partly to compensate sailors for the additional hazards they faced, and partly because the general price level rose as aggregate demand

---

[34]*BPP*, Trade and Navigation Accounts.

[35]*Ibid.*

[36]Barry E. Supple, *The History of the British Coal Industry. Volume 4: 1914-1946: The Political Economy of Decline* (Oxford, 1987), 48.

[37]Derek H. Aldcroft, "The Eclipse of British Coastal Shipping, 1913-1921," *Journal of Transport History*, 1st ser., VI, No. 1 (1963), 25, reprinted in Armstrong (ed.), *Coastal and Short Sea Shipping*, 163-177.

[38]Derek H. Aldcroft, *British Transport since 1914: An Economic History* (Newton Abbot, 1975), 18.

exceeded aggregate supply. As a result, the coaster's costs rose, and it had to adjust its freight rates upward to compensate. Isserlis calculated that for all tramp ships, coastal and blue-water, the freight rates charged in 1918 were about ten times those in 1913.[39] This is an extraordinary rise and more of a problem for coastal than overseas shipping, for the former faced competition from another mode, namely the railways. While "the coasting trade...remained free until the last year of war when a loose form of control was instituted,"[40] the railways were taken at an early date into direct government control. This had two significant effects. Firstly, the government directed some traffic which had previously been carried by coaster onto the railways because the risk of enemy action on the railways was tiny compared to seaborne trade. Secondly, the rates of carriage on the railways remained frozen at their 1914 level throughout the war. The government guaranteed to railway shareholders a return at least equivalent to that received before the war, so in effect the government was paying an immense subsidy to the users of the railway. Against this the coasters found it increasingly difficult to compete. Whereas before the war freight rates by sea were much cheaper than by rail, especially for long distances, Aldcroft estimates that "by the end of the war...coastal rates were from fifty to two hundred percent in excess of the comparable railway rates."[41] Thus, coastal shipping firms found it very difficult to compete with the railways, especially when it is borne in mind that the railways had a plethora of "exceptional rates" which were even lower than the published rates. As a result, many routes which the coaster operated became unprofitable and hence the economic rationale compounded the direct effects of hostilities to reduce the amount of work performed by the coastal ship.

Stagnation, 1918-1930

If the war had a deleterious effect on the coastal trade, it might be thought that once hostilities ceased and the abnormal circumstances were reversed, the coastal trade would bounce back to its prewar position. This was not the case, as table 1 shows. Although entrances of coasters carrying cargoes rose from 16.8 million tons in 1918 to 24.2 million tons in 1920, they then dipped in 1921, recovered in 1922 to a little above the 1920 level, and then for the rest of the 1920s stayed well below the 1922 peak. This was at best a miserable performance, but it is even worse when it is appreciated that the peak year of the 1920s, 1922, saw a level of coastal activity which was less than three-quarters of that achieved in 1913. Thus, the coastal trade throughout the

---

[39]Mitchell with Deane (comps.), *Abstract*, 224.

[40]Aldcroft, *British Transport*, 16.

[41]*Ibid.*, 19; and Aldcroft, "Eclipse," 28.

1920s, and indeed in the 1930s as well, never regained the position it had held in 1913. The interwar period was one of stagnation for the coastal trade. If the average tonnage of entrances for 1909-1913, 32.5 million tons, is compared to that for 1919-1930, twenty-two million tons, a reduction of about one-third is indicated, roughly 10.5 million tons.

Table 1
Net Registered Tonnage of Ships Entering UK Ports in the
Coasting Trade with Cargoes, 1913-1930 (million tons)

Year	Net Registered Tonnage
1913	34.759
1914	36.001
1915	27.468
1916	22.360
1917	19.201
1918	16.780
1919	19.901
1920	24.176
1921	20.949
1922	25.680
1923	21.407
1924	20.576
1925	20.627
1926	16.197
1927	22.160
1928	23.038
1929	24.021
1930	24.394

Source: Great Britain, Parliament, *Parliamentary Papers* (*BPP*), Trade and Navigation Accounts, various years.

Various explanations as to why the coaster did not regain its prewar position in the internal transport system of Britain have been advanced. One cause cited by a number of shipowners and captains can be dismissed fairly quickly. That is the complaint that foreign competition was forcing British ships and mariners out of the trade. In particular, the Dutch motor coaster was blamed. For instance, Jim Uglow, skipper of a sailing barge throughout the interwar period, complained that "by 1930 the Dutch motor coasters were a real menace on the British coast,"[42] while Hervey Benham, immensely knowledgeable on the activities of small ships, claimed that "the Dutchmen were the

---

[42]Jim Uglow, *Sailorman: A Barge Master's Story* (London, 1975), 84.

barges' chief competitors up to the Second World War."[43] However, this is unlikely to explain the stagnation of coastal shipping. The figures of entrances and clearances include foreign-owned ships, and if the Dutch captured a large market share this would still be included. As it was, foreign ships played only a small part in the British coastal trade. In the period 1920-1930 their average share of the market was 0.7 percent, and the peak year, 1920, recorded only 1.3 percent.[44] Thus, although the Dutch coaster may have been prominent in some trades it was not a significant player, and seeking government help to reserve the coastal trade to British ships was unlikely to have had much effect. It might be that the Dutch coaster had an indirect effect upon the British coastal trade in that it pushed freight rates so low that they became unremunerative. If this were so, one would have expected the Dutch to have captured a large share of the market, which was patently not the case.

The second argument that can be dealt with equally summarily is that the coaster had started a terminal decline from which it could not recover. The facts do not fit this interpretation. Although the coastal trade stagnated throughout the interwar period and declined even further during the Second World War, after the close of these hostilities it demonstrated rapid growth. In 1949 entrances exceeded those of the best interwar year, 1937, and in 1953 they passed the previous highest level achieved in 1914. The trend continued at least until the 1960s when it was twenty-five percent above the level of 1914.[45] Thus, the coastal trade grew steadily after the 1939-1945 war, which suggests that there were peculiar circumstances operating in the interwar period which caused it to stagnate.

Before 1914 the single most important commodity carried by the coaster was coal. As we have seen, in 1913 about 22.8 million tons were carried coastwise, of which about nine million went into London. Any reduction in coal consumption was likely to have had an adverse effect upon the coastal trade. The coal industry was severely depressed for most of the interwar period,[46] with output never reaching the 1913 peak, and average output for 1920-1929 was about 230 million tons per annum compared to 270 million in 1909-1913, a reduction of around fifteen percent.[47] Admittedly, 1921 and 1926 were

---

[43]Hervey Benham, *Last Stronghold of Sail: The Story of the Essex Sailing Smacks, Coasters and Barges* (London, 1948; reprint, London, 1986), 178.

[44]*BPP*, Trade and Navigation Accounts.

[45]*Ibid.*

[46]The sub-title of Supple's book on the interwar coal industry makes it clear: *The Political Economy of Decline.*

[47]Mitchell with Deane (comps.), *Abstract,* 116.

hard hit by major strikes, especially the latter when the coal strike lasted more than nine months (even if the General Strike was over in ten days), but this did not augur well for the coastal trade. However, another conventional truth of the decline in coal output between the wars is that much of it was attributable to a reduction in export demand. Whereas in 1909-1913 the average quantity of coal exported was sixty-five million tons per annum, between 1920 and 1929 it averaged forty-nine million, a reduction of about one-quarter.[48] Given that the output figures show a reduction of about forty million tons per annum and the export demand was down by about sixteen million, there is still plenty of reduction to account for.

Turning to coal carried coastwise, it is evident that whereas in the five years 1909-1913 on average twenty-one million tons of coal were conveyed as cargo plus 2.4 million tons for bunkers in the coasting trade, a total of about 23.5 million tons,[49] in 1923-1930 these figures, respectively, were fifteen million tons and 1.4 million tons for a total of 16.4 million tons.[50] Accordingly, there was a decrease of about seven million tons, equivalent to thirty percent. In other words, the reduction in the amount of coal carried by coastal ship was proportionately greater than the fall in export demand, a fact that has hardly ever been mentioned. The absolute reduction, averaging seven million tons of coal, could account for nearly as much shipping, say about five million registered tons, which was approximately half of the reduction previously noted in the activity level of coastal shipping in the interwar period compared to the prewar era. This decline in the coastal coal trade might have resulted from increased competition from the railways and the diversion of coastal coal traffic to the rail system. However, the railways did not increase the amount of coal they carried in the 1920s compared to before the war. Indeed, they carried less coal than previously, down by about twenty-one million tons per annum.[51] So it was not that the railways gained at the coaster's expense. Rather, the reduction in coal traffic experienced by the railways, at about ten percent, was much less than that experienced by the coaster. So the overall market was in decline, but the railway was retaining its market share to a greater extent than the coaster.

One symptom of the decline in demand for coastal shipping was the low level of remuneration earned. Many shipowners and masters bewailed the ruinously low freight rates being offered. Certainly there was a very rapid reduction in freight rates following the cessation of hostilities and the collapse of

---

[48]*Ibid.*, 121.

[49]*BPP*, Annual Reports on Mines and Quarries.

[50]Chamber of Shipping, Annual Reports, statistical tables.

[51]Munby, *Inland Transport Statistics*, 83.

the postwar shipping boom. The average freight rate in 1920 was about half that in the last year of the war, and that for 1921 was less than one-quarter of the 1918 rates.[52] However, there is some dispute as to how low freight rates sunk thereafter. Isserlis suggests that in the period 1919-1930 the rate was always above that obtained in 1913 and that even in the worst year, 1930, it was about thirty-seven percent above the 1913 levels.[53] Capie and Collins, basing their figures on *The Economist,* disagree. They suggest that for 1928-1930 the average freight rate was below that of 1913, in 1930 some twenty percent less.[54] These differences are quite important. The average of the Isserlis rates for 1921-1930 is eighty percent above that of 1913, whereas *The Economist* series suggests only ten percent. This is a large discrepancy. But these rates are all composites of a number of routes, mostly in the overseas trade. It would be preferable to use an index based solely on coastal routes, but to the best of my knowledge none exists.[55] We do know that the freight rates on coal from the Tyne to London fell drastically in the postwar slump. Between 1918 and 1921 they plunged from about 17s 3d to 6s 6d per ton, registering a fall in each year of at least twenty-five percent.[56] Then from 1921 to 1923 they dropped more slowly to about four shillings per ton. The course of coastal freights thereafter is less clear. Despite this uncertainty, it seems likely that rates in the mid-1920s were little more than the prewar level in money terms, where the average for 1911-1913 was about 3s 7d, and by the late 1920s the rates were below the prewar levels at, for instance, 3s 2d in 1927 and 2s 8d in 1930.[57]

---

[52]Mitchell with Deane (comps.), *Abstract,* 224.

[53]*Ibid.*

[54]Forrest Capie and Michael Collins, *The Inter-War British Economy: A Statistical Abstract* (Manchester, 1983), 84.

[55]The only coastal freight rate used by Isserlis to construct his overall tramp shipping index, coal from the Tyne to London (route 85), is very suspect as he has it *falling* from 1914 to 1918, until in the latter year it is about half that of the former: L. Isserlis, "Tramp Shipping Cargoes and Freights," *Journal of the Royal Statistical Society,* CI, Part 1 (1938), 120. This also conflicts with the "average" freights given for 1914-1923 in D.W. Lloyd and F.C. Swallow, *The North Country and Yorkshire Coal Annual for 1924* (Cardiff, 1924), 64, which show a rise from 1914 to 1917 when it is nearly five times the 1914 figure.

[56]Lloyd and Swallow, *North Country,* 64.

[57]This is based on an analysis of the freight rates appearing in the *Iron and Coal Trades Review* and *The Coal Merchant and Shipper.*

At the same time, most of the coaster's costs had risen, including wages and fuel. For example, the average cost of coal in 1921-1930 was nearly half as much again as in 1909-1913.[58] Whereas the able seamen and firemen on coastal colliers prewar were earning about thirty shillings per week,[59] in the 1920s it was more like fifty-six shillings, a rise of over eighty percent.[60] It is therefore clear that the freight rates earned by the coaster did not keep pace with increasing costs and, as a result, coasting became less remunerative. These low rates forced some owners to switch to overseas trade, or even to abandon shipping, leaving vessels rusting at anchor. Thus, a low level of demand and an excessive supply of coasters, especially in the non-liner trades, forced freight rates down to a barely remunerative level. This in turn forced some players out of the industry, prevented the renewal and modernization of the fleet and made it less efficient.

Of course, the coaster was not trading in a vacuum. Quite the contrary; as well as facing competition from the railways, the coaster faced potential competition from the upstart road transport industry. Motor lorries had made great strides in the First World War under the imperative of moving troops and munitions.[61] In theory, road haulage could have impinged on demand for the coaster's services. In fact, its impact in the 1920s was slight. Initially, lorries and vans were used for short-distance deliveries, from the railway station to the factory or from the port to the industrial estate, replacing horse-drawn carts and wagons.[62] In this role they complemented coasters rather than competing with them. In the 1920s the largest lorries were only rated at four tons and, although regularly overloaded, were nowhere near the capacity of the coaster.[63] Similarly, they were perceived as economical over relatively short distances by the coaster's standards, maybe up to sixty miles.

---

[58]Mitchell with Deane (comps.), *Abstract*, 483-484.

[59]Armstrong, "Late Nineteenth-Century Freight Rates."

[60]MacRae and Waine, *Steam Collier Fleets*, 138.

[61]James M. Laux, "Trucks in the West during the First World War," *Journal of Transport History*, 3rd ser., VI, No. 2 (1985), 64-70.

[62]T.C. Barker and Dorian Gerhold, *The Rise and Rise of Road Transport, 1700-1990* (Basingstoke, 1993), 85.

[63]*Ibid.*, 85-86; and T.C. Barker, *The Transport Contractors of Rye, John Jempson and Son: A Chapter in the History of British Road Haulage* (London, 1982), 20.

Roads were not constructed for motor traffic, signposting was poor,[64] and as a result journey times were slow: road haulage consequently posed no real threat to the coaster because the types of traffic it served were quite different. The same cannot be said of the railways.

The railways represented a very real threat to the coaster and, indeed, shipowners and captains believed an unfair one. In the immediate postwar years the railways remained under government control while "most of the limited control which had been exercised over coastal shipping was abandoned" in December 1918.[65] As a result, freight charges on the railways remained at the 1913 level until January 1920.[66] For this period the railway had a prodigious advantage over the coaster, whose costs had escalated by at least 200 percent, and its freight rates were largely uncompetitive against the government-subsidized railway rates. Even when the government raised rates on the railways – three times in 1920 alone[67] – and eventually restored them to private ownership, albeit now in four regional monopolies, the complaints from coastal owners did not cease. They objected to the habit of the railways charging "exceptional rates" which were below those officially quoted and never published or publicized. Some shipowners believed the railways made no profits on such rates and cross-subsidized them with higher charges on routes where competition from coastal ships was not possible.[68] In 1928 it was estimated that two-thirds of all goods were carried at exceptional rates,[69] and the proportion was probably not much less in the earlier 1920s. Certainly coastal shipowners perceived them as a significant problem.

The other complaint voiced by shipowners concerned charges levied by railways on short-haul traffic at the beginning and end of each voyage. It was very difficult for shipowners to avoid this cost except by employing road haulage, which was used later in this period. The shipowners felt that the railways held the whip hand in this traffic, being able to enforce high charges because no alternative existed.[70] As a result, the rate per ton mile on these short

---

[64]Barker, *Transport Contractors*, 20-23; and Barker and Gerhold, *Rise and Rise*, 86.

[65]Derek H. Aldcroft, "The Decontrol of British Shipping and Railways after the First World War," *Journal of Transport History*, 1st ser., V, No. 2 (1961), 97.

[66]Aldcroft, "Eclipse," 28.

[67]Aldcroft, "Decontrol," 93.

[68]*BPP*, 1921, XV, 344-345.

[69]Munby, *Inland Transport Statistics*, 96.

[70]*BPP*, 1921, XV, 344.

hauls could be three or four times as great as long hauls of the same commodity.[71] The railways might well retort that the realities of railway economics were such that there were economies in long hauls and diseconomies in short, but to the shipowner the charges seemed discriminatory and unfair. This was bolstered by their perception of the railway network after the 1921 reorganization as being an oligopoly and, worse still, effectively a collection of regional monopolies. Thus, the railways seemed to be in a strong competitive position, and shipowners, despite various attempts, could not assail this.

Another cause of stagnation in the coastal trade of the 1920s, according to the shipowners, was the poor condition of port facilities. They argued that where there was limited capital available for improving berths and unloading equipment, which first went to the overseas traders, leaving the coastal trade with obsolete, inadequate and inefficient facilities.[72] The result was that coasters often had to wait to get onto a berth, reducing revenue-earning time. When at a quay they were given low priority when labour was allocated. Docks were often tidal, lacked electricity, did not work at night and did not have adequate storage space.[73] The shipowners were also concerned because they smelled the whiff of unfair competition. They pointed out that over fifty British ports were owned by the railway companies and that, in these, facilities for coasters deteriorated while those for overseas trade were modernized.[74] They felt that this was deliberate policy on the part of the railways to disadvantage the coastal trade and so encourage the diversion of freight onto the railways. Certainly the coastal shipowners placed considerable blame on the poor quality of port provision. This was a common complaint, for there was always tension between shipowner and port authority. The former wished the most expensive facilities for the least cost, the latter to provide the cheapest facilities for the largest dues. It remains unclear how far the shipowners' accusations against port authorities and railway companies were designed to divert blame from their own failings.

The picture painted so far has been gloomy. However, there were some aspects which were outside the control of the coaster owner and even the British government, such as the establishment of the Irish Free State, with the reclassification of some trade as foreign, and the worsening relationship be-

---

[71]Savage, *Inland Transport*, 26; and *BPP*, 1929-1930, XVI, 365.

[72]Anon., *The Coastwise Trade of the UK Past and Present and Its Possibilities* (London, 1925), 64; and Couwper, "British Coasting Trade," 230.

[73]Philip S. Bagwell, *The Transport Revolution from 1770* (London, 1974; new ed., London, 1988), 281.

[74]Savage, *Inland Transport*, 28.

tween the two governments culminating in the "economic war."[75] Moreover, the coastal trade was not passive in this period. The coastal liner continued to offer a reliable service, so much so that it was used in 1919 to carry copies of Keynes's *Economic Consequences of the Peace* from the printers R. and R. Clark of Edinburgh to the publishers Macmillan in London.[76]

One of the long-term effects of the war on the railway industry was to bring about the mergers which created four regional monopolies. The logic of this was that during the war, when the government had operated the railway system as an integrated whole, it believed it to have run at a significantly higher level of efficiency. This view is endorsed by Peter Cain, who believed the average haul increased significantly and that further operating economies were extracted through the amalgamation of companies.[77] In this way the railway companies entered the 1920s with a more appropriate structure which made them more efficient. The coastal liner trade followed this precedent, and that of many other industries, and was subject to a significant number of mergers and amalgamations between the wars. In 1917 Coast Lines Ltd. was established, which was an amalgamation of three liner firms. In the next three years it absorbed a further ten liner companies, and in the 1920s another ten coastal shipping businesses were merged into the firm, by which time it had a large share of the trade.[78] Coast Lines itself was taken over by the Royal Mail Steam Packet Group in 1917, and the advantages that should have flowed from being part of a large group may well have been negated by the difficulties the group faced, which culminated in its crash and dismemberment in the early 1930s.[79] Thus, the merger mania may have brought some economies to the coastal liners, and, external to the individual firm, there continued to be a series of conferences and other agreements, both between coastal liner companies and between them and the railway companies. These aimed to restrict the excesses of competition and to seek co-operation on freight rates, frequencies of sailings, etc. In other words, just as the railways had become four regional monopolies,

---

[75]*Ibid.*, 26.

[76]Robert Skidelsky, *John Maynard Keynes: Hopes Betrayed, 1883-1920* (London, 1983), 381 and 393.

[77]Peter J. Cain, "Private Enterprise or Public Utility? Output, Pricing and Investment on English and Welsh Railways, 1870-1914," *Journal of Transport History*, 3rd ser., I, No. 1 (1980), 12, 13 and 20.

[78]Peter Mathias and Alan W.H. Pearsall, *Shipping: A Survey of Historical Records* (Newton Abbot, 1971), 37-39.

[79]Edwin Green and Michael S. Moss, *A Business of National Importance: The Royal Mail Shipping Group, 1902-1937* (London, 1982).

the coasters endeavoured to gain greater control of their market. The extent to which they achieved this still needs investigation, as does the relative degree of efficiency gain in the two modes of transport, but it seems likely that the railways gained more from this exercise than did coastal shipping.

The coastal fleet in the 1920s also demonstrated technical improvements. The number and tonnage of sailing ships continued to decline, and there was growth in both the number and tonnage of diesel ships and coastal tankers. The latter, for instance, may have mustered 800 registered tons in 1914; by 1929 it was about 5000.[80] The criticism that can be levelled against the British coastal fleet is not that it did not innovate but that it did so too slowly. However, with low freight rates, excess capacity and poor rates of return, it is not surprising that the firms were not eager to invest larger sums.

---

[80]Isserlis, "Tramp Shipping Cargoes," 96-97.

Chapter 15
The Shipping Depression of 1901 to 1911:
The Experience of Freight Rates in the
British Coastal Coal Trade[1]

The period 1900 to 1914 has been characterized as one of continuing depression in the international shipping industry. James considered 1901 to 1911 to have been an acute depression in shipping,[2] while Sturmey considered the period 1904 to 1911 to have been "the first truly international shipping depression."[3] More recently Aldcroft, examining ocean freight rates, the earnings of half a dozen British liner companies and the movement of earnings per ton of four British shipping firms, has concluded that "during the first decade of the twentieth century British shipping was more depressed than at any time during the last quarter of the nineteenth century."[4] Aldcroft admits the statistical basis of his conclusions "are very weak,"[5] and indeed he presents no new index of freight rates to support his case, relying on data from Angier and Isserlis.[6]

No cognizance has been taken of coastal freight rates in this period, and indeed not much is known generally about the movement of coastal freight rates. It has been established that coastal liner companies followed the practice of the railways in charging on an eight-fold classification based on the value of the good and the difficulty of handling it and that the coastal liner charged less

---

[1]This essay appeared originally in *Maritime Wales*, No. 14 (1991), 89-112.

[2]Frank Cyril James, *Cyclical Fluctuations in the Shipping and Shipbuilding Industries* (Philadelphia, 1927; reprint, New York, 1985), 24.

[3]S.G. Sturmey, *British Shipping and World Competition* (London, 1962), 50, note 1.

[4]Derek H. Aldcroft, "The Depression in British Shipping, 1901-11," *Journal of Transport History*, 1st ser., VII, No. 1 (1965), reprinted in Aldcroft, *Studies in British Transport History, 1870-1970* (Newton Abbot, 1974), 113.

[5]*Ibid.*

[6]*Ibid.*, 100-101.

than the railway for virtually all commodities and routes.[7] However, there has been no attempt to look at a series of coastal freight rates to see how they fluctuated over time. Freight rates charged by liner companies were only one aspect of the coastal trade. There was also the tramping coaster, not tied to any specific route but going wherever there was a cargo to be carried and working to no published schedule of sailings but departing as and when the cargo had been loaded. Although the liner companies carried the higher-valued cargoes, leaving the bulkier, lower-valued commodities to be hauled by tramps, liners were almost certainly a minority of the total number of ships in the coastal trade and carried a smaller aggregate tonnage of goods than the tramp ships. Thus, to concentrate only on liner freight rates is to ignore charges made on the major portion of coastal trade. The single most important cargo carried by the coaster in the nineteenth and early twentieth century was coal. The quantity of coal carried by coasters rose from about eleven million tons in 1870 to nearly eighteen million tons in 1900 and just under twenty-three million tons in 1913.[8] No other single commodity required seaborne transport to the same extent. The pig iron trade was probably the next largest but was nowhere near as significant. The coastal coal trade grew at over two percent per annum for more than forty years, a faster growth rate than the overall British economy.[9]

The nature of the coal trade was essentially movement from the sources of production to the large urban centres where it was consumed. There were two main coal-producing areas which used the coaster for distribution in the early twentieth century, the North East and South Wales. Of the various towns and cities served, London was the most important because of its large population, geographic size with concomitantly high domestic demand, and requirements for steam coal for factories, shipping, and gas and electricity generation usually based on, or near, the Thames.[10] The coal trade was not served by liners but by either specialized tonnage – the screw colliers largely having replaced the collier brigs on the east coast by the late nineteenth cen-

---

[7]John Armstrong, "Freight Pricing Policy in Coastal Liner Companies before the First World War," *Journal of Transport History*, 3rd ser., X, No. 2 (1989), 180-197, reprinted in Armstrong (ed.), *Coastal and Short Sea Shipping: Studies in Transport History* (Aldershot, 1996), 112-129.

[8]Great Britain, Parliament, *Parliamentary Papers* (*BPP*), Annual Reports on Mines and Quarries, Part III.

[9]Charles Feinstein, "Slowing Down and Falling Behind: Industrial Retardation in Britain after 1870," *Refresh*, No. 10 (1990), 6, reprinted in Anne Digby, *et al.* (eds.), *New Directions in Economic and Social History* (2 vols., Basingstoke, 1992), II.

[10]Robin Craig, *Steam Tramps and Cargo Liners, 1850-1950* (London, 1980), 45.

tury – or general purpose tramps which sought a return cargo of some low-value, bulky commodity such as cement or grain. Given that this was not a liner trade and was neither an urgent nor high-value commodity, there was little occasion for a conference or price-fixing arrangements as occurred quite commonly among coastal liner companies.[11] Freight rates were impersonally determined by the interplay of market forces, the supply of shipping compared to the amount of coal requiring transport to the urban centres.

Table 1
Tramp Freight Rates, Tyne to London Carrying Coal, 1900-1913

	Isserlis Route 85	Angier per Ton	Angier Index	Isserlis Overall Index
1900	NA	NA	NA	133
1901	100	4s 3d	100	100
1902	81	3s 5d	80	86
1903	81	3s 5d	80	86
1904	77	3s 3e	76	86
1905	95	4s	94	89
1906	83	3s 6d	82	91
1907	87	3s 8d	86	95
1908	71	3s	71	79
1909	74	3s 1d	73	81
1910	78	3s 1d	77	88
1911	88	3s 9d	88	102
1912	101	4s 3d	101	137
1913	95	4s	95	119

Sources: L. Isserlis, "Tramp Shipping Cargoes and Freights," *Journal of the Royal Statistical Society,* CI, part 1 (1938), 120; and E.A.V. Angier, *Fifty Years' Freights, 1869-1919* (London, 1920), 106-140.

The sheer size of the coastal coal trade was likely to influence the freight rates set for other commodities, for a general purpose coaster could as well load coal as china clay or pig iron or sand. The actual freight rates received would vary depending on the commodity, but the overall movement of the rates was likely to be in sympathy with those for coal cargoes. Thus, an index of coastal coal freight rates would probably serve as an indication of the general trend in coastal tramp freights. The aim of this article is to look at coastal coal freight rates from ports in South Wales in the period 1900-1914 to see how far they conform to "a depression," to compare this with the experience of the North East ports, the other major coal-producing region, and to ascertain the basis of tramp freight rates.

[11]John Armstrong, "Conferences in British Nineteenth-Century Coastal Shipping," *Mariner's Mirror,* LXXVII, No. 1 (1991), 55-65.

Table 2
Average Money Freight Rates for Coal from Cardiff, 1900-1913

		1900	1901	1902	1903	1904	1905	1906	1907	1908	1909	1910	1911	1912	1913
Holyhead	Steam	4s4d	4s1d	4s1d	4s3d	4s1d	4s1d	4s1d	4s4d	4s	3s9d	3s9d	4s9d	4s6d	4s4d
Portmadoc	Sail	3s3d	3s1d	2s10d	3s3d	3s3d	3s6d	3s6d	3s7d	3s	3s1d	3s4d	3s6d	3s10d	4s3d
Fowey	Steam	4s1d	3s10d	3s9d	3s9d	3s7d	3s6d	3s6d	3s7d	3s6d	3s6d	3s6d	5s	5s	4s7d
Fowey	Sail	4s7d	4s7d	4s4d	4s1d	3s10d	3s10d	3s10d	4s7d	3s10d	3s9d	3s10d	5s1d	5s4d	5s
London	Steam	4s9d	4s7d	4s4d	4s2d	4s1d	4s4d	4s4d	4s6d	4s3d	4s3d	4s1d	5s4d	4s9d	5s
Gravesend	Sail	6s7d	6s3d	6s	5s7d	5s3d	5s	NA	6s7d	NA	NA	NA	NA	NA	NA
Belfast	Steam	4s9d	4s7d	4s3d	4s2d	3s10d	3s7d	3s7d	4s4d	3s10d	3s8d	3s9d	4s10d	4s9d	4s7d
Waterford	Sail	5s6d	5s4d	4s7d	5s	4s6d	4s3d	4s4d	4s10d	4s6d	4s6d	4s9d	6s	6s1d	6s
Londonderry	Steam	5s1d	5s1d	4s7d	4s7d	4s4d	4s2d	4s4d	4s6d	4s4d	4s4d	4s6d	5s6d	5s6d	5s1d
Sligo	Sail	8s	7s10d	7s1d	7s3d	6s1d	6s3d	6s3d	7s	7s9d	7s6d	8s	9s	9s3d	

Source: Joseph Davies and C.P. Hailey (eds.), *South Wales Coal Annual for 1916* (Cardiff, 1916), 104.

One of the best known and most quoted indices of shipping freight rates is that of Isserlis, published in 1938.[12] Fortunately, as well as the overall index Isserlis gave individual indices for over 300 routes served by tramp ships. Of these, however, only one was a coastal rate, that from the Tyne to London with coal (route 85 in his system),[13] and this only commenced in 1901. The index numbers for this route are shown in table 1. The majority of Isserlis' freight rates were drawn, as he stated explicitly,[14] from the annual reports compiled by Angier Bros. and from 1909 by E.A.V. Angier. These gave only the lowest and highest freight rates for each route for each year, and Isserlis computed his figure by taking the mean of these two extremes. Obviously, this is not the ideal method. It would have been preferable to have weighted each separate freight rate by the number of times it was indexed or by the quantity of coal carried and so have achieved a truly representative figure. However, since this information no longer survives we can do no better than to take Isserlis' average figure.

It is possible to go back to the original source which Isserlis used to determine the actual freight rate (Isserlis gives only an index number but never mentions the money amount) and check his calculations. The figures produced in Angier's reports have been summarized and published[15] in the same form as Isserlis used them, that is, as minimum and maximum rates for the year. The simple average of these rates is given in table 1 as are the index numbers based on these money freight rates. An examination of table 1 shows that Angier's figures do move in very close harmony to the index figure derived by Isserlis and that, therefore, Isserlis' index was a competent calculation. More importantly, these figures demonstrate that coastal freight rates on the east coast were indeed depressed in the early years of the twentieth century. Compared to the figure for 1900 there were two nadirs, in 1904 and 1908, when prices were twenty-three and twenty-nine percent below the 1900 figure. These rates did not recover to the 1900 base until 1912 when there was a dramatic surge which was not fully sustained into 1913. Figures beyond this have not been quoted, although they are given in Angier, since the approach and actuality of war distorted normal peacetime rates.

However, this is only one rate for one route, and it might be argued that this was not necessarily representative of coastal coal freight rates as this trade has a high proportion of dedicated tonnage, the screw collier, which

---

[12]L. Isserlis, "Tramp Shipping Cargoes and Freights," *Journal of the Royal Statistical Society*, CI, part 1 (1938), 53-134.

[13]*Ibid.*, 106.

[14]*Ibid.*, 74.

[15]E.A.V. Angier, *Fifty Years' Freights, 1869-1919* (London, 1920).

could not be easily deployed into other trades. Hence, there was a relatively inelastic supply of shipping tonnage. Fortunately, there is another source which provides freight rates for coal carried from Cardiff in the years just before the Great War.[16] This gives annual freight rates from 1901 to 1913 for nine different ports, five rates for steamers and five for sailing ships. These figures are given in table 2. As with the Angier figures, the source gives simply the highest and lowest freight rates fixed in the year, with no indication of the distribution or number of rates. Therefore, the prices given in table 2 are the simple unweighted means of these two extremes. The figures for Gravesend are so patchy that they cannot be used to make comparisons and have been ignored from future conclusions. In table 3 these have been converted, with the exception of those for Gravesend, into index numbers with a base in 1901 to allow comparison with the figures for the Tyne-Thames trade in table 1.

Table 3
Average Freight Rates for Coal from Cardiff, 1900-1913
(Index Numbers)

	Holy-head	Port-madoc	Fowey	Fowey	Lon-don	Bel-fast	Water-ford	Lon-don-derry	Sligo
	Steam	Sail	Steam	Sail	Steam	Steam	Sail	Steam	Sail
1900	106	104	106	100	103	103	102	100	102
1901	100	100	100	100	100	100	100	100	100
1902	100	92	97	95	95	92	86	90	90
1903	103	104	97	89	90	90	93	90	92
1904	100	104	94	84	89	84	84	85	78
1905	100	112	90	84	95	77	79	81	79
1906	100	112	90	84	95	77	81	84	79
1907	106	116	94	100	97	95	91	88	89
1908	97	96	90	84	92	84	84	85	98
1909	91	100	90	81	92	79	84	85	95
1910	91	108	90	84	89	81	88	88	102
1911	115	112	129	111	116	105	112	107	114
1912	109	124	129	116	103	103	114	107	117
1913	106	136	119	108	108	100	112	100	114

Source: Joseph Davies and C.P. Hailey (eds.), *The South Wales Coal Annual for 1915* (Cardiff, 1915), 64.

A number of conclusions can be drawn. Firstly, it is apparent that coastal freight rates were not monolithic and that there were wide variations between routes as to the extent and timing of the changes in rates. For in-

---

[16]Joseph Davies (ed.), *South Wales Coal Annual* (Cardiff, 1903-1915). I am very grateful to Robin Craig for directing me to this source.

stance, the lowest figure recorded in the individual series varies between seventy-seven and ninety-two, showing a considerable difference, and the highest index numbers recorded in the individual series vary between 105 and 136, an even greater disparity. There are significant differences as to the timing of the nadir of this particular trade cycle as far as coastal shipping freights were concerned. The individual series bottom out in seven different years, and although 1905 was the mode for the lowest freight rate recorded, it scored only twice. It is not even that the nadir is represented by a cluster of dates: the years of lowest freight include nearly all even years and a couple of odd ones (1902, 1904, 1905, 1906, 1909 and 1910). There is, however, much more agreement as to when coastal freight rates peaked: it was towards the end of the period. Here there is a cluster around 1911 and 1912, with four individual series reaching their apogees in each of these years.

If table 3 is compared to table 1, the series of freights for the coal traffic from the Tyne to the Thames seem much more depressed than those for the South Wales coal trade. The low point of the Angier/Isserlis coastal index (seventy-one) is several points below the lowest of the Welsh series (seventy-seven), and the high point of the east coast coal index at 101 is below the lowest high point of the Welsh series at 105. Thus, the freight rates on the route from the North East to London appear to have fallen further and recovered less than was the case for the Welsh coal trade. There is little agreement about the timing of the depth of the depression when the east coast coal trade is compared to that from South Wales. The lowest point of the Tyne-Thames series came in 1908, which is different from all but one of the Welsh series. There is, however, a great measure of consensus over the timing of the peak of the rates, the east coast series reaching its zenith in 1912, the same date that half of the Welsh series peaked.

What do these rates suggest about the concept of a "depression" in British shipping in the early years of the twentieth century? Aldcroft stated that freight rates "began to decline towards the end of 1900,"[17] implying that 1900 was a reasonable year for freights. In the absence of any series of coastal freight rates for the last quarter of the nineteenth century, it is impossible to compare the Welsh series with anything other than its own internal evidence. It shows that freight rates began to fall by 1901, supporting the timing of the onset of the depression posited by Aldcroft. A measure of "depression" in coastal freight rates is needed. I shall take the simple idea that there should be more years in which the index number of freight rates is below the 1900 figure than the number of years in which the freight rate is equal to or above 1900. Using data for the period 1901-1910 to test this proposition lends convincing support for the idea of a depression. Of ninety possible observations, eighty-one are below the 1900 figure. If we repeat this operation using the 1901 fig-

---

[17]Aldcroft, "Depression in British Shipping," 100.

ure as the acid test and employing the period 1902-1910, the answer remains unchanged. Of eighty-one possible observations, sixty-six are below the 1901 figure, bearing out the notion that the period 1901-1910 saw depressed coastal freight rates. It is also straightforward to pinpoint the timing of the recovery from the depression. For the period 1911-1913, of twenty-seven observations all but one are equal to or more than the 1900 rate, indicating that the depression was over by 1911. This supports the timing suggested by both Aldcroft and Sturmey.

Table 4
Coal Freight Rates from Newcastle, 1900-1911
(per Ton)

	London	Plymouth	Dartmouth	Devonport	Dublin	Cork
1900	4s 1d	5s 8d	6s	5s 7d	6s 6d	5s 9d
1901	3s 3d	4s 6d	4s 7d	4s 9d	4s 10d	4s 5d
1902	3s 3d	4s 4d	4s 6d	4s 7d	4s 11d	4s 5d
1903	3s 2d	4s 2d	4s 4d	4s d	4s 7d	4s 4d
1904	3s 1d	4s 1d	4s 1d	4s 9d	4s 4d	4s 6d
1905	3s 3d	4s 3d	4s 5d	4s 7d	NA	NA
1906	3s 3d	4s 5d				
1907	3s 4d	4s 11d				
1908	2s 11d	4s 4d				
1909	2s 10d	4s 3d				
1910	2s 11d	4s 7d				
1911	3s 4d	5s 2d				

Source: Newcastle Central Library, L622.32, 49927SA, book of newspaper cuttings on the coal trade.

There is some additional, if partial, evidence on coal freight rates from Newcastle to six British ports. This is shown in table 4. Unfortunately, four of the series do not continue beyond 1905, but what is there broadly supports the view of a depression in coastal freight rates. 1900 was the peak year for these coastal freight rates, as was the case for the Welsh series of coaster freights, and there was a significant fall in 1901 and no real recovery before 1911. The impact of the depression was greater on Tyne freights than on those from South Wales, for the minimum fall of the North East series in 1900-1901 was sixteen percent, which was the maximum reduction for the Welsh ports for the period 1900-1902. Most of the coastal freights from the Tyne showed a reduction of over twenty percent in one year, whereas the average reduction for the Welsh freights for the same period was only ten percent (over two years). These data confirm the impression given by the Angier figures that the fall in freight rates was more immediate and more severe in the North East than in the South Wales' coal trade.

In both series the depths of the slump coincided, i.e., 1904 or 1905 and 1909. However, whereas freights from the Tyne to London had fallen to seventy-six and seventy percent, respectively, of the 1900 rate, that from South Wales had fallen on average to eighty-eight and eighty-nine percent, respectively. The decline was similar for the cost of transport from the Tyne to Plymouth, the figure for 1904 reaching seventy-two percent of 1900 and that for 1909, seventy-five percent. Both for the London and Plymouth trades the falls were greater than for virtually all of the series from South Wales' ports.

The same trend can be seen in the upturn in freight rates at the end of the period. From the second nadir of 1909 all series agree on a steady recovery in 1910 and 1911. However, there was a great difference in the degree of recovery between the Welsh series and the two from the Tyne. Whereas between 1909 and 1911 the freights from the North East to London and Plymouth increased by fifteen and twenty-three percent, respectively, those from the South Wales' ports increased on average by twenty-eight percent. As a result of this higher increase, and because the Welsh series had not fallen as steeply as those from the Tyne, whereas the actual rates in 1911 from South Wales exceeded those current in 1900, those from the Tyne still languished below their original level.

The explanation for this differential performance can be sought in the nature of the product and the market. Coal was not a homogenous product. By the early twentieth century most of the coal produced in South Wales was "steam coal," that is, it was particularly efficient at evaporating water to produce steam.[18] It was therefore in great demand for ships' bunkers as it minimized space and weight. Most of the Tyne coal was of lower efficiency for raising steam, having a lower carbon content than Welsh steam coal. Kirkaldy calculated that twelve percent more Newcastle coal was needed compared to the best Welsh steam coal for any given distance, and that the Welsh took only forty cubic feet to the ton compared to forty-five or forty-six for Northumberland coal,[19] a saving of twelve to fifteen percent. The demand for Welsh steam coal was buoyant at a time when British and world steamship tonnage was growing fast. In 1900 British registered steam tonnage was just over seven million tons. By 1913 this had risen to over eleven million.[20]

It was not only the merchant marine which was expanding rapidly in this period. So too was the Royal Navy. These were the years of the height-

---

[18]Martin J. Daunton, *Coal Metropolis: Cardiff, 1870-1914* (Leicester, 1977), 5-7.

[19]Adam W. Kirkaldy, *British Shipping: Its History, Organisation and Importance* (London, 1914; reprint, Newton Abbot, 1970), 455 and 611.

[20]B.R. Mitchell with Phyllis Deane (comps.), *Abstract of British Historical Statistics* (Cambridge, 1962; reprint, Cambridge, 1976), 219.

ened naval rivalry between Germany and Great Britain which was dubbed the "naval race," when each strove to out-build the other in terms of number and size of warships, and Britain introduced the Dreadnought and Super Dreadnought classes of battleships.[21] As part of his reforms in naval strategy, Sir John Fisher, when made First Sea Lord in 1904, concentrated the British fleet in home waters rather than having it dispersed around the globe, recognizing that the main threat now came from Germany.[22] Hence, there were an increased number of naval ships, and a larger proportion was in the North Sea, the Channel or the Atlantic. Until *Queen Elizabeth* (1912), the first of the Super Dreadnoughts, these were mainly coal-fuelled.[23] There was therefore a much greater demand for the best steam coal at the naval dockyards serving the home fleets, such as Portsmouth and Plymouth, both of which were easily served by coasters from South Wales.

The high demand for Welsh steam coal was reflected in its price, which was greater than the price for coal from the North East in the early years of the twentieth century. For example, in 1911, when the average price for large best steam coal f.o.b. at Cardiff was 17s 8d, large, screened Durham steam coal cost 11s 7d.[24] This represented a premium of over fifty percent in favour of Welsh coal, reflecting the greater demand and quality. In such buoyant demand conditions the price for transport of a much sought-after commodity was likely to be less affected by the trade cycle.

The other feature which may have influenced coastal freight rates was the level of competition for coal traffic experienced from the railways. To the NER and GNR, the two companies which jointly formed the "East Coast Joint Stock" to run trains from the North East to Kings Cross,[25] coal traffic was an important component of total tonnage, and these railways competed for it against the coastal ship. For instance, under George Gibb, appointed general manager in 1891, the NER concentrated on train loading as the key to profitability and began providing larger capacity wagons and running more in one

---

[21]Peter Padfield, *The Great Naval Race: Anglo-German Rivalry, 1900-1914* (London, 1974).

[22]*Ibid.*, 117-124.

[23]*Ibid.*, 270-271.

[24]Davies and Hailey (eds.), *South Wales Coal Annual for 1913*, 173; and *The North Country Coal and Shipping Annual for 1912* (London, 1912), 126-127.

[25]Charles H. Grinling, *The History of the Great Northern, 1845-1922* (London, 1905; rev. ed., London, 1966), 221.

train.[26] As a result of these improvements, the average wagonload of minerals on the NER rose by twenty percent between 1903 and 1913, the average train load of minerals doubled between 1900 and 1913 and the total minerals carried by the railway increased twenty percent between 1900 and 1912.[27]

That the importance of such economies of scale was appreciated by the GNR was demonstrated by Lord Allerton, its chairman, in a speech in February 1903. He emphasized that the railway had introduced twenty- and thirty-ton wagons to improve its ability to carry minerals and was also replacing older engines with much larger, more powerful types, capable of pulling longer, heavier trains.[28] The latter were designed by Henry Ivatt, chief mechanical engineer of the GNR from 1895 to 1911, who also designed the "Atlantic" crack passenger engines.[29] However, although the GNR was moving towards larger wagons, including some ordered in 1905 of thirty-five-ton capacity, the vast majority remained small, ten tons being the normal size.[30]

The GWR seems to have been less interested in coal traffic, for despite experimenting with larger coal wagons in the 1890s and 1900s, these were only used to haul the company's own locomotive coal.[31] The bulk of the coal was carried in privately owned wagons, and the GWR was unable to persuade its owners to change. This and the problem of returning numerous empty wagons led the GWR to give less priority to coal and hence less effective competition to the coaster than was the case on the east coast.

Unfortunately, the two series for the North East disagree in regards to timing, with Angier's suggesting that 1901 was the peak year for coastal freights in the very early twentieth century, whereas those in table 4 indicate 1900. We do not have a freight rate from Angier for 1900, so it is possible that the figure he would have produced would have been higher than that for 1901. A more likely explanation is that the two series are out of synchronization by one year and that they need to be adjusted. If the Angier series is advanced by

---

[26]Robert J. Irving, *The North Eastern Railway Company, 1870-1914: An Economic History* (Leicester, 1976), 213 and 221-222.

[27]*Ibid.*, 222, 241 and 293.

[28]Grinling, *History*, xi.

[29]Philip S. Bagwell, *Town of Trainmakers: Doncaster, 1853-1990* (Exeter, 1991), 30.

[30]*Ibid.*

[31]Edward T. MacDermot, *History of the Great Western Railway, Vol. 2: 1863-1921* (London, 1931; rev. ed., London, 1964), 305.

one year (Angier's 1901 rates become 1900) there is a remarkably close fit between the figures for London in the two series.

Ideally, rather than simply taking the individual rates for each port, it would be worthwhile to compile one overall freight index for Cardiff by weighting each of the freight rates by the quantity of coal sent from Cardiff to each port. The latter can be ascertained from the *South Wales Coal Annual.* In practice, however, this is not a very meaningful exercise. The quantity of coal taken by London is so huge that it totally dwarfs the amount taken by the other eight ports. For instance, in 1903 London took 795,794 tons of coal from Cardiff whereas the other eight ports in aggregate took only 88,471 tons.[32] This relationship held for the other years of the early nineteenth century. The result of this imbalance is that a weighted freight rate index from Cardiff becomes virtually the same as that for London.

The Welsh coastal freight rates can be compared to the overall index of tramp shipping freight rates constructed by Isserlis. The latter is shown in the last column of table 1. Although this claims to be valid for all tramp shipping, it should be borne in mind that of the more than 300 routes for which Isserlis collected freight rates, only one was a coastal route, and therefore his index is more properly for overseas freight rates only and not necessarily representative of coastal rates. Two points emerge from this comparison. The nadir of Isserlis' overseas index, at seventy-nine, is lower than the average low point for the nine Welsh series at eighty-four. Similarly, the apogee of Isserlis' index, 137, is much above the average highpoint for the Welsh coal series at 117.

This suggests that overseas freights were much more volatile than coastal freight rates. Deep-water tramp freights covered a wider range, reaching greater depths than coastwise but also achieving higher peaks. Thus, coastal freights were more reliable, moved less rapidly and were less exciting. This is supported by an examination of table 3 which shows that on a large number of occasions the freight rate for a particular coastal route remained unchanged from one year to the next, suggesting that customary rates prevailed and were slow to alter. A little more evidence can be adduced for greater volatility in foreign freight rates than those in the coasting trade. Aldcroft suggests that by 1901 "many rates had fallen to 50 percent below those ruling in the autumn of the previous year,"[33] indicating a sharp fall. By comparison, none of the Welsh coastal rates fell by anything approaching fifty percent, the maximum reduction being that for Waterford where the fall from 1900 to 1902 was about sixteen percent. This suggests that deep-water tramp rates were more volatile than this set of coastal tramp rates. This is also backed up by the sharp

---

[32]Davies (ed.), *South Wales Coal Annual for 1904.*

[33]Aldcroft, "Depression in British Shipping," 100.

fall in the overall Isserlis index for 1900 to 1902. The reduction is about thirty-five percent, approaching Aldcroft's figure and double the greatest fall in any of the coastal series for the same period. Thus, although the evidence is sparse and partial, there seem some grounds for suggesting that coastal freight rates were much less volatile than deep-water freights, reflecting the steady demand for coal – the single most common cargo – throughout the swings of the trade.

Table 5
Freight Rates for Coal from South Wales Ports, 1902
(Sailing Coasters, Winter Rates Compared to Summer)

	Winter	Summer	Difference	Difference %
		Freight per Ton		
Swansea to Plymouth	4s 10d	3s 10d	1s	+26
Swansea to Medina (IoW)	5s 6d	5s	6d	+10
Swansea to London	5s 6d	5s	6d	+10
Swansea to Dublin	5s	4s 6d	6d	+11
Swansea to Cork	5s	4s	1s	+25
Swansea to Grimsby	7s	6s 3d	9d	+12
Swansea to New Ross	4s 9d	4s 3d	6d	+12

Source: Davies (ed.), *South Wales Coal Annual for 1902.*

The South Wales' coal freights also allow an examination of the factors which determined the specific rate paid. One of these was the season of the year. Throughout the series from 1900 to 1913 there is a note that "[a]s a general rule the freights payable on these Small coasting Sailers rule lower during the summer months than during the winter months."[34] Two pieces of evidence back this up. For 1902 a number of freight rates are given for a range of ports by month for coasting sailing ships. Many ports have freights quoted in only one or two months. Where there are, however, a number of freights for both winter and summer, the freight rates for the winter months are invariably higher than those for the summer (see table 5). The difference, although only between six pence and one shilling per ton, is significant in percentage terms, ranging from ten to twenty-five percent extra for winter voyages than those in summer.

A similar result is obtained from an examination of the tables given in each year of the *Annual* from 1904 to 1915 for "average freights (sailers) from South Wales and Monmouthshire Ports to Coastwise Ports." Winter is defined in this context as October to March inclusive, and summer as April to September. This definition was followed when calculating average freights for table 5. In all of the tables in the *Annual,* for all ports and all years, the winter freight rate is higher than the summer freight rate. The usual range was between six

---

[34]Davies (ed.), *South Wales Coal Annual for 1910,* 277.

pence and 1s 6d per ton extra for winter sailings. Some examples for 1913 are given in table 6. The extra paid for winter sailings varies between fourteen and twenty percent. Thus, until the First World War it was the practice to pay sailing ships a higher freight rate for identical cargoes and routes in the winter than in the summer.

Table 6
Sample of Sailing Ship Freight Rates for Coal from South Wales' Ports in 1913
(per ton)

	Winter	Summer	Difference	Percent
Cardigan	6s 6d	5s 6d	1s	+18
Shoreham	7s 9d	6s 6d	1s 3d	+19
Lynn	7s 9d	7s	1s	+14
Medina (IoW)	7s	5s 9d	1s 3d	+22
Appledore	3s	2s 6d	6d	+20
Rochester	7s 3d	6s	1s 3d	+21
Plymouth	5s 6d	4s 3d	1s 3d	+29

Source: See table 5.

Table 7
Freight Rates on Coal from South Wales' Ports, 1903 and 1912,
Average Summer and Winter Rates (per ton by Steamer)

Port	Summer	Winter	Difference	
			Money Terms	Percent
1903				
Pembroke	3s 4d	4s 2d	10d	25
Devonport	3s 3d	3s 7d	4d	10
Portsmouth	3s 9d	3s 11d	2d	4
Dublin	3s 10d	4s 6d	8d	17
Cork	3s 10d	4s 1d	3d	6
Shoreham	4s 11d	5s 4d	5d	8
London	4s 1d	4s 3d	2d	4
1912				
Pembroke	3s 4d	4s 6d	1s 2d	35
Devonport	3s 9d	4s 6d	9d	20
Portsmouth	4s 5d	3s 9d	(8d)	(15)
Dublin	4s 6d	5s 1d	7d	13
Cork	5s 1d	6s	11d	18
Shoreham	5s 4d	6s 6d	1s 2d	22
London	5s 1d	6s 3d	1s 2d	23

Source: Davies (ed.), *South Wales Coal Annual*, 1903 and 1912.

It is more difficult to ascertain whether demand or supply factors were dominant in determining this winter premium, for two factors concatenate in the winter. The demand for coal rose in the winter months because of the extra demand for heating fuel on top of the normal demand for steam coal for boilers. Moreover, the winter was a more dangerous time for ships to be trading, with greater possibility of gales, fog or snow, and rougher and heavier seas. As a result the load line, required by the legislation of 1890, was lower for winter than summer to give the ship more reserve buoyancy. Thus, the quantity of cargo carried by each ship in winter was less than in summer, and as the costs of operating the vessel were not reduced, the unit cost rose and freight rates reflected this.

The payment of higher freight rates in winter than summer was not confined to coastal sailing ships. Another part of the *South Wales Coal Annuals* for 1903 to 1912 gives the range of freight rates each month for a number of ports for coasting steamers. Where there is material which allows comparison, the freight rates for the winter months were higher than those for the summer months. The average of the range of freight rates for summer and winter months for a number of ports which received regular cargoes from South Wales' ports are given in table 7. From these it will be seen that, with one exception, the winter rates were greater than the summer by between two pence and 1s 2d per ton. These differences represented a premium for winter voyages of between three and thirty-three percent. Steamers charged a premium for winter working for the same reasons as sailing ships. Rather remarkably, the range of this premium was greater in 1912 than it was in 1903. Admittedly, 1912 saw much higher freight rates on average than 1902, but the evidence suggests that there was no withering away of this seasonal premium as a result of technical progress. Both sailing ship and steamer expected, and received, a higher freight rate for braving the more dangerous and difficult coastal waters in winter than in summer and carrying less cargo, at least until the First World War. This was reflected by the higher price of coal to the customer in winter than in summer.

The aberrant freight rate for Portsmouth is explained by one consignment sent in September from Cardiff attracting the unusually high rate of 6s 6d per ton. The next highest freight rate in the summer months was 4s 9d. If the latter figure were used to calculate the simple arithmetic mean, it would give a summer freight figure of 3s 7d, nearly one shilling less than the actual figure and more in line with expectations. The precise reason for this one abnormally high freight rate is difficult to discern. Given that it was going to Portsmouth, a town with a large naval dockyard, it is conceivable that the Admiralty had pressing strategic reasons for a cargo of coal and was willing to pay over the odds to get the consignment there in especially rapid time.

Table 8
Freight Rates from South Wales Ports for Coal,
Steamer Compared to Sailing Ship (price per ton)

| | Steam | | Sail | |
	Winter	Summer	Winter	Summer
1903				
Pembroke	4s 2d	3s 4d	5s	4s
Devonport	3s 7d	3s 3d	6s 6d	4s 6d
Portsmouth	3s 11d	3s 9d	6s 6d	5s 3d
Dublin	4s 6d	3s 10d	6s 6d	4s 9d
Cork	4s 1d	3s 10d	6s 6d	4s 6d
1912				
Pembroke	4s 6d	3s 4d	5s	4s
Devonport	4s 6d	3s 9d	6s 6d	5s 3d
Portsmouth	6s 6d	5s 4d	7s 9d	6s 6d
Dublin	5s 1d	4s 6d	6s 6d	6s
Cork	6s	5s 1d	6s 6d	6s

Source: See table 7.

This source also throws up a strange fact of coaster freight rates. For a given route at a given time of year a sailing ship was paid a higher freight rate than a steamer. Some examples of this for 1903 and 1912 are given in table 8. It should be stressed that these are only a sample. In all cases where there was comparable data this relationship was found to be true. Sailing coasters earned a larger sum per ton than did steamers for traversing the same sea route. This is difficult to explain. It is easier to explain above-average freights for particular destinations because of the difficulty or danger of entering a harbour or unloading. For instance, Finch makes the point that unloading on an open beach, such as Brighton or Cromer, was sufficiently dangerous, if a strong east wind blew up, to need high reward.[35] Similarly, some ports were entirely shunned by steamers because of lack of depth of water, the need to lie on the bottom at low tide or poor facilities for unloading.[36] However, this does not explain why a sailing ship trading to the same port as a steamer should attract a higher freight, or indeed more to the point, why a merchant should be willing to pay a higher freight to use a sailing vessel which was likely to be slower and less reliable than a steamer. For some ports the sailing ship might incur some costs which the steamer could avoid, for instance, towage into or

[35]Roger Finch, *Coals from Newcastle: The Story of the North East Coal Trade in the Days of Sail* (Lavenham, 1973), 178.

[36]Basil Greenhill, *The Merchant Schooners* (2 vols., London, 1951-1957; 4th rev. ed., London, 1988), II, 103 and 168; and Roger Finch and Hervey Benham, *Sailing Craft of East Anglia* (Lavenham, 1987), 10.

out of harbour, or up narrow rivers to towns such as in Suffolk or Norfolk. Indeed, given that the sailing ship would normally be slower than the steamer, it might incur higher crew costs if the crew size was identical and it was paid by the day or week. However, many coastal sailing ship crews were paid "on shares," i.e., a proportion of the freight rate, irrespective of the time taken.[37] Even where crews were on time rates and hence wages had to be paid for a longer period than on the equivalent steamer, since there was no greater value to the shipper of using the sailing ship, the normal reaction of sailing ship owners was to reduce the crew members to a minimum and cut wages to compensate. The extra freight rate paid to sailing ships must have equalled some greater utility to the customer, but it is difficult to see what. I leave the solution of this particular riddle to others with more knowledge and wit.

Table 9
Freight Rates per Ton-Mile from Cardiff for Coal in 1900 and 1913

	Freight Rate		Distance (miles)	Ton-Mile Rate (pence per ton-mile)	
	1900	1913		1900	1913
Holyhead, steam	4s 4d	4s 4d	221	0.24	0.24
Portmadoc, sail	3s 3d	4s 3d	212	0.18	0.24
Fowey, sail	4s 7d	5s	220	0.25	0.27
Fowey, steam	4s 1d	4s 7d	220	0.22	0.25
London, steam	4s 9d	5s	596	0.10	0.10
Gravesend, sail	6s 7d	NA	577	0.14	NA
Belfast, steam	4s 9d	4s 7d	332	0.17	0.17
Waterford, sail	5s 6d	6s	178	0.37	0.40
Londonderry, steam	5s 1d	5s 1d	396	0.15	0.15
Sligo, sail	8s	9s	517	0.19	0.21

Source: Davies (ed.), *South Wales Coal Annual*, 1900 and 1913; and Harold Fullard (ed.), *The Mercantile Marine Atlas* (London, 1959), 8-10.

It is also interesting to investigate whether there were "economies of distance" in coastal shipping. It is usually argued that a high proportion of a coaster's costs were terminal charges, incurred at the beginning and end of a

---

[37]Greenhill, *Merchant Schooners*, II, 127.

voyage.[38] These comprised pilotage (and towage for sailing ships), dock or harbour dues, the cost of loading and unloading, etc. If this hypothesis is valid, it might be expected that freight rates per ton-mile for long-distance routes would be lower than for shorter ones. Some freight rates have been calculated on this basis in table 9. If the rates are divided into steam and sail and then rearranged in order of distance from Cardiff, as presented in table 10, it will be seen that there is an inverse relationship between distance and unit freight rate. As distance rose, the unit freight rate fell. However, the relationship was not linear, as is indicated by the correlation coefficients for the four series, but there is significant support for the belief that there were reductions in unit cost for longer-distance routes, thus reinforcing the belief that coastal shipping had a distinct economic advantage for long-distance routes.

Table 10
Freight Rates per Ton-Mile from Cardiff for Coal in 1900 and 1913

	Distance	Ton-Mile Rate (Pence)	
		1900	1913
Sail			
Gravesend	577	0.14	-
Sligo	517	0.19	0.21
Fowey	220	0.25	0.27
Portmadoc	212	0.18	0.24
Waterford	178	0.37	0.40
Steam			
London	596	0.10	0.10
Londonderry	396	0.15	0.15
Belfast	332	0.17	0.17
Holyhead	212	0.24	0.24
Fowey	220	0.22	0.25

Source: See table 9.

It might be argued that freight rates were affected not only by economies of distance but also by economies of scale. Ports receiving large quantities of coal might attract larger, more efficient ships which could be employed more regularly in the coal trade and that therefore the freight rates would be lower to reflect this steady employment. Ports receiving smaller quantities, where ship employment was more irregular and unpredictable, would attract higher rates per ton-mile. This can be tested on the ton-mile rates already calculated. The relationship does not hold, as can be seen from table 11. There is

[38]John Armstrong and Philip S. Bagwell, "Coastal Shipping's Relationship to Railways and Canals," *Journal of the Railway and Canal Historical Society*, XXIX, part 5 (1988), 215.

no support for the hypothesis that freight rates per ton-mile fell as the quantity shipped to a port rose. Indeed, for the sailing ships the opposite happened for the first three observations.

Having calculated the ton-mile rates, it might be possible to test the hypothesis advanced elsewhere that coastal tramp freight rates were on average lower than those for coastal liners.[39] However, this would be fairly meaningless, since the liners did not carry coal and we would not be comparing like commodities. There is, therefore, no evidence for or against this hypothesis.

Table 11
Freight Rates on Coal from Cardiff in 1903 Compared to Quantity Shipped

	Quantity Shipped (tons)	Freight Rate (d per ton-mile)
Sail		
Waterford	35,312	0.34
Fowey	19,900	0.22
Gravesend	4527	0.12
Sligo	1410	0.17
Port Madoc	850	0.18
Steam		
London	795,794	0.08
Fowey	19,900	0.20
Belfast	12,270	0.15
Londonderry	7705	0.14
Holyhead	6497	0.23

Source: Davies (ed.), *South Wales Coal Annual for 1903.*

Since the alternative method of moving coal from the pits to the centres of consumption was by railway, one ought to see how freight rates compared between the two competing forms of cargo carrier – the train and the coaster. It is notoriously difficult to ascertain the freight rates the railways actually charged. The maximum rates were published, but as one journal commented, "the actual charges are in the majority of cases considerably lower."[40] The problem is that there were the published rates and then the special rates which were negotiated on a customer-by-customer basis.[41] Very few records of these rates survive, presumably because they were of a less official nature, even in the voluminous material in the National Archives. However, one book

---

[39]Armstrong, "Freight Pricing Policy," 196.

[40]Davies (ed.), *South Wales Coal Annual for 1905,* 100.

[41]Peter J. Cain, "Traders Versus Railways: The Genesis of the Railway and Canal Traffic Act of 1894," *Journal of Transport History,* New ser., II, No. 2 (1973), 66.

has survived for the GWR which gives the freight rates on coal for a range of South Welsh collieries to a number of London stations.[42] It is undated, but internal evidence suggests it applied to the period 1902-1908. The rate for coal traffic from South Wales to London stations varied between 7s 5d and 7s 10d for owners' wagons with a minimum of four tons. A simple comparison with the rates in table 2 shows that the railway was charging much more than the coastal steamer, the rates for which varied between 4s 1d and 5s 4d.

The land distance from the South Wales ports to London is between 160 and 190 miles. This gives a ton-mile rate for the railway of 0.49d to 0.58d compared to 0.3d to 0.4d for the steam coaster to London. In other words, the coastal rate was between sixty and sixty-eight percent of the equivalent railway charge. However, it should be borne in mind that the ton-mile rate for the ship is entirely notional, since it had to take a much more circuitous route to reach the capital and in so doing travelled nearly 600 miles. Thus, the actual coaster ton-mile rate was much lower than the notional one calculated above, between 0.08d and 0.11 per ton-mile. Given that the coaster could charge such a rate and at least break even or perhaps make a small return on its capital, the implication is that the costs of the coaster were a fraction of those of the train, perhaps less than twenty percent per ton-mile of the railway rate. Nor could it be argued that the railways were greedily insisting on higher returns on their capital than the coaster, for in the late nineteenth and early twentieth centuries the yield earned on the railways had dropped considerably and was a cause of complaint among shareholders. Aldcroft calculated that net return on capital fell from about 4.5 percent in 1870 to 3.5 percent between 1900 and 1912.[43]

To summarize, the coaster was charging a much lower rate than the railway for the journey to London, somewhere around two-thirds of the train rate. For a cargo such as coal, where the price of transport rather than speed or regularity was the main determinant, the coaster thus had a great advantage. The railway still carried coal to the metropolis because some final destinations for coal were distant from the port of London or the river. To reach these via coaster involved expensive horse-and-cart or primitive motor lorry carriage or transhipment onto the railway for the last few miles. For some such destinations it was logistically simpler and perhaps no dearer to send the coal all the way by rail. That said, the actual costs of the coaster per ton-mile were a fraction of those of the railway on long-distance routes.

---

[42]Great Britain, National Archives (TNA/PRO), RAIL 253/212.

[43]Aldcroft, *Studies in British Transport History*, 33.

Conclusions

From the evidence gathered on coastal coal freight rates, it is clear that coastal shipping suffered from the same depression of the early twentieth century which affected overseas shipping. There was no uniformity in terms of the degree of reduction in freight rates on different routes, nor of the timing of the low points. There is general agreement on the timing of the recovery. In general, the freight rates for coal from the North East fell further and recovered less than those from South Wales. This may be explained by the demand for Welsh coal being more buoyant than that from Tyneside and hence, as shipping space is a derived demand, the demand for coastal ships was reflected in less depressed freight rates. The reason for the buoyancy of demand for Welsh coal was that much of it was high-quality steam coal used for ship's bunkers. One method of cutting costs and increasing cargo space in a shipping depression is to switch to the most efficient, least bulky fuel – this was best Welsh steam coal. Compared to overseas shipping, coastal freight rates appear less volatile and more staid. Perhaps the mundane nature of their cargoes made them less subject to rapid fluctuations in demand compared to more exotic overseas trade. Hence, coastal rates were less exciting but more reliable.

  In this period coastal coal freight rates seem to have fallen in real terms, for the vast majority of the coastal rates rose less than the retail price index. Feinstein has recently suggested that the cost of living was rising between 1899 and 1913 by just under one percent per annum.[44] A re-examination of table 1 shows that freight rates from the North East did not rise in money terms and therefore fell in real terms by about eleven to twelve percent. Similarly, from table 3 it can be ascertained that the majority of the rates from South Wales' ports, although they rose more than the series from the North East, did not keep up with the general rise in the cost of living. Using 1900 as a base, the rise in average prices by 1911 would have been about 10.7 percent, 11.6 percent in 1912 and 12.6 percent in 1913. Of the twenty-seven observations for these three years, only eight exceeded the rise in the general cost of living, whereas fifteen were less and four roughly equal. Thus, at a period when the average cost of living was rising, the real costs of moving coal by coaster were falling, suggesting increasing cost efficiency.

  A number of other points emerge from this study about coastal coal tramp freight rates. Even as late as 1913, for both sailing and steam ships freight rates were higher in the winter months than in the summer. Sailing ships received a higher rate, for the same route and time of year, than did steamers. Both sailing vessels and steamships enjoyed economies in the long

---

[44]Charles H. Feinstein, "What Really Happened to Real Wages? Trends in Wages, Prices, and Productivity in the United Kingdom, 1880-1913," *Economic History Review*, 2nd ser., XLIII, No. 3 (1990), 344.

haul in that unit freight rates tended to fall as the length of journey increased. This is more explicable in terms of a high proportion of total costs being incurred as "terminal" costs than in terms of economies of scale.

Chapter 16
The Coastal Trade of Connah's Quay in the Early Twentieth Century: A Preliminary Investigation[1]

The coastal trade of individual ports has been little researched. Where ports have been studied it is often with reference to their technical development, engineering works or role in the more exotic overseas trades.[2] Coasters called at ports more frequently than deep-water ships because their voyages were shorter compared to the longer overseas journeys and thus had a greater impact on harbour activity. With a few honourable exceptions there has been little research on the trade or fortunes of individual ports and even less on small ports.[3] This is probably as much due to a dearth of records as a lack of interest. By good fortune one source has survived which records the trade of Connah's Quay from 1905 to the First World War, ironically among the records of a railway company. Connah's Quay was a small port on the Dee Estuary about eight miles downstream from Chester. It no longer operates as a port. The Wrexham, Mold and Connah's Quay Railway (WMCQR) was approved by Parliament in 1862 and was built by 1865.[4] The railway company developed the wharves at Connah's Quay. It was absorbed by the Manchester Sheffield and Lincolnshire Railway (MSL) in 1890, and by 1897 the MSL was part of the Great Central (GC). In 1904 the GC formally took over the WMCQR.[5] It

---

[1]This essay appeared originally in the *Flintshire Historical Society Journal*, XXXIV (1996), 113-133 (with David Fowler).

[2]Gordon Jackson, *The History and Archaeology of Ports* (Tadworth, 1983); and Jackson, "The Ports," in Michael J. Freeman and Derek H. Aldcroft (eds.), *Transport in Victorian Britain* (Manchester, 1988), 177-208.

[3]John H. Farrant, *The Harbours of Sussex, 1700-1914* (Brighton, 1976); David F. Gibbs, "The Rise of the Port of Newhaven, 1850-1914," *Transport History*, III, No. 3 (1970); and H.C. Brookfield, "Three Sussex Ports, 1850-1950," *Journal of Transport History*, 1st ser., II, No. 1 (1955), 35-50.

[4]George Dow, *Great Central, Vol. 3: Fay Sets the Pace* (London, 1965), 48-50; K. Davies, "The Growth and Development of Settlement and Population in Flintshire, 1851-91," *Flintshire Historical Society Journal*, XXVI (1973-1974), 150; and James I.C. Boyd, *The Wrexham, Mold and Connah's Quay Railway* (Oxford, 1991).

[5]Dow, *Great Central*, 39 and 69-70.

required the port of Connah's Quay to keep a register of all ships entering and clearing the port, including their registered tonnage, the type and quantity of cargoes they were carrying, their captain's name and their port of despatch or destination.[6] This source can be used to shed light on the nature of the trade of this tiny port.

The port register was kept from 1904 to 1922 and appears to have recorded all ships which used the wharves in chronological order. Ships calling at the other small ports of the Dee Estuary were not included, and the temptation is to see the register as a record from which invoices for port charges were compiled. Given the number of pieces of data and the long time period, an initial project was formulated to extract all the information for four years: 1905, 1906, 1912 and 1913.[7] This was entered on a computer database which could then be analyzed using SPSS.[8] All statements which follow based on this source were determined in this way. As with any manuscript source there were a number of omissions, misspellings, uncertainties and mistakes. Where possible these have been corrected or inserted by extrapolation from other material in the register. The number of such occurrences is very small, and their exclusion would make no difference to the overall trends or patterns.

As a check, it was possible to compare entries for some ships with the detail of voyages given in their "crew agreements," where they are lodged in the National Archives. For instance, *T&EF*, the cryptically named schooner owned in Millom, appears seven times in the register for 1906, on six occasions departing for Duddon with coal and once arriving from the same port with iron ore. Its registered tonnage was recorded as fifty-seven and its captain as "Humphreys." The crew agreement for 1906 confirms many of these details, including the tonnage and captain, and that among its thirty-nine coastal trips that year, six were from Connah's Quay to Duddon and one from Duddon to Connah's Quay.[9] This sort of concordance justifies using the register as a reliable source for ship movements at the port in the early twentieth century.

I

Some writers have suggested that the port of Connah's Quay came into being as a result of the canalization of the Dee in 1737 which led the channel to the

---

[6]Great Britain, National Archives (TNA/PRO), RAIL 226/110.

[7]1904 was excluded as it was not a complete year.

[8]Our thanks to Linda Brown who solved all the technical software problems calmly and efficiently and to the Polytechnic of West London for the use of computer space and software.

[9]TNA/PRO, Board of Trade (BT) 99/2469.

southern side of the previously wide estuary and hence favoured Connah's Quay while disadvantaging Mostyn, Bagillt and Flint.[10] Hawkes, however, suggested that the name Connah's Quay did not come into being until a little before 1784, probably based on no particular individual but on the fact that a number of Connahs had lived in the area since at least the sixteenth century.[11] Connah's Quay then developed as a port for the despatch of coal, aided by the Irish Coal Company which was based there, as well as sending Buckley pottery to London, Ireland and South Wales. Buckley ware was cheap everyday earthenware which thrived until the third quarter of the nineteenth century when it went into a steady but terminal decline.[12] However, the red marl which was associated with the coal measures of Flint and Denbigh was found to be ideal as fireclay for tiles, bricks and other necessities of the building trade, and from 1865 onwards there was a thriving export trade in these products[13] after the construction of the WMCQR linked the port to the clay and coal areas of Flint and Denbigh.[14] By 1872 Connah's Quay had become the most important port on the Dee Estuary in the eyes of the Commissioners of Customs, for they had their Principal Coast Officer move there from Flint.[15] The WMCQR had developed the docks and wharves at the port from 1865 and further extended them in 1886. By 1900 Connah's Quay had ceased to be the most important port of the Dee Estuary, for Mostyn and Point of Ayr both cleared a larger tonnage of ships.[16] However, by then there was a thriving community based on the railway-owned docks with chemical and engineering works as well as shipbuilding, ships' chandlers and agencies. Indeed, the inflow of English from across the river fused with the native Welsh to create a community which

---

[10]Arthur H. Dodd, *The Industrial Revolution in North Wales* (Cardiff, 1933; 2nd ed., Cardiff, 1951), 127; and George Lloyd, "The Canalization of the River Dee in 1737," *Flintshire Historical Society Publications,* XXIII (1967/1968), 39.

[11]G.I. Hawkes, "The Founder of the Port of Connah's Quay," *Flintshire Historical Society Journal,* XXXII (1989), 177-179.

[12]Dodd, *Industrial Revolution,* 192-193; and Joseph E. Messham, "The Buckley Potteries," *Publications of the Flintshire Historical Society,* XVI (1956), 31-49.

[13]Dodd, *Industrial Revolution,* 193.

[14]G.I. Hawkes, "Shipping on the Dee: The Rise and Decline of the Creeks of the Port of Chester in the Nineteenth Century," *Maritime Wales,* No. 11 (1987), 121.

[15]*Ibid.,* 123.

[16]Derrick Pratt, "Sidelights on the Dee Navigation, 1892-1912," *Maritime Wales,* No. 6 (1981), 63.

looked seaward for most of its income and employment. Families like the Coppacks and Reneys were typical of the local enterprising families.[17]

Across the water, from the 1890s the iron and steel works of John Summers was beginning to take shape. It contributed to the traffic on the Dee for it "built up its own fleet of ships to ply between the works and the ocean-going ships at Liverpool."[18] However, little of this traffic came into the Connah's Quay register as it went directly to Shotton.

II

In the last few years before the First World War the port of Connah's Quay was in decline. Whereas in 1905 and 1906 an average of 450 ships of an aggregate registered tonnage of about 35,000 entered and cleared the port with cargoes, by 1912 and 1913 only 276 ships arrived and departed with a registered tonnage of just under 22,000. There was a similar reduction in the quantity of cargo loaded and unloaded in the port. In 1905 and 1906 it averaged over 65,000 tons per annum, but by 1912 and 1913 it was less than 43,000 tons. Table 1 gives the details of port activity. This was not in line with national trends in shipping. In fact, it was contrary to the movement in coastal shipping entrances and clearances. Between 1905 and 1913 the total tonnage of ships entering British ports in the coastal trade with cargoes rose by six percent, from nearly thirty-three million to almost thirty-five million registered tons (see table 2). The "shipping depression" which has been identified as occurring between 1901 and 1911[19] meant that there was no growth in the aggregate figures between 1906 and 1911, but in the last two prewar years there was a strong recovery in which Connah's Quay did not share. The national figures show a reduction of about nine percent in the number of ships entering coastwise, from 185,000 to 170,000. This was due to a decline in the number of sailing ships in the coastal trade, while there continued to be growth in the number of steamers, which on average were larger; hence, the average size of ship rose by about sixteen percent.

---

[17]Tom Coppack, *A Lifetime with Ships: The Autobiography of a Coasting Shipowner* (Prescot, 1973).

[18]Brian Redhead and Sheila Gooddie, *The Summers of Shotton* (London, 1987), 105.

[19]Derek H. Aldcroft, "The Depression in British Shipping, 1901-11," *Journal of Transport History*, 1st ser., VII, No. 1 (1965), reprinted in Aldcroft, *Studies in British Transport History, 1870-1970* (Newton Abbot, 1974); and John Armstrong, "The Shipping Depression of 1901 to 1911: The Experience of Freight Rates in the British Coastal Coal Trade," *Maritime Wales*, No. 14 (1991), 89-112.

Table 1
Connah's Quay: Port Activity, 1905-1913

	Ship Movements Numbers			Registered Tonnage			Cargo Weight Tons		
	Arrivals	Departures	Total	Arrivals	Departures	Total	Inward	Outward	Total
1905	144	275	419	11,528	21,072	32,600	19,378	40,756	60,134
								(274)	(418)
1906	205	276	481	16,520	21,857	38,377	26,211	44,492	70,703
					(274)	(479)	(200)		(476)
Average	175	275	450	14,024	21,436	35,462	22,794	42,624	65,418
1912	95	184	279	8102	14,262	22,364	14,521	28,952	43,473
				(93)					
1913	118	156	274	9309	11,657	20,966	18,213	24,259	42,472
Average	107	170	276	8705	12,960	21,665	16,367	26,605	42,972

Note: The figures in brackets indicate the number of voyages for which information is available where it is less than the total shown in the first three columns.

Source: Derived from Great Britain, National Archives (TNA/PRO), RAIL 226/110.

Table 2
Number and Tonnage of Vessels Entering the Ports of the
UK, with Cargoes Only, in the Coastwise Trade

	No. (thousands)	Tons (millions)	Average Size of Ships (Tons)
1905	185	32.7	176.8
1906	182	32.9	
1907	170	31.2	
1908	165	31.0	
1909	166	31.6	
1910	167	32.2	
1911	162	31.3	
1912	167	33.1	
1913	169	34.8	205.9

Source: Great Britain, Parliament, *Parliamentary Papers* (*BPP*), Trade and Navigation Accounts, 1905-1913.

Table 3
Number and Net Tonnage of British and Foreign Sailing and Steam Vessels,
Including Repeated Voyages, Entering and Clearing the Port of
Chester with Cargo, 1905-1913

	Entered		Cleared	
	No.	Tons	No.	Tons
1905	879	64,218	1290	92,069
1906	967	71,859	1234	89,474
1907	834	60,728	1460	106,434
1908	734	52,923	1336	98,281
1909	603	43,686	1315	96,484
1910	534	39,965	1257	92,884
1911	466	34,716	1066	78,523
1912	572	42,074	1162	85,177
1913	634	45,770	1366	100,850
Average				
1905-1906	923	68,038	1262	90,771
1912-1913	603	43,922	1264	93,013
Percent change	-35	-35	+ <1	+2

Source: See table 2.

The decline in Connah's Quay did not mirror the regional situation either. The kingdom was divided up into sections of coast with each designated by the name of its major port. Connah's Quay did not rate its own separate section but was included within the Chester district. The figures for the coastal trade of that "port" between 1905 and 1913 are given in table 3. From this it will be seen that there were contrary trends. The number and volume of ships arriving in the regional coastal trade declined drastically, by about one-third, whereas the number and volume departing with cargoes increased by a very small percentage. Thus, Connah's Quay, though moving in the same direction as the region in arrivals, was doing worse than the regional trend, and in departures was bucking the trend and declining sharply when the regional trend was a slight rise. This suggests that we cannot look solely for national or regional trends, such as the shipping depression or the changing importance of the North West, to explain this decline. Rather, we must look for explanations peculiar to the local conditions of Connah's Quay. The downside of this is that Connah's Quay cannot be taken as representative of other ports. It is not typical of the general trend and may be unique in losing traffic when most other ports were gaining it. Thus, in discovering why Connah's Quay was in decline we may be able to isolate factors which might cause the decline of any coastal port, but we cannot presume them to apply at this time to any other port.

The port of Connah's Quay has been characterized as essentially a port for the coasting trade. If foreign trade was ever extensive from this location, which seems unlikely, then by the early twentieth century it was a very small proportion (see table 4). In most years and for most categories, foreign trade made up less than five percent of total activity. In 1912 foreign arrivals made up an eighth of the tonnage of ships arriving, for ocean-going ships tended to be larger than coasters, but even then the proportion of cargo tonnage was only just over five percent, so that as movers of goods the deep-water ships were never significant to Connah's Quay. Hence, for this period it is fair to characterize Connah's Quay as essentially a coastal trade port.

Table 4
Connah's Quay: Foreign Trade (Percent of Total Trade)

| | Ships | | | | | | Cargo | | |
| | Number | | | Registered Tons | | | (Tons) | | |
	In	Out	Overall	In	Out	Overall	In	Out	Overall
1905	2.8	1.5	1.9	7.0	2.4	4.0	5.8	2.2	3.3
1906	1.9	2.2	2.1	4.7	3.1	3.8	1.9	3.5	2.9
1912	3.1	0.5	1.4	12.8	1.0	5.3	5.5	0.9	2.4
1913	0	0	0	0	0	0	0	0	0

Source: TNA/PRO, RAIL 226/110.

III

The normal explanation for the decline of the port of Connah's Quay is couched in geographical terms. The Dee Estuary for more than a century had been renowned as a treacherous stretch of water. Its channel was narrow, tidal,

twisting, shallow and continually changing.[20] Despite the work of the River
Dee Company, or perhaps because of it, by the 1880s the maximum draught
between Mostyn and Connah's Quay was thirteen feet, and the distance from
Mostyn to Connah's Quay had increased from thirteen to thirty miles.[21] The
channel was subject to silting, and sandbanks appeared and disappeared unpre-
dictably. As a result, expert knowledge was required to navigate it in safety,
and even old hands were occasionally caught out.

Table 5
Connah's Quay: Average Size and Capacity of Ships

	Ship Size (registered tons)			Cargo (tons)			Deadweight/ Registered Ratio		
	In	Out	Overall	In	Out	Overall	In	Out	Overall
1905	80	77	78	135	149	144	1.69	1.94	1.85
1906	81	80	80	131	161	148	1.62	2.01	1.85
1912	85	77	80	153	157	156	1.8	2.04	1.95
1913	79	75	76	154	155	155	1.95	2.07	2.04

Source: Derived from PRO, RAIL 226/110.

The result of the restrictions on draught was to prohibit the use of
large ships. The largest ship to call at the port in the four years studied was the
SS *Nerva* of 424 registered tons which brought wood pulp from Poireau, but it
was only partly laden to reduce the depth of water needed. On average, the
size of ships using Connah's Quay was much smaller. Table 5 shows that the
average was just under eighty registered tons and did not alter significantly
over the period. There was little difference between ships entering and departing, the consistently slightly larger size of ships entering probably being attributable to a small number of large foreign-owned ships bringing in timber
products. The mean size of ship entering Connah's Quay was also much
smaller than the national average for the coastal trade, which was 177 registered tons in 1905 and rose steadily to 206 in 1913.[22] Thus, whereas the UK
trend was to increased size, that for Connah's Quay was stagnation. Because of
the difficulties of navigation and the small population, there was no liner trade
to Connah's Quay. By the early twentieth century coastal steam liners were

[20]Davies, "Growth," 146.

[21]*Ibid.*, 156.

[22]Calculated from Great Britain, Parliament, *Parliamentary Papers* (*BPP*),
Trade and Navigation Accounts, various years.

often as large as 2000 tons and could not negotiate the shallow channel.[23] Moreover, the port's small size and restricted hinterland were unlikely to generate sufficient passengers or goods traffic to justify a regular visit by a large liner.

The national trend in coastal shipping was a decline in the number and tonnage of sailing ships and a steady growth in the absolute and relative importance of steamships. Whereas in 1905 sailing ships made up twenty-nine percent by number and eleven percent by registered tonnage of all coastwise entrances, by 1912 these proportions had slipped to twenty-four and eight percent, respectively.[24] The proportion of sailing ships in the trade of Connah's Quay was much higher than the national average (see table 6). This is explained partly by the deteriorating geographical conditions of the Dee Estuary. As the average size of steam coasters rose, only the smaller steamer or sailing ship could easily navigate the channel, and throughout the period 1830-1913 the average size of a coastal sailing ship was much less than that of a steamer. The disadvantage of this for Connah's Quay was that the sailing ship did not enjoy the speed or regularity which could be achieved by the steamer, and because sailing ships remained small they were not able to extract the economies of scale that benefited larger steamers. Thus, sailing ships were pushed into the least valuable cargoes on runs which were not attractive to steamships and hence earned low freight rates. In many ways Connah's Quay was becoming marginalized, and the evidence of the port register supports the view that the difficult conditions combined with the increasing average size of coastal ships conspired to put Connah's Quay's seaborne trade in jeopardy. However, the emphasis on sailing ships was not solely a function of geography. It was partially dictated by the nature of the largest single commodity shipped from Connah's Quay, namely bricks. Steamships were considered less suitable than sailing ships for bricks, tiles, clay pipes and other such goods which were frangible and liable to chip, crack or crush. The combination of vibration from the engines plus the exaggerated pitching of screw propulsion made the cargo more likely to move in the hold, thus leading to a higher proportion of damaged goods. The sailing ship, being less capital intensive, was willing to spend more time at the quay loading carefully whereas the steamer wanted a more rapid turnaround and faster loading.[25] This is borne out by the experience at Connah's Quay where between eighty-seven and ninety-one percent of all brick

---

[23]For instance, in 1909 the Carron Co. took delivery of *Carron IV* of 2354 gross registered tons for its Edinburgh-to-London service: Ian Bowman, "The Carron Line," *Transport History*, X, Nos. 2-3 (1979), 210.

[24]*BPP*, Trade and Navigation Accounts.

[25]Roy S. Fenton, *Cambrian Coasters: Steam and Motor Coaster Owners of North and West Wales* (Kendal, 1989), 70.

cargoes were carried in sailing ships in the four sample years. Bricks were never a significant cargo for steamers but comprised between fifty-eight and sixty-seven percent of all trips made by sailing ships.

Table 6
Connah's Quay: Breakdown of Ship Activity, Steam and Sail

| | 1905 | | 1906 | | 1912 | | 1913 | |
	Sail	Steam	Sail	Steam	Sail	Steam	Sail	Steam
Ships Calling (numbers)	140	23	136	29	92	18	76	17
Percent	86	14	82	18	84	16	82	18
Trips Made (numbers)	326	82	377	107	235	53	205	72
Percent	80	20	78	22	82	18	74	26
Cargo (tons)	48,138	11,478	53,343	16,222	34,563	9770	29,287	13,025
Percent	80	19	77	23	78	22	69	31

Source: Derived from TNA/PRO, RAIL 226/110.

It has already been pointed out that Connah's Quay could boast no liner trade. However, this did not mean a stochastic pattern of trade. There were a number of ships which were regular traders in all four years of the period studied. For instance, SS *Maggie* in 1906 made twenty-six trips into Connah's Quay, all from Liverpool and all carrying grain. In 1913 *Elizabeth Hyam* made ten outward journeys from Connah's Quay, all with bricks as a cargo, to a variety of mostly Irish ports.[26] In 1912 *T&EF* made ten trips from Connah's Quay to Duddon with coal and brought return cargoes of iron ore on four occasions. Thus, a number of ships had a regular pattern of trade based on the port and some, like SS *Maggie,* were liner-like in the regularity of their movements, shuttling between the two ports.

There was a degree of specialization evident even among such small ships as served Connah's Quay. For instance, many of the ships which brought grain from Liverpool were dedicated to that role, i.e., they only brought grain from Liverpool and neither took back loads nor deviated into other types of cargo. For instance, in 1913 SS *Pelican* made twelve such voyages, while SS *Penrhyn* made seven; in 1906 SS *Maggie* made twenty-six and SS *Chili* fifteen. Nor was it that there were no cargoes available to act as back hauls, for in all four years substantial quantities of bricks were sent to Liverpool. However, most of the ships specializing in the grain trade from Liverpool were steamers

---

[26]The ship appears as *E. Higham* in the register, almost certainly a misspelling as the registered tonnage of this ship matches that *of Elizabeth Hyam*; see *Alphabetical List of British Registered Sailing Vessels* (London, 1908).

and, as has been mentioned previously, steamers were considered less suitable than sailing ships for bricks and tiles because of breakages and chipping of the cargo. This led some ships to specialize in the brick trade. For instance, in 1912 *Lota* made seven trips to Liverpool with cargoes of bricks, and in 1906 made eight such voyages; in 1913 *Menra* made five trips to Belfast carrying bricks. Some ships specialized in bulk cargoes which were less vulnerable to breakage. For instance, *T&EF* in both 1912 and 1913 carried coal to Duddon and iron ore back.

IV

The trade of Connah's Quay was also vulnerable because it was so narrowly based. Relatively few commodities were involved in the seaborne trade. Tables 7 and 8 show the narrow range of goods in which there was any substantial trade. Table 9 shows the degree of dependence of the port on the top three commodities. In all four years just three products made up over seventy percent of traffic by weight of cargo and number of ship movements. Similarly, the most important single cargo, bricks, accounted for over half the total trade except in the last prewar year. Thus, trade was heavily concentrated on a few commodities. Although this degree of concentration decreased from the very high levels of 1905, even in 1913 three commodities still made up three-quarters of all trade. Thus, any changes in demand or in the nature of the supply structure for these products which put North Wales at a competitive disadvantage had a disproportionate effect on Connah's Quay, which relied on the derived demand for shipping services arising from the sale of just a few goods.

Table 7
Connah's Quay: Main Cargoes Departing

	Bricks		Coal		Manure	
	No.	Weight	No.	Weight	No.	Weight
1905	245	37,502	25	2718	1	49
1906	250	41,089	22	2794	0	0
Average	247	39,295	23	2756	0.5	24
1912	155	24,934	16	2026	8	1014
1913	130	20,539	14	2175	5	599
Average	142	22,736	15	2100	6.5	807

Source: TNA/PRO, RAIL 226/110.

A similar degree of concentration is evident in the number of ports with which Connah's Quay traded. Table 10 shows that the total number of ports involved in the trade each year appears quite large. There were over sixty ports in 1905, a little over seventy in 1906, and close to fifty in the last two prewar years. However closer examination of the pattern of trade shows that one port (Liverpool) accounted for about a quarter of all ship movements, and the top three ports (Liverpool, Dublin and Belfast) accounted for roughly half. Although the degree of dependence on a few ports declined, in all four years over one-fifth of all trade was with just one other port, and the top five ports provided well over half of all ship movements in all four of the years examined. This reinforces the findings relating to the small number of commodities, for not only was the trade of Connah's Quay dependent on a narrow range of commodities but also it was with a small number of ports. Thus, it was very vulnerable to any changes in the nature of these few trades. The narrow range of ports meant the risk was not spread, as a good share portfolio should be. Rather most of the eggs were in a very few baskets and thus vulnerable. As figure 1 shows, it was essentially a regional rather than a national trade.

Figure 1: Connah's Quay and Its Most Important Trading Partners, 1905-1913

Source: Courtesy of the author.

Table 8
Connah's Quay: Main Cargoes Arriving

	Grain		Iron Ore		Pig Iron		Scrap Iron		Timber		China Clay	
	No.	Weight	No.	Weight	No.	Weight	No.	Weight	No.	Weight	No.	Weight
1905	89	11,422	9	1317	6	1192	10	1088	12	1094	5	896
1906	85	11,680	18	2843	11	1616	26	2948	37	2904	4	739
Average	87	11,551	13	2080	8	1404	18	2018	25	1999	4	817
1912	32	4353	14	2132	11	1932	-	-	-	-	5	973
1913	45	6220	26	4831	8	1693	-	-	-	-	4	754
Average	38	5286	20	3481	9	1812	-	-	-	-	9	863

Source: PRO, RAIL 226/110.

Table 9
Top Three Cargoes (Percent of Total Trade)

		No.	Tons	Rank	No.	Tons
1905	Bricks	58.3	62.4	1	58	62
	Grain	21.1	19.0	2	79	81
	Coal	5.9	4.5	3	85	86
1906	Bricks	51.6	58.1	1	52	58
	Grain	17.6	16.5	2	69	75
	Scrap Iron	5.4	4.2	3	75	79
1912	Bricks	55.5	57.3	1	55	57
	Grain	11.5	10.0	2	67	67
	Iron Ore	5.0	4.9	3	72	72
1913	Bricks	47.7	48.4	1	47	48
	Grain	16.4	14.6	2	64	63
	Iron Ore	9.5	11.4	3	73	74

Source: Derived from TNA/PRO, RAIL 226/110.

Table 10
Ports with which Connah's Quay Traded

	1905	1906	1912	1913
Number of different ports,	62	71	49	46
of which coastal	57	63	46	46
Most important ports by number of ship movements:				
Liverpool	123	122	60	63
Dublin	51	69	39	22
Belfast	42	52	38	40
Duddon	23	28	27	18
% of total ship movements made up by:				
Top port	29.3	25.4	21.5	23.0
Top 3 ports	51.4	50.5	49.1	45.6
Top 5 ports	61.9	61.3	59.5	52.9

Source: See table 9.

This problem is reinforced when the distribution of trade between individual ships is examined. Table 11 shows that there was much less reliance on a few ships than there was on a few commodities and ports. The three ships which called most frequently never contributed more than seventeen percent of all trade, and even the top ten ships at the peak contributed just over a third – a much smaller percentage of aggregate traffic than was the case with the top ports or cargo types. However, table 11 shows that the proportion of total traffic made up by any given number of ships was rising steadily over time. Thus, whereas in 1905 the top five ships accounted for sixteen percent of the total number of trips and fourteen percent of the total cargo weight, by 1913 these had grown to twenty-three percent each. In other words, Connah's Quay was becoming increasingly dependent on a small number of regular traders. Rather than spreading its risk, the port was becoming more narrowly based and hence more vulnerable.

Table 11
Trade Made up by the Most Frequently Calling Ships

	Trips (No.)	Cargo (tons)	As a Percent of Total Trade	
			No.	Cargo
1905				
1	17	2440	4.0	4.1
3	45	5168	10.7	8.6
5	66	8174	15.7	13.6
10	111	13,987	26.4	23.3
1906				
1	26	3818	5.4	5.4
3	61	8670	12.6	12.3
5	90	12,200	18.6	17.2
10	141	18,430	29.1	26.1
1912				
1	14	2346	4.9	5.4
3	36	5024	12.5	11.6
5	55	7780	19.1	17.9
10	94	13,903	32.6	32.0
1913				
1	15	3120	5.5	7.3
3	42	7120	15.3	16.8
5	64	9936	23.4	23.4
10	102	15,351	37.2	36.1

Source: See table 9.

Finally the vulnerability of the trade at Connah's Quay can be further stressed by looking at the imbalance between inward traffic and outward voyages. In all four years many more outward cargoes were available than inward, suggesting that ships found it difficult to obtain a cargo to bring into Connah's Quay and thus, *ceteris paribus,* would prefer another port where both an inward and outward cargo was likely to give them the opportunity of two freight-earning voyages rather than one in ballast for which no remuneration was earned.

V

As table 7 showed, the single most important commodity despatched from Connah's Quay, and indeed the good with the largest weight to pass through the port, was bricks. These were being produced inland based on the convenient proximity of appropriate clay deposits and coal measures which could be used as fuel to bake them. The area of Buckley Mountain was renowned for its cheap, everyday pottery from the eighteenth century,[27] and there is evidence to suggest the clay was used to make pots as early as the fifteenth century.[28] By 1900 the competition from cheap white china, glass and tinware had killed off the pottery industry at Buckley.[29] However, the manufacture of bricks, tiles and firebricks, which had commenced by the middle of the eighteenth century, continued at least until the 1930s.[30] Most of the brickmakers had their own sidings into the WMCQR and packed their bricks 252 at a time into wooden boxes which were equipped with castors and lifting eyes. These were then rolled straight on to flatbed railway wagons and lifted by steam crane into the ship's hold[31] – an early form of containerization.

The brick industry of North Wales suffered two problems in the early twentieth century. The first was that the construction of houses in the country as a whole was declining. Thus, the demand for both bricks and firebricks, which derived in large part from new construction, was depressed. Although there is some discrepancy in the figures given by various experts, the trends are similar. Brinley Thomas suggests that whereas the volume of building rose by about thirty percent between 1894 and 1903, between 1899 and 1908 it rose by only five percent, and in the period 1904 to 1913 it fell by twenty-two percent.[32] Mitchell and Deane concur in the trend, suggesting house construction peaked in 1903 at about 157,000 completions and then fell steadily to 1912 when only 53,000 new houses were completed. This represents a fall of two-

---

[27]Messham, "Buckley Potteries," 31-49.

[28]Joseph E. Messham, "A Ewloe Bailiffs' Account and the Origins of the Buckley Pottery Industry," *Flintshire Historical Society Journal*, XXXII (1989).

[29]Messham, "Buckley Potteries," 55.

[30]Dodd, *Industrial Revolution*, 192-193.

[31]Boyd, *Wrexham*, 64, 79 and 319.

[32]Brinley Thomas, "Migration and International Investment," in Alan R. Hall (ed.), *The Export of Capital from Britain, 1870-1914* (London, 1968), 48.

thirds.[33] Another series shows a similar reduction, from 155,000 to 54,009 houses completed between 1903 and 1913, a fall only fractionally less steep.[34] The causes of this building cycle do not concern us here, but its impact does. With such a reduction in national housebuilding, the demand for bricks plummeted. This affected the number of bricks despatched from North Wales and hence the large reduction in outward brick cargoes shipped from Connah's Quay.

Without the internal records of the North Wales' brickmaking firms, we cannot be sure of the precise causes of the decline of brick shipments, but as well as the fluctuations in demand resulting from national patterns of housebuilding, it was likely to be affected by competition from other manufacturers. The brickmaking industry was going through radical changes in the late nineteenth and early twentieth century which brought new producers and new processes into the industry.

Perhaps the most important change was the development of the fletton brickmaking industry.[35] This was based on outcrops of Oxford clay around Bedford and Peterborough. It had a number of advantages over conventional "stock" brickmaking, namely that the clay itself contained a high proportion of carbonaceous matter, and this ignited to give substantial fuel savings. The clay could be dry-pressed, not needing the addition of water, so reducing processing and saving on fuel for drying. The fletton brick is lighter than the non-fletton, making transport a little cheaper. Much of the fletton industry developed close to the railway lines and employed private sidings to keep costs down.

Simultaneously, particularly in the fletton works, there was a search for ways of reducing labour costs through mechanization. This resulted in the Hoffman continuous kiln, in use from the 1880s but subject to continuous small-scale improvements, so that by 1898 the London Brick Co. had a version which held 50,000 bricks at any one time. Mechanical methods of removing clay were explored in the first years of the twentieth century using a steam-navvy excavator and other forms of soil removal and transfer. There were also improvements in brick-pressing methods, such as the Whittaker machines. The result of these increases in mechanization was to reduce the cost of brickmaking, from about twenty-two shillings per thousand in 1896 to about twelve shillings per thousand in 1905. This presented the North Wales brick industry

---

[33]B.R. Mitchell with Phyllis Deane (comps.), *Abstract of British Historical Statistics* (Cambridge, 1962; reprint, Cambridge, 1976), 239.

[34]London and Cambridge Economics Service, *The British Economy – Key Statistics, 1900-1970* (London, 1970), 13.

[35]This section draws heavily on Richard Hillier, *Clay that Burns: A History of the Fletton Brick Industry* (London, 1981).

with greatly increased competition. At the same time, there was a tendency among the southern producers to combine into trade associations to try to maintain prices and regulate output. For instance, in 1890 the Fletton Brick-makers Association was formed, in 1900 the Peterborough and District Brick Manufacturers Association and in 1909 the Pressed Brick Makers Association Limited. These changes, plus the proximity of the fletton manufacturers to the markets which were growing fastest – London, the Midlands and the South East – when compared to Buckley, marginalized the trade from North Wales and confined it to the Irish market and some local regions. This is evident in table 12. Over time the proportion of the brick trade from Connah's Quay go-ing to Ireland rose, while in all other English and Welsh markets it declined. These other mainland markets could be serviced by the fletton manufacturers much more easily than the Irish market, where the necessity for sea transport and the greater distance of the fletton makers from their final customers re-duced their competitive advantage. The market closest to the fletton makers was obviously the one which was most likely to benefit from their growth, and table 12 bears this out. The region designated as the South East, including London and Kent, was most easily served by Bedford and Peterborough, and this market showed the largest proportionate decline in brick shipments from Connah's Quay in 1912 and 1913 compared to 1905 and 1906. Thus, the Buckley brickworks were being forced out of the mainland markets and were forced to rely much more on the Irish market for their sales. The trade of Connah's Quay consequently also became dependent on the Hibernian market.

Table 12
The Distribution of Outward Brick Cargoes from Connah's Quay by Region

| | 1905 | | 1906 | | 1912 | | 1913 | |
	No. of Cargoes	Cargo (tons)	No. of Cargoes	Cargo (tons)	No. of Cargoes	Cargo (tons)	No. of Cargoes	Cargo (tons)
Irish								
No.	130	19,354	133	20,307	105	16,772	90	13,533
%	53.1	51.6	53.2	49.4	67.7	67.3	68.7	65.9
South East								
No.	30	7087	31	8323	12	2612	11	2689
%	12.2	18.9	12.4	20.2	7.7	10.5	8.4	13.1
North West								
No.	37	4588	45	5564	21	2919	15	2065
%	15.1	12.2	18.0	13.5	13.5	11.7	11.4	10.0
South West								
No.	34	4377	21	2792	10	1428	9	1516
%	13.9	11.7	8.4	6.8	6.4	5.7	6.9	7.4

Source: Derived from TNA/PRO, RAIL 226/110.

VI

The single most important commodity entering the port of Connah's Quay in all four years was grain (see table 8). It came from Liverpool and was almost certainly foreign grain being unloaded over the side from deep-water ships and brought to Connah's Quay for milling into flour.[36] This conformed to the national trend to rely increasingly on imported rather than home-grown wheat. Whereas in the 1870s about forty percent of British milled flour was from imported wheat, by 1900-1904 this had risen to sixty percent, and by 1905-1909 to seventy percent.[37] Imported wheat was cheaper than that produced at home and was harder and more suited to bread production. The grain was consigned in the port register to "CM Co.," which stood for the Cobden Mills at Wrexham.[38] From Connah's Quay the sacks of grain were carried by the WMCQR to the mill's own siding. The trade was large and regular enough for Cobden Mills to possess its own twelve-ton, long-wheelbase open wagon for this trade.[39]

However, market forces were working against the small-scale corn miller in the early twentieth century. Wheat consumption per capita was stagnant from the 1870s to the 1900s at about 1.05 lbs.[40] Although flour consumption per capita rose a little, from about 200 lbs. per annum in 1880 to 220 lbs. per annum in the early 1900s,[41] one expert said of bread that "since 1900 it has become 'an inferior' good"[42] because consumption fell as real incomes rose. This stagnant per capita consumption was not indicative of any drop in the standard of living, quite the contrary. From the 1870s, when prices of primary

---

[36]Coppack, *Lifetime with Ships*, 53 and 66, indicates that this was a normal trade for their ships.

[37]Richard Perren, "Structural Change and Market Growth in the Food Industry: Flour Milling in Britain, Europe and America, 1850-1914," *Economic History Review,* 2nd ser., XLIII, No. 4 (1990), 425.

[38]Coppack, *Lifetime with Ships*, 41.

[39]Boyd, *Wrexham*, 318.

[40]Mitchell with Deane (comps.), *Abstract*, 358.

[41]W. Hamish Fraser, *The Coming of the Mass Market, 1850-1914* (London, 1981), 27.

[42]E.J.T. Collins, "The 'Consumer Revolution' and the Growth of Factory Foods: Changing Patterns of Bread and Cereal-Eating in Britain in the Twentieth Century," in Derek J. Oddy and Derek S. Miller (eds.), *The Making of the Modern British Diet* (London, 1976), 27.

products fell, the concomitant increase in real income was spent on a wider variety of foodstuffs, particularly more nutritious and interesting products, such as dairy products, fresh fruit and vegetables, and also the wider range of manufactured foods beginning to come from the new food factories – jam, chocolate, meat extract and custard.[43] This meant that the miller faced a depressing demand situation where the market was growing only inasmuch as the population grew; it was not a deepening market.

At the same time as in the brickmaking industry, there were changes in techniques, scale and location which worked against small rural millers.[44] In the 1880s roller milling using steel cylinders was beginning to replace stone milling. This allowed the gradual separation of the constituent parts of the wheat and was more suited to the hard North American wheat which was making up a larger part of total raw materials. Roller milling was more capital-intensive but gave economy in labour costs. However, imported wheat, especially from the less-developed parts of the world, such as India or Russia, contained a higher proportion of impurities. Therefore, more sophisticated filtering and screening apparatus was required, which increased the capital cost. To compensate for the increased degree of capital intensity, firms needed to use the machinery intensively and hence to have a larger throughput. Thus, the national trend in flour milling was to fewer, more capital-intensive plants located near deep-water terminals owned by public companies such as Ranks or Spiller. This tendency left the small-scale Welsh miller with high costs and declining demand.

Thus, one might expect a reduction in the coastal trade in grain from Liverpool to Connah's Quay. However, unlike the brick trade, where a number of firms were involved, the grain trade into Connah's Quay was entirely for the one firm, Cobden Mills, and all came from Liverpool. In all four years all grain landed at the quay was consigned to "CM Co." As a result, this trade depended on the fortunes of a single firm. This made the grain trade even more vulnerable than the brick trade. Any fluctuation in the fortunes of the Wrexham firm hit the shipping trade at Connah's Quay. In 1916 the worst occurred when the Cobden Mills closed[45] and hence terminated this aspect of shipping activity.

---

[43]Charles Wilson, "Economy and Society in Late Victorian Britain," *Economic History Review,* 2nd ser., XVIII, No. 1 (1965), 183-198.

[44]Howard F. Gospel, "Product Markets, Labour Markets, and Industrial Relations: The Case of Flour Milling," *Business History,* XXXI, No. 2 (1989), 86.

[45]Boyd, *Wrexham,* 168.

VII

In conclusion, this paper has shown the nature and scale of a small coastal port in the years before the Great War. Connah's Quay experienced declines in the number of ships arriving and departing, their aggregate registered tonnage and the weight of cargo involved of about forty percent between the years 1905-1906 and 1912-1913. The goods involved in the trade were entirely high bulk, low-value commodities such as bricks, grain and coal. No high-value consumer goods were mentioned in the port register. This is not surprising since there was no liner trade to the port, and such finished products needed the regularity and speed of the scheduled liner.[46] The trade of Connah's Quay was entirely performed by steam tramps and sailing ships, though some displayed a regular pattern and specialized in certain trades.

The weakness of the seaborne trade of Connah's Quay was that it lacked resilience because it was very narrowly based in terms of commodities carried and ports served. Any alteration to the pattern of trade in a handful of industries hit the demand for shipping severely. Thus, changes in the national economy which affected the demand for bricks and flour were reflected in a reduction in the demand for shipping space at Connah's Quay. This, combined with the navigational difficulties of the Dee Estuary, depressed the coastal trade.

This study confirms some aspects of the coastal trade advanced elsewhere. It has been suggested that compared to the canal or railway the coastal ship in the late nineteenth and early twentieth centuries had a long average haul: 250 miles has been postulated, compared to the railway at fifty miles and the canal at even less.[47] The average journey made by ships trading with Connah's Quay was 224 miles in 1905, 284 in 1906, 199 in 1912 and 190 in 1913. Two points arise from this. Firstly, these figures are not grossly out of line with the statistic advanced in the 1987 study: they average out at well over 200 miles. They also show significant variation from year to year, and the difference between the average of the two pairs of years is quite large. The former works out at just over 250, the latter pair at less than 200. It might be tentatively postulated that this decline in average haul was indicative of the trade of Connah's Quay becoming increasingly regional – serving the Irish Sea and the

---

[46]John Armstrong, "Freight Pricing Policy in Coastal Liner Companies before the First World War," *Journal of Transport History*, 3rd ser., X, No. 2 (1989), 180-197, reprinted in Armstrong (ed.), *Coastal and Short Sea Shipping: Studies in Transport History* (Aldershot, 1996), 112-129.

[47]John Armstrong, "The Role of Coastal Shipping in UK Transport: An Estimate of Comparative Traffic Movements in 1910," *Journal of Transport History*, 3rd ser., VIII, No. 2 (1987), 176, reprinted in Armstrong (ed.), *Coastal and Short Sea Shipping*, 148-162.

west coast of Britain – rather than any pretensions at a national trade. Figure 1 shows that Connah's Quay traded mostly in this region, though there were always some examples of long voyages, such as to London, Rochester, Plymouth or Limerick. In the last two prewar years, Connah's Quay had become even more reliant on this regional trade, as was noted above when discussing the destination of brick cargoes.

The relationship between the registered and deadweight tonnage of a ship can be calculated from the port register. These are given in table 5. They have been characterized elsewhere as "carrying coefficients," and it was suggested that in 1910 nationally for coastal ships they varied between 2.0 and 2.8 depending on the type of ship, year of construction and type of cargo.[48] This study shows that the "carrying coefficients" for ships trading with Connah's Quay were rather lower, ranging between 1.6 and 2.1 with an average of about 1.9. The difference need not concern us unduly, since the ships trading with Connah's Quay were below average size, and the ratio between deadweight and registered tonnage increased as ship size rose. They suggest that a national figure of well over two would not be unreasonable.

Given the paucity of studies of individual small ports in this period, it is hoped that this article may spur others to attempt to answer similar questions for other locations. At the moment it looks as though Connah's Quay was atypical of the national trend, but if trade was becoming increasingly concentrated at a smaller number of large ports to the detriment of the fortunes of the smaller ports, this need not be the case. Hopefully, further examination of this port register may yield additional detail on the trade of this small port.

[48]*Ibid.*, 172-175.

Chapter 17
The Cinderella of the Transport World:
The Historiography of the British Coastal Trade[1]

Derek Aldcroft once described the British coastal trade as "the Cinderella of the transport world."[2] He was referring to the lack of provision made for it by the ports compared to overseas trade just before and during the First World War. His appellation might also be applied to its position in the research hierarchy in that it has attracted relatively little scholarly time and attention. One indication of this is that the standard textbook on the coasting trade in early modern England remains that written by T.S. Willan in 1938.[3] Although a number of articles have been published adding to and amending the picture portrayed by Willan, there has been no attempt to incorporate this recent research into a new book-length synthesis. For the modern period the situation is worse. There is no book dealing with this aspect of British transport history. Worse still, with one honourable exception, those textbooks which do exist on British transport history devote a tiny proportion of their often impressive bulk to the topic. Apart from a few articles, many of them in the *Journal of Transport History*, and a couple of chapters in two collections of essays, the coastal trade remains largely overlooked. It is not surprising, then, with this degree of neglect by professional scholars that the neophyte coming fresh to the subject, having made reasonable endeavours to locate a body of knowledge, might conclude that the coastal ship was never of any great significance to the trade and growth of the British economy and to the modernization of society.

---

[1]This essay appeared originally in John Armstrong (ed.), *Coastal and Short Sea Shipping: Studies in Transport History* (Aldershot, 1996), ix-xxiv.

[2]Derek H. Aldcroft, "The Eclipse of British Coastal Shipping, 1913-1921," *Journal of Transport History*, 1st ser., VI, No. 1 (1963), 32, reprinted in Armstrong (ed.), *Coastal and Short Sea Shipping*, 163-177. This is probably based on the use of the same phrase, "the Cinderella of the transport world," in Anon., *The Coastwise Trade of the UK Past and Present and Its Possibilities* (London, 1925), 39 and 64, because of the attitudes of the port authorities. The phrase was echoed by Sir Alfred Read, chairman of Coast Lines Ltd., though his reason was because it received so little government help or encouragement; Alfred Read, "Coastal Shipping in Relation to Transport Planning," *Journal of the Institute of Transport*, XVIII, No. 1 (1936), 11.

[3]T.S. Willan, *The English Coasting Trade, 1600-1750* (Manchester, 1938; reprint, Manchester, 1967).

This essay tries to correct this view. More specifically it attempts three things. It puts forward a little more evidence to sustain the contention that coastal shipping deserves the sobriquet of the Cinderella trade. It then endeavours to explain why there has been such a profound neglect. The third section sketches, albeit very briefly, the main outlines of what we now know about the economics and impact of the coastal trade and, in so doing, provides a context for the other articles in this volume.

I

Let us then reinforce the view that the coastal trade has remained a relatively under-researched topic, relative, that is, to most other forms of transport. One piece of evidence may support this view. Recently a bibliography of the British coastal trade was published.[4] It claimed to cover the whole period from medieval to modern, and the spectrum from enthusiast to academic. It contained approximately 300 items. By comparison, Ottley's classic bibliography of British railway history contains just under 13,000 items.[5] A similar impression would be gained if comparisons were made between the coverage, on the one hand, of the British coastal trade and, on the other, of British overseas trade and shipping. The output of the latter outweighs the former by a massive multiple.

If we turn to textbooks on the topic of transport history, we discover that most were written some years ago and thus are in need of new editions in order to incorporate recent scholarship. Also, the coastal trade received scant or no coverage when compared to most other modes of transport, even those that have existed for much shorter time spans, such as railways, motor transport and air travel. For instance, the largest and most comprehensive text on British transport, written by H.J. Dyos and Derek H. Aldcroft, reached a large audience by virtue of being published as a Pelican paperback and devoted a mere two pages specifically to coastal shipping.[6] Similarly, a textbook initiated by Christopher Savage, and whose latest edition was compiled by Theo Barker, though more restricted in its scope inasmuch as it only dealt with "inland" transport and hence did not cover overseas shipping or air travel,

---

[4]John Armstrong, "An Annotated Bibliography of the British Coastal Trade," *International Journal of Maritime History*, VII, No. 1 (1995), 117-192.

[5]George Ottley (comp.), *A Bibliography of British Railway History* (London, 1965; 2nd ed., London, 1983); and Ottley (comp.), *A Bibliography of British Railway History Supplement* (London, 1988).

[6]H.J. Dyos and Derek H. Aldcroft, *British Transport: An Economic Survey from the Seventeenth Century to the Twentieth* (Leicester, 1969; reprint, Harmondsworth, 1974), 208-210.

managed to find enough material to fill only two pages.[7] The most recently published of the standard texts, Bagwell, has the fullest coverage. It devotes one whole chapter and several sections of chapters to the coastal trade and addresses the period from 1770 to 1985.[8] By comparison, the previous two texts mention coastal shipping in the nineteenth century and then ignore it, as though it was a short-lived mushroom that popped up in the Victorian period and then rapidly disintegrated. However, as good as it is to see a good proportion of the pages devoted to the coastal trade, Bagwell was originally published in 1974 and the second edition merely reprinted the early chapters and did not contain any assessment of later research.

It might be argued that these three textbooks should not be taken to task too severely for their imperfect coverage of the coastal trade. They were all written and produced about three decades ago, before there was much published research on the topic. However, more recent general works on transport history have not always devoted much greater attention to the significance of coastal shipping. For instance, Simon Ville's brave attempt to summarize the impact of transport developments on European economies for the entire period of rapid industrialization, though eminently successful in other respects, devotes very little space to coastal shipping and does not engage in any extended discussion of its role.[9] More reprehensible is the recent monograph by Rick Szostak which makes the case that Britain was the first country to industrialize because of its superior endowment of natural and man-made transport facilities.[10] The monograph hardly mentions coastal shipping. His work can be criticized for seeking simplistic monocausal explanations since, rather than there being one cause, it was the interaction of a number of factors which brought about Britain's industrialization. Also, to confine his analysis only to France and Britain is too narrow given the economic performance of Belgium, The Netherlands and Prussia. For the purposes of this chapter, the neglect of coastal shipping, given the great extent of Britain's coastline and navigable rivers, seems perverse, as it was probably the feature of early modern transport in which Britain had the largest differential of advantage over other European countries.

---

[7]T.C. Barker and Christopher I. Savage, *An Economic History of Transport in Britain* (London, 1974), 70-72.

[8]Philip S. Bagwell, *The Transport Revolution from 1770* (London, 1974; new ed., London, 1988).

[9]Simon P. Ville, *Transport and the Development of the European Economy, 1750-1918* (London, 1990).

[10]Rick Szostak, *The Role of Transportation in the Industrial Revolution: A Comparison of England and France* (Montréal, 1991).

Two volumes deserve honourable mention as they each devote a chapter to the coastal trade. These are the two books edited by Derek Aldcroft and Michael Freeman on transport during the Industrial Revolution and in Victorian Britain, respectively.[11] These are useful syntheses of knowledge that was current t the time, although both are a little dated now. Overall, however, they are rare beacons in a fog of neglect and ignorance. The general pattern has been to pay no attention to the coastal trade and by so doing transmit the signal that it is not worth studying.

II

We now need to determine why there has been such widespread neglect of this topic. It might be suggested that this is a result of the sources. Historians can only be as good as the records they have to work with, and the sources available for the coastal trade, it might be argued, are few and far between and difficult to use. We can demolish the former argument fairly quickly. There are a large number of different sources available for the study of the British coastal trade, including the shipping registers, which give information on ownership;[12] the crew agreements, which contain data on crews, wages and voyage patterns;[13] the bills of entry, which for some ports and some periods give details of commodities and merchants involved, as well as ships and routes;[14] local and specialist trade papers;[15] and records of companies involved in the coastal trade, both records of the companies themselves and those lodged with the Registrar of Joint Stock Companies.[16] Thus, it is difficult to sustain the argument that there is a lack of primary material relevant to the coastal trade.

---

[11]Derek H. Aldcroft and Michael J. Freeman (eds.), *Transport in the Industrial Revolution* (Manchester, 1983), chap. 5; and Freeman and Aldcroft (eds.), *Transport in Victorian Britain* (Manchester, 1988), chap. 5.

[12]Rupert Jarvis, "Ship Registry to 1707," *Maritime History*, I, No. 1 (1971), 29-45; Jarvis, "Ship Registry, 1707-86," *Maritime History*, II, No. 2 (1972), 151-167; and Jarvis, "Ship Registry, 1786," *Maritime History*, IV, No. 1 (1974), 12-30.

[13]Keith Matthews, "Crew Lists, Agreements, and Official Logs of the British Empire, 18631913," *Business History*, XVI, No. 1 (1974), 78-80.

[14]Edward Carson, "Sources for Maritime History (1): Customs Bills of Entry," *Maritime History*, I, No. 2 (1971), 176-189.

[15]These include publications such as *Fairplay, Shipping and Mercantile Gazette, Daily Freight Register,* and *Nautical Magazine*.

[16]Great Britain, National Archives (TNA/PRO), Board of Trade (BT) 31 and 44.

This is reinforced by comparison with sources for deep-water activity. Virtually all of the types of documents available to scholars for the coastal trade are also available to those studying overseas trade. Thus, if the latter can be examined in depth, why cannot the former? This commonality of sources is also relevant to the debate about the difficulty of using those records which do exist on the coastal trade. Undoubtedly, some of them are intractable to use, some series of data are not continuous or are recorded on a different basis, and many of the manuscripts are so voluminous that even with electronic databases a great deal of time is required to input the mass of material. However, if these sources can be used to shed light on overseas trade, they are obviously not totally unwieldy, and the techniques used for that sector could be applied to the coastal. Although each record has its weaknesses and biases, it is not an absence of sources nor their total intractability which has led to the neglect of the coastal trade.

Difficulties of definition have sometimes been advanced as an explanation for the low quantity of research on the coastal trade. The official figures at times compound the home trade with the coasting, where the home trade included voyages to the near continent, but this is not really a problem since the figures can usually be disaggregated. Another complaint is that it is difficult to determine what constitutes a "coastal" ship since many were deployed flexibly, conducting whatever trade offered the best freight rates or most regular flow of cargoes.[17] There is undoubtedly truth in this. Some ships did ply in both the coastal and foreign trades; for example, some ships of the General Steam Navigation Co.[18] and the Wilson Line[19] sailed to the Baltic area as well as on purely coastal voyages. However, the notion of flexibility can be pushed too far. Some companies, such as the Aberdeen Steam Navigation Co.,[20] only ran coastal voyages, thus all their ships were coasters. Other trades had specialized vessels which were difficult to deploy outside their appointed trades,

---

[17]Simon P. Ville, "The Deployment of English Shipping: Michael and Joseph Henley of Wapping, Shipowners 1775-1830," *Journal of Transport History*, 3rd ser., V, No. 2 (1984), 16-33.

[18]Sarah R. Palmer, "'The Most Indefatigable Activity:' The General Steam Navigation Company, 1824-50," *Journal of Transport History*, 3rd ser., III, No. 2 (1982), 1-22.

[19]For example, TNA/PRO, BT 99/1509 and BT 99/1779.

[20]Clive H. Lee, "Some Aspects of the Coastal Shipping Trade: The Aberdeen Steam Navigation Company, 1835-80," *Journal of Transport History*, 2nd ser., III, No. 2 (1975), 94-107, reprinted in Armstrong (ed.), *Coastal and Short Sea Shipping*, 90-103.

such as the "flat iron" colliers built for up-river deliveries on the Thames.[21] At the other extreme, some ships, such as Atlantic passenger liners, deep-draught cargo carriers and tea clippers, were intended only for oceanic voyages and were inappropriate for coastal traffic. To cite flexible deployment as an excuse for not studying the coastal trade is to prejudge the issue. We do not know the extent of vessel crossover between coastal and overseas trade. It might be marginal, with most vessels plying steadily in one aspect and only a few working both the coastal and near continental trades. This issue requires more detailed research, and its study would enhance our understanding of both sectors.

It should also be stressed that we are primarily discussing the coastal "trade," not the ships themselves. Once the voyage is known it is quite clear which trade is under discussion. It is more difficult to designate specific vessels as either "coastal" or "overseas," but this is less important when we are trying to understand trade flows, the economics of operation and their impact on growth. The situation on the mainland of Europe might be more problematic. There, with adjoining land borders and contiguous coastlines a voyage following the coast might well mean a venture into overseas trade, but this does not apply to the UK before 1922 when any voyage along the coast necessarily meant staying in internal trade. This situation might help to explain the relative lack of research on coastal trade in the countries of mainland Europe, but it applies less to the UK.

Perhaps one of the prime reasons for the neglect of the coastal trade can be found in one of John Masefield's poems.[22] The coaster was unattractive compared to the image of foreign trade. The cargoes in coasters were mundane – coal, ironware, pig lead and cheap tin trays – compared to ivory and apes, or sandalwood and wine, brought in from abroad. However, the reality of foreign trade did not match this image. The bulk of imports comprised timber, grain, iron ore and raw cotton, rather more prosaic than the contents of Masefield's quinquereme. Similarly, the sound of ports such as Port Said, Tokyo, or Bombay is much more romantic and exotic than South Shields, Runcorn or Preston. In fact, of course, many of these "exotic" destinations were unhealthy, unsafe, overcrowded slums where port efficiency was low and mariners spent time kicking their heels and feeling unwell, but the glamorous image persists.

It may also be that the theoretical literature in economics and economic history has laid more stress on overseas than on internal trade as an engine of growth. The literature on ocean freight rates and their impact, via reduced prices, in promoting specialization and a greater exchange of goods is

---

[21]J.A. MacRae and Charles V. Waine, *The Steam Collier Fleets* (Wolverhampton, 1990), 54.

[22]See John Masefield, "Cargoes," in Masefield, *Ballads* (London, 1903).

vast and spread over a long time period.[23] The impact of trade in opening less-developed countries and bringing them into the world economy by trading their surplus agricultural or extractive commodities is also well known. By comparison, less has been written on the role of internal trade and how it promoted specialization and greater efficiency within a country. Hence, the apparent glamour of overseas trade combined with the seemingly greater theoretical part it can play in world economic growth may help to explain the comparative neglect of the British coastal trade. There has been similar neglect of the importance of the coaster in promoting economic growth. Whereas there have been two econometric studies of the part played by railways in Britain in fostering economic growth during the nineteenth century, there has been no such attempt to calculate the coaster's contribution.[24]

The view of the British coastal trade handed down by writers on deep-water marine activity has sometimes been patronizing and has downplayed its role. The notion that the coasting trade was the "nursery" of seamen suggests a kindergarten for immature sailors who would eventually graduate to a higher form of education, presumably the blue-water trades. It implies that the coasting trade was no more than a training ground, certainly not an end in itself, and that no mature sailor would want to do more than pass through it, in the process gaining experience. The suggestion has been made that following a coastline required quite different and inferior skills to the real sailoring of deep-water voyages. That there were differences in some of the skills required is not doubted. Whether the skills required to work a coaster safely and expeditiously were inferior to those needed for ocean voyages is open to debate, especially in the treacherous waters round Britain's coastline, and still needs testing. As Bruce shows, the record of technological innovation in the short-sea trades is impressive, and pioneering on these routes was normal.[25] However, this impression of inferiority may have led scholars to dismiss the coastal trade as unworthy of their attention and hence help to account for the low level of research output.

---

[23]See John Armstrong, "Late Nineteenth-Century Freight Rates Revisited: Some Evidence from the British Coastal Coal Trade," *International Journal of Maritime History*, VI, No. 2 (1994), 45-50, for a summary of this literature.

[24]G.R. Hawke, *Railways and Economic Growth in England and Wales, 1840-1870* (Oxford, 1970); and Wray Vamplew, "Railways and the Transformation of the Scottish Economy," *Economic History Review*, 2nd ser., XXIV, No. 1 (1971).

[25]J. Graeme Bruce, "The Contribution of Cross-Channel and Coastal Vessels to Developments in Marine Practice," *Journal of Transport History*, 1st ser., IV, No. 2 (1959), 65-80, reprinted in Armstrong (ed.), *Coastal and Short Sea Shipping*, 57-72.

III

Although the output of research on the coastal trade remains disappointing, a clear view is emerging of the role it played, the economics of the coaster and how it competed with other forms of transport, and the commodities which it carried and trades in which it was active. This section attempts an outline of current thinking on the coastal trade and situates the contributions within it.

There seems little doubt that from very early times coastal shipping was a normal and ubiquitous form of transport.[26] It was a natural extension of riverine and estuarial traffic and, initially, navigation in sight of land was easier than trans-oceanic voyages. In the medieval and early modern periods coastal shipping remained an essential part of the transport network. Its strengths lay in the size of the coaster, capable of carrying cargoes of ten to a hundred tons when road transport using packhorse, cart or wagon could move between a few hundredweight and a few tons.[27] To move fifty tons, an average size for an eighteenth-century coaster, required approximately ten to twelve large wagons, or about thirty two-wheeled carts or around 400 packhorses. Each of the carts and wagons required a driver, and the number of packhorses that could be guided in one string was limited to perhaps twenty. Thus, the labour costs of moving large quantities were enormous compared with the coaster which was manned by a handful of sailors. In addition, whereas the wind came largely free, the horses were very expensive to run, consuming huge amounts of fodder even when not working, and if made to pull or carry extra loads, to move at greater speeds or to tackle steep gradients required the number of horses and their consumption of oats and hay to rise alarmingly. Of course, the wind for the sailing coaster was not free of cost. Sails blew out and spars broke, as did masts occasionally. However, the repair of sails and ropes was one of the skills required of sailors in their less busy periods, as was painting and general maintenance, and repair work was needed on carts and wagons as well as on spars and masts.

The capital cost of the wooden sailing coaster was probably less than that of wagons relative to carrying capacity, i.e., the initial cost of a fifty-ton coaster was about half that for ten wagons and forty horses. Hughes estimated

---

[26]J.B. Blake, "The Medieval Coal Trade of North East England: Some Fourteenth Century Evidence," *Northern History*, II (1967), 1-25; and K. Lloyd Gruffydd, "Maritime Dee during the Later Middle Ages," *Maritime Wales*, No. 9 (1985), 7-31.

[27]This and the next paragraph rely heavily on John Armstrong, "The Significance of Coastal Shipping in British Domestic Transport, 1550-1830," *International Journal of Maritime History*, III, No. 2 (1991), 63-94; and Dorian Gerhold, "Packhorses and Wheeled Vehicles in England, 1550-1800," *Journal of Transport History*, 3rd ser., XIV, No. 1 (1993), 1-26, reprinted in Dorian Gerhold (ed.), *Road Transport in the Horse-Drawn Era: Studies in Transport History* (Aldershot, 1996), 139-164.

that coasters during the eighteenth century cost about £7 per ton to build and equip.[28] Turnbull thought the capital cost of a wagon was approximately £35 and that draught horses cost about £10 apiece.[29] Four of the latter could pull about four or five tons to give a total cost of about £75 for a wagon and team. Thus, to haul a fifty-ton load would cost about £350 for a coaster and about £750 for ten sets of wagons. However, whereas a wagon horse was lucky to have a working life of five or six years, a coaster, barring tragedies, could survive for twenty years or more. The annual depreciation – though not couched in these terms by contemporaries – was therefore a much smaller percentage for the coastal ship than for the horse.

As a result of lower manning costs per ton, lower depreciation and no necessity for fodder, the coastal ships' costs per ton-mile were much lower than horse-drawn road transport. Following on from this, the prices charged for moving goods by sea were much cheaper than by land. It might be argued that the sailing ship was less reliable than horse-drawn transport because it was at the mercy of the fickle wind which might be too strong, too light or from the wrong direction, forcing the coaster to remain in harbour for days or to lie disconsolately becalmed at sea. There is much truth in this.

However, three points need to be borne in mind. First, until the improvements in surfaces and routes of the eighteenth century, road transport could be delayed by climatic conditions such as floods or ice on inclines. Secondly, the extent of such delays, in both road transport and coastal shipping, is open to debate and has never been investigated systematically. What percentage of journeys was so affected and what was the average addition to the journey time are questions which still need answering. Was there a great deal of difference in these figures for road and coastwise travel? Thirdly, the attitude toward punctuality was more relaxed before the advent of national time and strict timetables, which were largely products of the early nineteenth century and the railways. A few hours made little difference. Wagons and coasters were scheduled to depart on a particular day and sometimes in the morning or afternoon, but rarely at a specific time. Thus, merchants expected less precise times of departure and arrival. Also, if speeds are compared for the sailing coaster and horse-powered road transport it is very difficult to say which was quicker. The horse could plod at two miles per hour for ten or more hours a day. More work than this was likely to increase its fodder consumption or shorten its life. Twenty miles per day was not a bad average. The coaster, on the other hand, could bowl along at up to twice the speed of the horse for vir-

---

[28]Edward Hughes, *North Country Life in the Eighteenth Century. Vol.II: Cumberland and Westmoreland, 1700-1830* (Oxford, 1965), 187.

[29]Gerard L. Turnbull, *Traffic and Transport: An Economic History of Pickfords* (London, 1979), 17.

tually the whole day given favourable wind conditions and could achieve up to 150 miles in a twenty-four-hour period – in theory. In practice, the average was likely to be much less, but just how road and sail compared on speed is a much more open question than might at first appear and is another area needing in-depth research. Perhaps the crucial advantages of road transport over the coaster were reliability and frequency of service; the wagon's time of arrival could be much more closely predicted than that of the coaster.

The characteristics of the coaster made it particularly suitable for some commodities: those that needed to be moved in large quantities; those that were of low unit value and, therefore, where the freight rate was a higher proportion of total cost, and the capital tied up in the goods was not excessive; and those that did not deteriorate rapidly. Thus, goods such as corn, coal, ores and building materials could not travel any distance by cart or wagon as the cost of transport added such a large proportion to the total cost. If they were to move any distance at all it had to be by coastal ship or inland water. The value of the coastal ship was enhanced by its ability to penetrate some distance inland by using the main navigable rivers, for the average-sized coaster was little larger than a nineteenth-century canal barge.

Many of these points are demonstrated in John Chartres's essay on Wiggins Key.[30] He shows that Wiggins Key had a large trade in corn in the late seventeenth century which came mostly from Kentish and East Anglian ports, though some was shipped from places much further afield, such as Hull and Swansea. The ships carrying the corn were mainly hoys, small, single-masted, square-sailed craft. The vessels ranged from nine to ninety-five tons with a mean of about forty tons. All came several miles up river to deliver. Chartres shows that voyage frequency was related to distance travelled: those coming from Kent made eighteen or twenty trips per annum, whereas the average was more like five or six. He also demonstrates that these ships frequently returned in ballast, having found no suitable back cargo and that the peak period for arrivals was February to May. This suggests that coasters could navigate even in winter months if the commodity was available and needed to be shipped. It downplays the significance of bad weather in preventing coasters from operating. The importance of Wiggins Key is emphasized in that Chartres estimates it handled about twenty percent of London's seaborne grain trade. The importance of Chartres' article is that he used manuscript sources to demonstrate the extent of this trade, and the source was robust enough to allow some quite sophisticated statistical analysis. When Roseveare examined the

---

[30]John Chartres, "Trade and Shipping in the Port of London: Wiggins Key in the Later Seventeenth Century," *Journal of Transport History*, 3rd ser., I, No. 1 (1980), 29-47, reprinted in Armstrong (ed.), *Coastal and Short Sea Shipping*, 1-19.

records on which Chartres's article was based,[31] he was able to add much information on the role of the Key in overseas trade and to demonstrate that its coastal trade was not the most important aspect of income generation for the owners, but he did not question Chartres' conclusions on the role the coaster played. This reinforces the value of Chartres' findings.

In the eighteenth century the relative position of road transport improved thanks to innovations in highway construction methods, the better breeding of haulage horses, the improved layout of road routes which reduced gradients and economies of scale through larger firms.[32] These lessened horse wear and tear and perhaps speeded up their journey times. However, road haulage was still expensive and slow – if a little more reliable than the sailing coaster. The goods carried long distances were those of high value relative to weight, such as bullion, specie and textiles where the carriage cost was a small percentage of total value. An article by Turnbull[33] reprinted in *Road Transport in the Horse-drawn Era* demonstrates that road transport was used for the long-distance carriage of linen cloth. It also draws out some of the weaknesses of the coastal ship in time of war, namely the fear of attack by enemy ships, whether official or privateers, and the increase in costs as a result of insurance premiums rising to allow for this additional risk and wage advances. He also stresses the possibility of adverse weather, such as storms, causing a total loss of ship and cargo, which was a highly unlikely eventuality in land carriage. How frequently this occurred at sea and, therefore, its probability is unknown and needs researching. It could be that shipwrecks, being great tragedies, attracted much attention but were actually very infrequent. Turnbull makes the point that although coastal transport normally was cheaper than land haulage by a factor of three or four, "normal" times were rare. In wartime the cost of insurance could cancel out this cost advantage. Even in conditions of peace the greater regularity and reliability of road transport caused it to be used, especially for the higher-value linens. However, when road transport routes were monopolized and charges were raised, there was a scramble to revert to the cheaper coastal routes.

Road transport rarely carried bulky, low-value goods more than a few miles because the cost was prohibitive. Given that the twin trends of industrialization and urbanization required the movement of masses of bulky raw ma-

---

[31]Henry Roseveare, "'Wiggins' Key' Revisited: Trade and Shipping in the Late Seventeenth-Century Port of London," *Journal of Transport History*, 3rd ser., XVI, No. 1 (1995), 1-20.

[32]Gerhold, "Packhorses."

[33]Gerard L. Turnbull, "Scotch Linen, Storms, Wars and Privateers: John Wilson and Son, Leeds Linen Merchant, 1754-1800," *Journal of Transport History*, 3rd ser., III, No. 1 (1982), 47-69, reprinted in Gerhold (ed.), *Road Transport*, 75-97.

terials such as coal, corn, iron ore, china clay, bricks, tiles, timber and lime-stone, it is difficult to conceive of the Industrial Revolution taking place in the absence of a vigorous coastal shipping industry. This is especially true of Britain where the variegated coastline and abundance of rivers ensured that few locations were more than fifteen miles from navigable water.[34] An article by Simon Ville[35] brings out the importance of the east coast coal trade not only to heat, light and power London but also as an export to the near continent and to keep the royal dockyards fuelled. He draws upon the rich Henley archive to show that in time of war *Freedom*, a very large coaster for the period, was continuously employed, mostly carrying coal from the North East. This oc-curred despite the drawbacks, outlined by Turnbull, such as the greater costs of operation due to higher wage rates and insurance premiums. It is an impor-tant article because, like the pieces by Chartres and Turnbull, it draws on pri-mary sources and demonstrates in detail how a shipowner deployed his vessel, the problems encountered, the actual rates of pay and time taken on voyages which, together, allow some estimate of profitability.

Canal construction was not a threat to the coaster for it did not bring competition but rather a complementary form of transport, since the canals were very rarely built parallel to the coastline or to cut off a protruding piece of coast. Where they did, like the Caledonian and Crinan canals, coasters could often use them. However, the majority of canals were built as extensions of the river system between two inland towns, or between ports and inland cities. In many ways canals extended the range of the coastal ship by lengthen-ing the all-water route which was usually the cheapest. Some coasters could enter some canals themselves and hence offer their service to a larger number of customers. Where this was not possible, coasters could often unload over the side directly into barges or lighters which then performed the river or canal part of the journey.[36] This form of transfer was much cheaper and quicker than unloading onto a wharf or quay and minimized the chance of pilferage. In this way the coaster worked with river and canal traffic to provide cheap transport of bulky materials which were crucial to the progress of the Industrial Revolu-tion.

The nineteenth century has been seen in transport terms as the age of the railway, which is rather ironic as steam propulsion was first applied com-

---

[34]Dyos and Aldcroft, *British Transport*, 84.

[35]Simon P. Ville, "Wages, Prices and Profitability in the Shipping Industry during the Napoleonic Wars: A Case Study," *Journal of Transport History*, 3rd ser., II, No. 1 (1981), 39-52, reprinted in Armstrong (ed.), *Coastal and Short Sea Shipping*, 21-34.

[36]Runcorn acted in this way; see H.F. Starkey, *Schooner Port: Two Centuries of Upper Mersey Sail* (Ormskirk, 1983; 2nd ed., Bebington, 1998).

mercially to shipping in 1812 on Henry Bell's *Comet*,[37] nearly two decades before the Rainhill Trials and the opening of the Liverpool and Manchester Railway. An article by Freda Harcourt[38] shows how one pioneer of steam navigation operated. Williams was instrumental in forming two companies to ply between Ireland and England. The article points out the large part played by Irish capitalists in financing the early steamship enterprises, which is explained by the different basis of Irish corporate law compared to English and by the realization on the Irish side that efficient cross-channel links were essential if their goods were to reach the large, rich and rapidly growing market that Britain represented. The article hints at the importance of this cross-channel trade to both the Irish and British economies and that success in this short-sea trade led to investment in steamers for more distant trades. In this way the coastal trade pioneered so that deep-water shipping could follow in its footsteps.

The article by Irvine takes up the story of the trade across the Irish Channel in the early days of steam.[39] It details the services available – suggesting that in 1846 there were over 100 services each week between Ireland and Britain – and the fares charged, showing that intense competition reduced prices and that fares could reach totally unremunerative levels at times of cutthroat competition. It also explores the quality of the accommodation offered, especially for the seasonal migrant who could afford only deck or steerage travel. The inadequacy of provision for this type of passenger led to legislation to improve safety and hygiene standards. Finally, Irvine attempts an estimate of the aggregate number of passengers who crossed the Irish Sea, suggesting a third of a million in the mid-1830s, rising rapidly in the 1840s when the Great Famine struck. The article is important because it pioneered the analytical study of this aspect of cross-channel trade, and it remains important because there has been little systematic work to supplement or refute it.

The impression sometimes given that the railways were the only form of long-distance freight transport in the nineteenth century is quite incorrect,[40]

---

[37]H. Philip Spratt, *The Birth of the Steamboat* (London, 1958), 87-88; and Brian D. Osborne, *The Ingenious Mr. Bell: A Life of Henry Bell (1767-1830), Pioneer of Steam Navigation* (Glendaruel, 1995).

[38]Freda Harcourt, "Charles Wye Williams and Irish Steam Shipping, 1820-50," *Journal of Transport History*, 3rd ser., XIII, No. 2 (1992), 141-162, reprinted in Armstrong (ed.), *Coastal and Short Sea Shipping*, 35-56.

[39]H.S. Irvine, "Some Aspects of Passenger Traffic between Britain and Ireland, 1820-1850," *Journal of Transport History*, 1st ser., IV, No. 4 (1960), 225-241, reprinted in Armstrong (ed.), *Coastal and Short Sea Shipping*, 73-89.

[40]See, for instance, John R. Kellett, *Railways and Victorian Cities* (London, 1979), 351: "long distance transport virtually was railway transport."

as is the notion that the railways caused the coastal trade to decline. The registered tonnage of coasters entering British ports with cargoes grew steadily throughout the period 1830-1913.[41] The growth was not spectacular, around two percent per annum, but the amount of work performed did increase.[42] In addition, a dense network of coastal liner routes was developed so that by the late nineteenth century all major port towns were connected by a regular, reliable, frequent and fast service of powerful, well-appointed ships which carried both cargo and passengers.[43] For instance, between 1860 and 1863, when the railway companies monitored the number of passengers travelling between Glasgow and Liverpool, they found the steamboats were carrying two-thirds of the total.[44]

An article by Lee[45] explains how one of these coastal liner companies operated. This article was significant at the time because so little was known about the way such firms performed, and it was based on the relatively rich records that survive of the business activities. It demonstrates that co-operation with the railways on through rates was an important strategy, as was some collusion with both the railways and other coastal liner companies to fix prices and avoid head-on competition. These strategies ensured that the firm was reasonably profitable, and this is demonstrated by its longevity. Lee's article can be supplemented by one by Geoffrey Channon, written a few years earlier, which discusses in depth the competition between the railway and the steamboat in carrying cattle and meat from Aberdeen to London. Lee demonstrates that the two modes of transport were evenly matched in the 1850s and 1860s, and that the coaster was able to compete with the railway.[46] Although Lee's

---

[41]Great Britain, Parliament, *Parliamentary Papers* (*BPP*), Trade and Navigation Accounts, various years.

[42]Philip S. Bagwell and John Armstrong, "Coastal Shipping," in Freeman and Aldcroft (eds.), *Transport in Victorian Britain*, 171.

[43]John Armstrong, "Freight Pricing Policy in Coastal Liner Companies before the First World War," *Journal of Transport History*, 3rd ser., X, No. 2 (1989), 180-197, reprinted in Armstrong (ed.), *Coastal and Short Sea Shipping*, 112-129; Ian Bowman, "The Carron Line," *Transport History*, X, Nos. 2 and 3 (1979), 143-170 and 195-213; and A.M. Northway, "The Tyne Steam Shipping Co: A Late Nineteenth-Century Shipping Line," *Maritime History*, II, No. 1 (1972), 69-88.

[44]TNA/PRO, RAIL 727/1, minutes 152, 172 and 256.

[45]Lee, "Some Aspects."

[46]Geoffrey Channon, "The Aberdeenshire Beef Trade with London: A Study in Steamship and Railway Competition 1850-1869," *Transport History*, II, No. 1 (1969), 1-24.

article stops in 1880, the firm continued trading until after the Second World War. This article remains important for its insight into the strategies of a coastal liner company in the mid-nineteenth century.

An attempt to determine the economics of the coastal liner companies is to be seen in an article by John Armstrong.[47] He demonstrates that the liner companies were able to charge much less than the railways for freight on long-distance hauls, even where the coastal route was significantly longer than the more direct rail journey. The article shows that the coastal companies monitored the railway's rates and methods of charging to ensure they stayed competitive, and used various marketing ploys to attract custom, such as long-term contracts, through rates and discounts for large quantities. It suggests that the speed and reliability of the scheduled liner services were not inferior to the service provided by the railways where freight traffic had a low priority and frequent shunting and marshalling meant low speed and uncertainty about the location of a particular consignment and its times of arrival. This article is significant because it demonstrates the basis of the coastal liner's competitiveness *vis-a-vis* the railways.

We have already alluded to the fact that steam services on coastal ships preceded steam traction on railways, and that the steamers across the Irish Sea led to steam being applied to deep-water routes. Bruce's article[48] spells out systematically the range of innovations that were pioneered in coastal or cross-channel ships and were then taken up in the ocean trades. These included means of propulsion such as paddle and screw, types of engine (such as compound, triple-expansion and turbine), the use of water ballast, the utilization of iron and steel for hulls and many others. The article is important because it demonstrates that the short-sea trades were not at all stagnant but rather continuous pioneers in technological innovation. This helps explain the improvements in speed, reliability and size of the coasters, as well as the productivity of both crew and capital. As a result of these improvements and the continuous shift from sail to steam, the coaster was able to compete with the railway, especially on long-distance routes, such as London to Scotland, and for bulky products, such as coal from South Wales and the North East into London and the South East.

Another of Armstrong's articles[49] was path-breaking because it suggested that the coaster was far from declining in the early twentieth century. It

---

[47]Armstrong, "Freight Pricing Policy."

[48]Bruce, "Contribution."

[49]John Armstrong, "The Role of Coastal Shipping in UK Transport: An Estimate of Comparative Traffic Movements in 1910," *Journal of Transport History*, 3rd ser., VIII, No. 2 (1987), 164-178, reprinted in Armstrong (ed.), *Coastal and Short Sea Shipping*, 148-162.

went on to argue that the coaster might be doing as much work as the railway, if measured in ton-miles, and then explained the basis of this competitive position. This article suggested that the coaster had particular strengths in long-distance routes and that its average haul was much greater than on the railways or canals, around 250 miles compared to forty miles on the railways. Thus, the coaster carried a smaller quantity than the railways over a much longer average distance. To date, there has been no challenge to this interpretation.

However, the coaster was not confined to long-distance trades. There were some locations where it had more direct routes than land transport. The article by Stanier[50] demonstrates that the Bristol Channel was one area where the coaster had the direct line whereas land transport from Cornwall or Devon to South Wales had a much lengthier journey. The cargoes involved – copper ore being carried for smelting to South Wales and a return cargo of coal to fuel the steam engines used to pump the mines dry – were precisely those suited to coastal transport. They were bulky goods capable of filling the 100-ton deadweight ships, initially brigs and later schooners, and the average haul was around 120 to 125 miles – lower than the average estimated by Armstrong but several times that of the railway. The article is based on locally generated archives and brings out well the nature of this unattractive working trade which was so important to Britain's industrialization.

There were also some routes where the ship was important because good roads were rare and the railways arrived late. One such area was studied by Wheeler[51] – the county of Sutherland in Scotland where shipping services were commenced by local worthies to provide a means of marketing agricultural produce. However, the lack of any industrial base to generate traffic, the seasonal nature of the demand to ship livestock, the lack of a return cargo and sparse density of population meant there was not sufficient regular traffic to support the service and it failed to establish itself. Thus, the sporadic attempts to establish a steamboat line were unsuccessful because of the nature of the local economy. This article is significant because it explains some of the features that were needed if a coastal steamer line were to be successful, stresses the role local capitalists played in promoting transport improvements, and demonstrates the long hauls required of the coaster. As in the case of Stanier's article, despite being decades old, no further detailed research has been carried out to supersede this viewpoint.

---

[50]Peter H. Stanier, "The Copper Ore Trade of South West England in the Nineteenth Century," *Journal of Transport History*, 2nd ser., V, No. 1 (1979), 18-35, reprinted in Armstrong (ed.), *Coastal and Short Sea Shipping*, 130-147.

[51]P.T. Wheeler, "The Development of Shipping Services to the East Coast of Sutherland," *Journal of Transport History*, 1st ser., VI, No. 2 (1963), 110-117, reprinted in Armstrong (ed.), *Coastal and Short Sea Shipping*, 104-111.

If Armstrong's estimate is correct, and the coastal ship was performing as much work as the railways, measured in ton-miles, the First World War saw a dramatic reversal of its fortunes, as Aldcroft demonstrates.[52] The short-term effects were bad enough, with large numbers of ships being sunk by mines, U-boats and shells, and other coastal ships being requisitioned by the government, but those still plying had to spend longer in port awaiting coastal convoys and escorts. As a result of the threat to shipping, the government diverted traffic onto the railways where security was unimpaired. The consequences of this were that the tonnage of shipping entering Britain's ports in the coastal trade with cargoes in 1918 was less than half the figure for 1913. This might be written off as merely the immediate impact of wartime conditions which would be reversed when the temporary emergency of the war was over. However, this did not occur. As Aldcroft shows, in 1921 the volume of coastal entries was only sixty percent of the 1913 level. He could have taken this analysis further. Although there was some improvement in this figure, except during 1926 when the General Strike took a heavy toll, even in 1937, the best interwar year for coastal shipping, entries in the coastal trade reached less than eighty-five percent of the 1913 figure.[53] Thus, not only was there no growth beyond the 1913 level of activity in the interwar period, but it was never approached. There was a long-term stagnation. However, this was not the beginning of an inevitable secular decline – the last days of coastal shipping – for after a dip in the Second World War, the tonnage of cargo-carrying entries rose again in the late 1940s, so that in 1952 the 1913 figure was exceeded and by 1964 was one-third higher than in 1913. Thus, there were circumstances peculiar to the interwar period which account for the poor performance of coastal shipping.

Two problems alleged in the interwar years can be dismissed quickly. The decline was not caused by competition from foreign-owned coasters, despite complaints about Dutch competition.[54] Throughout the interwar period foreign-owned tonnage never made up two percent of all entries or clearances and in most years was less than one percent. Nor was the advent of a new form of transport – motor vehicles – likely to have had much impact. Lorries were of small capacity, a few tons for a large lorry, slow and little used for

---

[52]Aldcroft, "Eclipse."

[53]*BPP*, Trade and Navigation Accounts, various years.

[54]For instance, "By 1930, the Dutch motor coasters were a real menace on the British coast;" Jim Uglow, *Sailorman: A Barge Master's Story* (London, 1975), 84.

long-distance journeys, especially in the 1920s.[55] The bulk cargoes being moved long distances by coasters were not likely to shift to road haulage. The explanation probably lies in increased competition from the railways as they began to lose short-distance traffic to motor vans and lorries, compounded by their unscientific[56] method of setting freight rates which allowed them to undercut the coaster while making no profit themselves. The stagnation and decline of Britain's staple industries, such as coal, iron and steel, and shipbuilding, also had a great impact on the coaster for it was in precisely the commodities of these industries, such as coal, iron ore, limestone, pig iron and china clay, that its bulk trade had resided. Domestic industrial stagnation combined with the stagnation of Britain's foreign trade to reduce the demand for coastal shipping. Before the war the coastal fleet had been heavily engaged in assembling British manufactures for export and distributing imported goods from the main ports. This role was much reduced in the interwar period.

As has already been mentioned, British coastal shipping took another dive in activity during the Second World War[57] for much the same reasons as in the First World War. There was a slow recovery afterwards with steady growth through to the 1960s to a new all-time peak,[58] but with a parallel steady growth of the proportion of foreign tonnage plying in the British coastal trade, which reached 13.5 percent in 1963. However, this is a largely un-researched area and we cannot comment in detail. Suffice to say that coastal shipping continued to play an important part in domestic freight transport after the Second World War.

IV

In conclusion, this chapter has endeavoured to contextualize the literature by showing some of it fits into the broad sweep of the history of the coastal trade over the last few hundred years. Inevitably this has been brief and has many limitations, and much of the detail has been omitted to allow a broad-brush depiction of the main trends. Given the small amount of scholarly research on the British coastal trade, the articles discussed stand out as beacons illuminating an otherwise dark-encompassed landscape. Not only were they original and illuminating when published but their findings and conclusions have not been

---

[55]T.C. Barker and Dorian Gerhold, *The Rise and Rise of Road Transport, 1700-1990* (London, 1993), 85-88; and Derek H. Aldcroft, *British Transport since 1914: An Economic History* (Newton Abbot, 1975), 37-38.

[56]Aldcroft, *British Transport*, 39.

[57]Christopher I. Savage, *Inland Transport* (London, 1957), *passim*.

[58]*BPP*, Trade and Navigation Accounts, various years.

reversed or amended by subsequent research. What is clear from them is that for a very long period the coastal trade has played an important and multifaceted role in British internal transport. Its special strengths lay in the long-distance carriage of bulky, low-value commodities. The coastal trade achieved its greatest degree of sophistication in the late nineteenth century when it divided the market into at least four broad segments.[59] The first was the coastal liner companies, offering speedy scheduled delivery of passengers and higher-value cargoes in small lots at a premium price. Then there were the regular traders, like the steam colliers, built for a specific trade or commodity, staying in that one trade and achieving speed and reliability. Third came the steam tramps, unscheduled, needing a full cargo to be profitable and with a steady if unspectacular speed, reasonably reliable and charging a low price. Finally, the coastal sailing ship offered the lowest freight rates for bulky, low-value products which were not required in a hurry or needed at a specific date, because these ships were still at the mercy of wind direction and strength, tides and currents. In this way the coastal ship offered a range of services at a variety of rates to cater for most goods and customer needs.

Without the coaster carrying coal from the North East and South Wales to London and the South East it is difficult to envisage how urban communities could have kept themselves warm in winter, lit their homes and streets and powered their steam engines. Similarly, large agglomerations of people required corn, livestock, beer and bacon. The coaster provided these foodstuffs as well as the low-value bulky building materials which were needed to construct the cities in the first place. Industrialization and urbanization depended on the coastal trade. It is a pity that it has been so under-researched. Perhaps this volume will spur scholars to delve into this topic in the future.

---

[59]See John Armstrong, "Coastal Shipping: The Neglected Sector of Nineteenth-Century British Transport History," *International Journal of Maritime History*, VI, No. 1 (1994), 182-185.

Bibliography of Writings by John Armstrong

"Coastal Shipping." In Aldcroft, Derek H. and Freeman, Michael J. (eds.). *Transport in the Industrial Revolution*. Manchester, 1983, pp. 142-176 (with Philip Bagwell).

"John Lawson Johnston" and "Lord Luke." In Jeremy, David J. (ed.). *Dictionary of Business Biography*. Vol. 3. London, 1985, pp. 510-521.

Directory of Corporate Archives. London, 1985.

"Hooley and the Bovril Company." *Business History*, XXVIII, No. 1 (1986), pp. 18-34. Reprinted in Davenport-Hines, R.P.T. (ed.). *Speculators and Patriots: Essays in Business Biography*. London, 1986, pp. 18-34.

"The Role of Coastal Shipping in UK Transport: An Estimate of Comparative Traffic Movements in 1910." *Journal of Transport History*, 3rd ser., VIII, No. 2 (1987), pp. 164-178. Reprinted in Armstrong, John (ed.). *Coastal and Short Sea Shipping: Studies in Transport History*. Aldershot, 1996, pp. 148-162.

Business Documents: Their Origins, Sources and Uses in Historical Research. London, 1987 (with Stephanie Jones).

"Coastal Shipping." In Freeman, Michael J. and Aldcroft, Derek H. (eds.). *Transport in Victorian Britain*. Manchester, 1988, pp. 171-217 (with Philip Bagwell).

"Coastal Shipping's Relationship with Railways and Canals." *Journal of the Railway and Canal Historical Society*, XXIX, Part 5 (1988), pp. 214-221 (with Philip S. Bagwell).

"Business History and Management Education." In Business Archives Council. *Proceedings of the Annual Conference 1988*. London, 1989, pp. 60-85.

"Freight Pricing Policy in Coastal Liner Companies before the First World War." *Journal of Transport History*, 3rd ser., X, No. 2 (1989), pp. 180-197. Reprinted in Armstrong, John (ed.). *Coastal and Short Sea Shipping: Studies in Transport History*. Aldershot, 1996, pp. 112-129.

"Transport and Trade." In Pope, Rex (ed.). *Atlas of British Social and Economic History since 1700*. London, 1989, pp. 96-133.

"The Rise and Fall of the Company Promoter and the Financing of British Industry." In Van Helten, Jean Jacques and Cassis, Youssef (eds.). *Capitalism in a Mature Economy: Financial Institutions, Capital Exports and British Industry, 1870-1939*. Aldershot, 1990, pp. 115-138.

"The Shipping Depression of 1901 to 1911: The Experience of Freight Rates in the British Coastal Coal Trade." *Maritime Wales*, No. 14 (1991), pp. 89-112.

"Railways and Coastal Shipping in Britain in the Later Nineteenth Century: Co-operation and Competition." In Wrigley, Chris and Shepherd, John (eds.). *On the Move: Essays in Labour and Transport History Presented to Philip Bagwell*. London, 1991, pp. 76-103.

"The Development of British Business and Company Law since 1750." In Turton, Alison (ed.). *Managing Business Archives*. London, 1991, pp. 27-59.

"Conferences in British Nineteenth-Century Coastal Shipping." *Mariner's Mirror*, LXXVII, No. 1 (1991), pp. 55-65.

"An Introduction to Archival Research in Business History." *Business History*, XXXIII, No. 1 (1991), pp. 7-34.

"The Significance of Coastal Shipping in British Domestic Transport, 1550-1830." *International Journal of Maritime History*, III, No. 2 (1991), pp. 63-94.

"Scholarly Sesame: A Plea for Academic Access to the Archives of Business." In Business Archives Council. *Access and Income Generation*. London, 1992, pp. 5-16.

"Road and Rail in the Inter-war Period: The Case of the Park Royal Industrial Estate." *Journal of the Railway and Canal Historical Society*, XXX (1992), pp. 459-468.

"The English Coastal Coal Trade, 1890-1910: Why Calculate Figures When You Can Collect Them?" *Economic History Review*, 2nd ser., XLVI, No. 3 (1993), pp. 608-611.

"The Golden Decade: The Business Archives Council, 1984-94." *Business Archives*, No. 66 (1993), pp. 1-15.

"Coastal Shipping: The Neglected Sector of Nineteenth-Century British Transport History." *International Journal of Maritime History*, VI, No. 1 (1994), pp. 175-188.

"Introduction." In Proctor, Margaret (ed.). *Transport on Merseyside: A Guide to Archive Resources*. Liverpool, 1994, pp. vii-xvi.

"Late Nineteenth-Century Freight Rates Revisited: Some Evidence from the British Coastal Coal Trade." *International Journal of Maritime History*, VI, No. 2 (1994), pp. 45-81.

"An Annotated Bibliography of the British Coastal Trade." *International Journal of Maritime History*, VII, No. 1 (1995), pp. 117-192.

"Forty Years of Transport History: Achievements and Prospects." *European University Institute Working Papers*, HEC No. 95/2 (1995), pp. 7-11.

Inland Navigation and Economic Development in Nineteenth-Century Europe. Mainz, 1995 (edited with Andreas Kunz).

"Inland Navigation and the Local Economy." In Kunz, Andreas and Armstrong, John (eds.). *Inland Navigation and Economic Development in Nineteenth-Century Europe*. Mainz, 1995, pp. 307-310.

"The Development of the Park Royal Industrial Estate in the Interwar Period: A Re-examination of the Aldcroft/Richardson Thesis." *London Journal*, XXI, No. 1 (1996), pp. 64-79.

"The Coastal Trade of Connah's Quay in the Early Twentieth Century: A Preliminary Investigation." *Flintshire Historical Society Journal*, XXXIV (1996), pp. 113-133 (with David Fowler).

Coastal and Short Sea Shipping: Studies in Transport History. Aldershot, 1996 (editor).

"Introduction: The Cinderella of the Transport World: The Historiography of the British Coastal Trade." In Armstrong, John (ed.). *Coastal and Short Sea Shipping: Studies in Transport History*. Aldershot, 1996, pp. ix-xxiv.

"Business History." In Butler, Larry J. and Gorst, Anthony (eds.). *Modern British History*. London, 1997, pp. 244-264.

"Management Response in British Coastal Shipping Companies to Railway Competition." *The Northern Mariner/Le Marin du nord*, VII, No. 1 (1997), pp. 17-28.

"Utilities, Transport and Other Services." In Goodall, Francis; Gourvish, Terry; and Tolliday, Steven (eds.). *International Bibliography of Business History*. London, 1997, pp. 375-473.

"Liverpool to Hull – by Sea?" *Mariner's Mirror*, LXXXIII, No. 2 (1997), pp. 150-168 (with Julie Stevenson).

"Coastal Shipping." In Simmons, Jack and Biddle, Gordon (eds.). *The Oxford Companion to British Railway History*. Oxford, 1997, pp. 440-441.

"Freight Transport by Rail and Sea: How Did the Coaster and Railway Co-exist?" *Journal of the Railway and Canal Historical Society*, XXXII, Part 4 (1997), pp. 242-251.

"Climax and Climacteric: The British Coastal Trade, 1870-1930." In Starkey, David J. and Jamieson, Alan G. (eds.). *Exploiting the Sea: Aspects of Britain's Maritime Economy since 1870*. Exeter, 1998, pp. 37-58.

"British Business Archives as a Research Resource for European Business History." *European Yearbook of Business History*, I (1998), pp. 157-170.

"An Estimate of the Importance of the British Coastal Liner Trade in the Early Twentieth Century." *International Journal of Maritime History*, X, No. 2 (1998), pp. 41-63 (with John Cutler and Gordon Mustoe).

"Transport History, 1945-95: The Rise of a Topic to Maturity." *Journal of Transport History*, 3rd ser., XIX, No. 2 (1998), pp. 103-121.

"The Crewing of British Coastal Colliers, 1870-1914." *The Great Circle*, XX, No. 2 (1998), pp. 73-89.

"Government Regulation in the British Shipping Industry, 1830-1913: The Role of the Coastal Sector." In Andersson-Skog, Lena and Krantz, Olle (eds.). *Institutions in the Transport and Communications Industries: State and Private Actors in the Making of Institutional Patterns, 1850-1950.* Canton, MA, 1999, pp. 153-171.

"Transport and Communications." In Schulze, Max-Stephan (ed.). *Western Europe: Economic and Social Change since 1945.* London, 1999, pp. 213-233.

"London's Railways – Their Contribution to Solving the Problem of Growth and Expansion." *Japan Railway and Transport Review*, No. 23 (2000), pp. 4-13 (with Terry Gourvish).

"Transport." In Waller, Phillip (ed.). *The English Urban Landscape.* Oxford, 2000, pp. 233-268.

"From Shillibeer to Buchanan: Transport and the Urban Environment." In Daunton, Martin (ed.). *The Cambridge Urban History, Vol. 3.* Cambridge, 2000, pp. 229-257.

"The British Coastal Fleet in the Eighteenth Century: How Useful Are the Admiralty's Registers of Protection from Impressment?" *American Neptune*, LX, No. 3 (2000), pp. 235-251 (with John Cutler).

Coastal Shipping and the European Economy, 1750-1980. Mainz, 2002 (edited with Andreas Kunz).

"British Coastal Shipping: A Research Agenda for the European Perspective." In Armstrong, John and Kunz, Andreas (eds.). *Coastal Shipping and the European Economy, 1750-1980.* Mainz, 2002, pp. 11-23.

Companion to British Road Haulage History. London, 2003 (edited with John Aldridge, Grahame Boyes, Gordon Mustoe and Richard Storey).

"The Steamboat, Safety and the State: Government Reaction to New Technology in a Period of *Laissez Faire*." *Mariner's Mirror*, LXXXIX, No. 2 (2003), pp. 167-184 (with David M. Williams).

"Coastal Transport," "Freight Transport," "Public Transport," "Railways,"
    "Road Transport" and "Sea Travel in the Twentieth Century." In
    Loades, David (ed.). *Reader's Guide to British History.* 2 vols. New
    York, 2003, pp. 280-281, 544-545, 1078-1080, 1090-1091, 1124-
    1125 and 1171-1172.

"Coastal Shipping and the Thames." In Owen, Roger (ed.). *Shipbuilding on
    the Thames and Thames-Built Ships.* West Wickham, 2004, pp. 146-
    156.

"The Role of Short-sea, Coastal, and Riverine Traffic in Economic Develop-
    ment since 1750." In Finamore, Daniel (ed.). *Maritime History as
    World History.* Gainesville, FL, 2004, pp. 115-129.

"Frederick Everard," "John Lawson Johnston" and "Lord Luke." In Mat-
    thews, Colin and Harrison, Brian (eds.). *Oxford Dictionary of Na-
    tional Biography.* Oxford, 2004, XVIII, pp. 781-782; XXX, pp. 361-
    363; and XXX, pp. 376-77.

"The Steamboat and Popular Tourism." *Journal of Transport History,* 3rd
    ser., XXXVI, No. 1 (2005), pp. 61-77 (with David M. Williams).

"Writing for an Academic Journal." *Journal of the Railway and Canal Histori-
    cal Society,* XXV, Part 3 (2005), pp. 150-154.

"The Strange History of the Ranterpike in British Coastal Trade." In Harlaftis,
    Gelina (ed.). *Proceedings of the Fourth International Congress of
    Maritime History.* Corfu, 2005 (Cd-rom).

"The Thames and Recreation, 1815-1840." *London Journal,* XXX, No. 2
    (2005), pp. 25-39 (with David M. Williams).

"Shipping, Coastal." In McCusker, John J. (ed.). *History of World Trade
    since 1450.* 2 vols. Farmington Hills, MI, 2005, pp. 661-663.

"The New History of the Steamboat." *International Journal of Maritime His-
    tory,* XVII, No. 2 (2005), pp. 241-247.

"Steam Shipping and the Beginnings of Overseas Tourism: British Travel to
    Northwestern Europe, 1820-1850." *Journal of European Economic
    History,* XXXV, No. 1 (2006), pp. 125-148 (with David M. Wil-
    liams).

"Some Aspects of the Business History of the British Coasting Trade." *International Journal of Maritime History*, XVIII, No. 2 (2006), pp. 1-15.

"The Privatisation of NFC." In Roads and Road Transport History Association. *Private or Public? Conference Papers.* Droitwich, 2007, pp. 19-21.

"Mail Services: Overland Mail," "Shipping Companies: Coastwise Cargo Companies" and "Trade Routes: Loss of Routes." In Hattendorf, John B. (ed.). *The Oxford Encyclopedia of Maritime History.* 4 vols. New York, 2007, III, pp. 446-448; III, 642-644; and IV, 160-162.

"The Steamship as an Agent of Modernisation, 1812-1840." *International Journal of Maritime History*, XIX, No. 1 (2007), pp. 145-160 (with David M. Williams).

"The Revolutionary Impact of New Technology: The Early Steamship, 1812-1840, An Interdisciplinary Study." *International Journal of Interdisciplinary Social Sciences*, II, No. 3 (2007), pp. 343-350 (with David M. Williams).

"The Perception and Understanding of New Technology: A Failed Attempt to Establish Transatlantic Steamship Liner Services, 1814-1828." *The Northern Mariner/Le Marin du nord*, XVII, No. 4 (2007), pp. 41-56 (with David M. Williams).

"Crisis and Response in the British East Coast Coal Trade to London, 1850-1900." In Scholl, Lars U. and Williams, David M. (eds.). *Crisis and Transition: Maritime Sectors in the North Sea Region, 1790-1940.* Bremerhaven, 2008, pp. 48-61 (with Roy Fenton).

"Promotion, Speculation, and Their Outcome: The 'Steamship Mania' of 1824/5." In Association for Information Management. *Proceedings*, LX, No. 6 (2008), pp. 642-660 (with David M. Williams).

"Technological Advance and Innovation: The Diffusion of the Early Steamship in the UK, 1812-1834." *Mariner's Mirror*, XCVI, No. 1 (2010), forthcoming.